T0370466

Intelligent Computing Applications for COVID-19

Innovations in Health Informatics and Healthcare: Using Artificial Intelligence and Smart Computing

Series Editors: Rashmi Agrawal, Manav Rachna International Institute of Research and Studies, and Mamta Mittal, G.B. Pant Government Engineering College

The aim of this series is to publish reference books and handbooks that will provide conceptual and advanced reference material centered around Health Informatics and Healthcare using AI and Smart Computing. There are numerous fields within the healthcare sector where these technologies are applicable including successful ways of handling patients during a pandemic time. Large volumes of data, data analysis, smart computing devices like IoT for sensing health data have drastically changed the way the healthcare sector functions. The scope of the book series is to report the latest advances and developments in the field of Health Informatics with the use of the latest technologies. Each book will describe in detail the use of AI, Smart Computing, Evolutionary Computing, Deep Learning, and Data Analysis in the field of Health Informatics and the books will include real-life problems that focus on the Healthcare System.

Intelligent Computing Applications for COVID-19
Predictions, Diagnosis, and Prevention
Edited by Tanzila Saba and Amjad Rehman Khan

For more information on this series, please visit: https://www.routledge.com/ Innovations-in-Health-Informatics-and-Healthcare-Using-Artificial-Intelligence- and-Smart-Computing/book-series/CRCIHIHUAISM

Intelligent Computing Applications for COVID-19

Predictions, Diagnosis, and Prevention

Edited by
Tanzila Saba
Amjad Rehman Khan

CRC Press
Taylor & Francis Group
Boca Raton London New York

CRC Press is an imprint of the
Taylor & Francis Group, an **informa** business

First edition published 2021
by CRC Press
6000 Broken Sound Parkway NW, Suite 300, Boca Raton, FL 33487-2742

and by CRC Press
2 Park Square, Milton Park, Abingdon, Oxon, OX14 4RN

© 2022 selection and editorial matter, Tanzila Saba and Amjad Rehman Khan; individual chapters, the contributors

CRC Press is an imprint of Taylor & Francis Group, LLC

Library of Congress Cataloging-in-Publication Data
Names: Saba, Tanzila, editor. | Khan, AR (Amjad Rehman), editor.
Title: Intelligent computing applications for COVID-19 : predictions, diagnosis,
and prevention / edited by Tanzila Saba and A.R. Khan.
Description: First edition. | Boca Raton : CRC Press, 2021. |
Includes bibliographical references and index.
Identifiers: LCCN 2021007042 (print) | LCCN 2021007043 (ebook) | ISBN 9780367692476 (hardback) |
ISBN 9780367692483 (paperback) | ISBN 9781003141105 (ebook)
Subjects: LCSH: COVID-19 (Disease)–Epidemiology–Data processing. | COVID-19 (Disease)–
Epidemiology–Simulation methods. | COVID-19 (Disease)–Diagnosis–Data processing. |
COVID-19 (Disease)–Diagnosis–Simulation methods. | Artificial intelligence–Medical applications.
Classification: LCC RA644.C67 I578 2021 (print) | LCC RA644.C67 (ebook) |
DDC 614.5/924140285–dc23
LC record available at https://lccn.loc.gov/2021007042
LC ebook record available at https://lccn.loc.gov/2021007043

ISBN: 978-0-367-69247-6 (hbk)
ISBN: 978-0-367-69248-3 (pbk)
ISBN: 978-1-003-14110-5 (ebk)

DOI: 10.1201/9781003141105

Typeset in Times
by SPi Technologies India Pvt Ltd (Straive)

Contents

Preface

The novel Coronavirus (COVID-19), originating in Wuhan, Hubei province, China, in December 2019, has infected hundreds of millions of people and caused millions of deaths globally. Additionally, the global dissemination of virus shut down major countries. Numerous mathematical and statistical models are being developed to forecast the future of the COVID-19 pandemic.

Accurate estimation, diagnosis and prevention of COVID-19 represent the current major global challenge for most of healthcare organizations and ministries. Innovative measures are required to tackle this challenge by introducing and implementing artificial intelligence (AI) and mathematical modeling applications. The main objective of this book is to provide a complete insight into the state of the art and recent advances of applications, statistical methods, and mathematical modeling in the healthcare industry and all related sectors. The tradeoffs between smart health applications, cost, and skills have become key challenges that must be addressed. In particular, the intense applications of AI and statistical methods require further research on emerging smart equipment, monitoring systems, remote caring centers, early prediction, diagnosis, prevention, regression analysis, and correlation of this pandemic to other factors.

This book also covers state-of-the-art and emerging applications of AI, statistical methods and mathematical modeling for COVID-19. It provides comprehensive information on the fundamental concepts, current applications, algorithms, protocols, new trends, challenges, and research results of AI, statistical methods, and mathematical modeling in the COVID-19 pandemic. It is observed that a powerful combination of AI, statistics, and mathematical applications assist in the early prediction, diagnosis, and prevention of COVID-19. These smart applications help to avoid unplanned downtime, increase operating efficiency, enable new medical products and services, and enhance risk management. The review of these applications is an invaluable resource, giving knowledge on the core and specialized issues of COVID-19, making it highly valuable for the health industry, for both new and experienced researchers, and for medical practitioners in this field.

Acknowledgment

This work was supported by the Artificial Intelligence & Data Analytics Lab (AIDA), CCIS, Prince Sultan University, Riyadh, Saudi Arabia. The editors are grateful for this support.

Notes on the Editors

Tanzila Saba earned her PhD in document information security and management from the Faculty of Computing, UniversitiTeknologi Malaysia (UTM), Malaysia in 2012. She won the best student award in the School of Computing UTM for 2012. Currently, she is serving as an Associate Chair of the Information Systems Department in the College of Computer and Information Sciences, Prince Sultan University Riyadh, KSA. Her primary research focus in recent years has been on medical imaging, pattern recognition, data mining, MRI analysis, and soft-computing. She has above 200 publications that have above 5,000 citations. Her publications are mainly in biomedical research published in ISI/SCIE indexed. Due to her excellent research achievements, she is included in Marquis Who's Who (S & T) 2012. Currently, she is an editor and reviewer of reputed journals and on the panel of TPC of international conferences. She led several funded research projects as a PI. She has full command of a variety of subjects and has taught several courses at graduate and postgraduate levels. On the accreditation side, she has ABET& NCAAA quality assurance. She is a senior member of IEEE. Dr.Tanzila is the leader of the Artificial Intelligence & Data Analytics Research Lab at PSU and an active professional member of ACM, AIS and IAENG organizations. She is the PSU WiDS (Women in Data Science) ambassador at Stanford University and Global WomenTech Conference. She earned the Best Researcher award at PSU for four consecutive years. She has been nominated as a Research Professor at PSU since September 2019.

Amjad Rehman Khan is a Senior Researcher in the Artificial Intelligence & Data Analytics Lab, Prince Sultan University, Riyadh, Saudi Arabia. He earned a PhD & Postdoc from the School of Computing, Universiti Teknologi Malaysia, Malaysia specialization in Forensic Documents Analysis and Security with honors in 2010 and 2011, respectively. He received the Rector award 2010 for the best student in the university. His keen interests are in data mining, health informatics, and pattern recognition. He is the author of more than 200 ISI journal papers and conferences, and is a senior member of IEEE. Currently, he is PI in several funded projects and has also completed projects funded from MOHE Malaysia, Saudi Arabia.

Contributors

Erum Afzal
National University of Sciences and
    Technology (NUST)
Islamabad, Pakistan

Dalyah Ajmal
Artificial Intelligence and Data
    Analytics Lab (AIDA)
CCIS Prince Sultan University
Riyadh, Saudi Arabia

Shahzad Akbar
Riphah College of Computing
Riphah International University
Faisalabad, Pakistan

Hind Alaskar
Artificial Intelligence and Data
    Analytics Lab (AIDA)
CCIS Prince Sultan University
Riyadh, Saudi Arabia

Noor Ayesha
School of Clinical Medicine Zhengzhou
    University
Zhengzhou, Henan, RP China

Muhammad Waqas Aziz
Riphah College of Computing
Riphah International University
Faisalabad, Pakistan

Shima Zarrabi Baboldasht
Department of Computer Engineering,
    Najaf Abad Branch
Islamic Azad University
Isfahan, Iran

Chukwudinma C. Okoli
Medical Laboratory Services
Hospital Management Board,
    HHSS
Abuja, Nigeria

Dipankar Das
Department of Computer Science and
    Engineering
Jadavpur University
Kolkata, West Bengal, India

Monika Dhariwal
Research Scholar
Amity University
Gurgaon, India

Nkereuwem S. Etukudoh
Office of the Provost/CEO
Federal School of Medical Laboratory
    Science
Jos, Nigeria

Nahid Fatima
Associate Professor
Prince Sultan University
Riyadh, Saudi Arabia

Aayush Gadia
Department of Information Technology,
    Department of Engineering and
    Technological Studies
University of Kalyani
Kalyani, West Bengal, India

Sahar Gull
Riphah College of Computing
Riphah International University
Faisalabad, Pakistan

Hemalatha Gunasekaran
IT Department
University of Technology and Applied
    Sciences
Oman

Israr Hanif
Department of Computer Science
Bahauddin Zakariya University
Multan, Pakistan

Majid Harouni
Faculty of Computer Engineering,
    Najafabad Branch
Islamic Azad University
Isfahan, Iran

Syed Ale Hassan
Riphah College of Computing
Riphah International University
Faisalabad, Pakistan

Sajid Iqbal
Department of Computer Science
Bahauddin Zakariya University
Multan, Pakistan

Sogand B. Jaferi
Department of Computer Engineering,
    Najaf Abad Branch
Islamic Azad University
Isfahan, Iran

G. Maria Jones
Department of Computer Engineering,
    Dolatabad Branch
Islamic Azad University
Isfahan, Iran

Mohsen Karimi
Department of Computer Engineering,
    Dolatabad Branch
Islamic Azad University
Isfahan, Iran

Muhammad Kashif
Department of Computer Science and
    Software Engineering
International Islamic University
Islamabad, Pakistan

Manmeet Kaur
Department of Microbiology
Punjab Agricultural University
Ludhiana, India

Amjad Rehman Khan
Artificial Intelligence and Data
    Analytics Lab CCIS Prince Sultan
    University
Riyadh, Saudi Arabia

Rahemeh Ramazani Mahounaki
Faculty of Computer Engineering
Najafabad Branch, Islamic Azad
    University
Isfahan, Iran

Kashif Mehmood
Department of Computer Engineering
University of Engineering and
    Technology
Taxila, Pakistan

Zahid Mehmood
Department of Computer Engineering
University of Engineering and
    Technology
Taxila, Pakistan

Palash Nandi
Department of Computer Science and
    Engineering
Jadavpur University
Kolkata, West Bengal, India

Afrooz Nasr
Department of Computer Engineering,
    Isfahan (Khorasgan) Branch
Islamic Azad University
Isfahan, Iran

Zahra Nourbakhsh
Computer Engineer
Azad University of Isfahan (Khorasgan)
Isfahan, Iran

Shakiba Khadem Olghoran
Department of Computer Engineering
Sheikh Bahaei University
Isfahan, Iran

Shadi Rafieipour
Department of Computer Engineering
Najaf Abad Branch
Islamic Azad University
Isfahan, Iran

Arash Raftarai
Faculty of Computer Engineering,
    Najafabad Branch
Islamic Azad University
Isfahan, Iran

Ziafat Rahmati
Department of Computer Engineering,
   Khoram Abad Branch
Islamic Azad University
Lorestan, Iran

K. Ramalakshmi
Alliance College of Engineering and
   Design
Alliance University
Bengaluru, Karnataka, India

Shalini Ramanathan
Research Scholar
National Institute of Technology
Tiruchirapalli, India

Tanzila Saba
Artificial Intelligence and Data
   Analytics Lab (AIDA)
CCIS Prince Sultan University
Riyadh, Saudi Arabia

Tariq Sadad
Department of Computer Science and
   Software Engineering
International Islamic University
Islamabad, Pakistan

A. George Maria Selvam
Department of Mathematics
Sacred Heart College
Tirupattur, India

Jaskanwar Singh
Production Department
Guru Nanak Dev Engineering College
Ludhiana, India

Mohan Singh
Computer Networking Department
Assiniboine Community College
Brandon, Manitoba, Canada

Nakisa Tavakoli
Department of Computer Engineering,
   Farsan Branch
Islamic Azad University
Chaharmahal Bakhtiari, Iran

Uchejeso M. Obeta
Department of Medical Laboratory
   Management
Federal School of Medical Laboratory
   Science
Jos, Nigeria

R. Venkatesan
IT Department
Karunya Institute of Science and
   Technology
Coimbatore, India

D. Vignesh
Department of Mathematics
Sacred Heart College
Tirupattur, India

S. Godfrey Winster
Department of Computer Science and
   Engineering
SRM Institute of Science and
   Technology
Chengalpattu, India

1 Deep Learning for COVID-19 Infection's Diagnosis, Prevention, and Treatment

Amjad Rehman Khan, Kashif Mehmood, and Noor Ayesha

CONTENTS

1.1 INTRODUCTION

A novel pneumonia-infected case was reported in Wuhan, Hubei Province, China in late December 2019, SARS-CoV-2 or COVID-19 (Wang et al., 2020a, 2020b). The main origin of COVID-19 was Huanan Seafood Wholesale Market where several cases were reported and which was closed to the public on December 31, 2019 (Siordia, 2020). The World Health Organization (WHO) declared the SARS-CoV-2, COVID-19 or coronavirus pandemic globally on March 11, 2020 (Ludvigsson, 2020).

The clinical manifestation of infected patients from bronchoalveolar lavage samples reported SARS-CoV-2 as a serious threat. Research studies show that COVID-19 associated 88% similarity with bat-related (bat-SL-CoVZC21 and bat-SL-CoVZC45) coronaviruses. It is as yet unknown whether COVID-19 transmission occurs directly from bats or another host. Human spread may occur through contact, direct or aerial transmission (cough, sneeze, droplet inhalation). Medical staff and flight attendants were also reported in close contact with COVID-19 patients. It mainly infects and initiates from lung disease in humans, leading to serious respiratory infections (Umakanthan et al., 2020). Around 61,299,371 confirmed COVID-19 cases, including 1,439,784 deaths, were reported up to November 28, 2020.

The virus affected people slightly, while some elderly and old had comorbidities such as pneumonia, acute respiratory distress syndrome (ARDS), and multi-organ dysfunction. The predicted fatality rate is 2%–3%. Distinct molecular assessments were undergone during diagnosis for the virus in respiratory secretions (Singhal, 2020). Imaging modalities such as computed tomography (CT), chest X-ray (CXR), and ultrasound (US) were used during the diagnosis procedures along with examining the nasopharyngeal swab, bronchoalveolar lavage, or tracheal aspirate samples and possible contact history (Lomoro et al., 2020). The key symbol of SARS-CoV-2 is the ground-glass opacities' (GGOs) two-sided distribution with or without consolidation in the lungs (Wang et al., 2020; Chung et al., 2020). In the analysis of CT imaging of lung injury-related infections due to COVID-19, several features were observed, such as crazy-paving pattern, airway variations, and reversed halo sign, for precise diagnosis (Afza et al., 2020; Abbas et al., 2018a, 2018b, 2019a, 2019b).

Researchers implemented numerous procedures and techniques for COVID-19 detection and diagnosing earlier with accurate classification (Khan et al., 2021). Several medical screening modalities, such as CXR, CT scans, and lung ultrasound (LUS), used deep learning (DL) for detection and diagnosing of infection (Amin et al., 2019c, 2019d). The analysis reported CT scan modality to be more effective, due to virus detection from patients' images (Amin et al., 2018). The development of computer-aided diagnosis (CAD) procedures started to perform efficiently and assist doctors in the examination process during the 1980s. Numerous machine learning (ML) models were developed and reported efficacy in medical imaging, such as support vector machines (SVMs), naïve Bayes (NB) classifiers, the K-nearest neighbors (KNN) algorithm, and decision trees (Adeel et al., 2020; Afza et al., 2019; Ejaz et al., 2018a, 2019, 2020). DL techniques, such as neural networks (Saba, 2021), and convolutional neural networks (Rehman et al., 2021b) were developed, and reported better performance than ML techniques in the case of big data. In this chapter, lung infection occurrence due to COVID-19 is analyzed and discussed comprehensively, evaluated using DL techniques.

1.2   LUNG INFECTIONS OVERVIEW

The CT reported variations of lung-related irregularities in patients having a history of fever or respiratory indications or both, while faintness, headache, nausea, vomiting and diarrhea are reported as low common symptoms (Wang et al., 2020; Khan et al., 2019; Saba, 2019).

COVID-19 infects any organ due to the existence of the ACE2 receptor in the human body. Mainly, the ACE2 receptor showed in lungs, gut enterocytes, and alveolar epithelial (type II) cells. ACE2 amino acid transport function is associated with gastrointestinal (GI) tract infections, and SARS-CoV2 may ease infection through microbiota. Superfluous lung indications in the GI tract appeared in some COVID-19 patients (Rehman et al., 2021).

Explicit microbial development occurred due to severe variations throughout lung pathology. Plenty of Porphyromonas, Neisseria, Haemophilus, and Fusobacterium was revealed in lung infection, asthma, and bronchoscopy samples. Previous indications reported minor asthma allergic irritation due to oral or intranasal microbiota management. Several types of bacteria are observed, such as Achromobacter and Pseudomonas, in different lung diseases and lung cancer. The reports also showed a connection between lung microbiota and pneumonia. Fungi, bacteria or others activate pneumonia by alveoli infection in the lung (Saba et al., 2019).

Cytokine storm (CS) is identified due to other positive coronavirus disease collaboration in patients. In COVID-19, respiratory failure, C-reactive protein (CRP) and IL-6 association have also been reported (Polidoro et al., 2020). During COVID-19 disease, IL-6 plays a significant role in lung injury activation. Constant boost of IL-6 leads to immune-facilitated lung injury and macrophage activation syndrome (MAS) in patients possibly connected with coronavirus (Mayor-Ibarguren & Robles-Marhuenda, 2020). The comparison reports between LUS and CT based on the detection of pleural and parenchymal regions and the screening capacity of different treatments showed better LUS due to fast detection and diagnosis of influenza during the avian influenza A(H7N9) epidemic in 2013. The disease was reported in a large geographic area in China, with 40% mortality rate. LUS also helps in COVID-19 disease detection and diagnosis. It is better than CT in detecting deep lung features in infected patients. In COVID-19, infection detection and diagnosing processes in patients, air bronchograms, lung inner and outer regions, B-lines, inaccurate pleural lines, consolidations, and more were screened through US. As more features are required for severe pneumonia variations, just like CT, it was frequently unsuccessful for lung injury detection throughout the influenza A(H1N1) epidemic in 2009. The patients' condition deteriorated rapidly into complex pneumonia. The average sensitivity, specificity, positive predictive value (PPV), and negative predictive value (NPV) of 94%, 89%, 86% and 96%, respectively, were reported for LUS, showing it to be more effective than CT (Convissar et al., 2020).

A total of 62 COVID-19-positive patients, comprising 39 males and 23 females with a mean age of 52.8 (range: 30–77), were studied by analyzing medical and CT data from China. The spread, CT lesions indications and degree of CT contribution score were assessed and evaluated by two radiologists using the Mann-Whitney U test. Early CT indications and complex coronavirus infection association were evaluated by chi-square test. The results reported lymphocyte amount reduction, improved erythrocyte sedimentation rate, high sensitivity CRP level evaluation, and more. CT results in infected patients showed GGO, consolidation, fibrotic streaks, vacuolar sign, GGO + reticular pattern of 40.3%, 33.9%, 56.5%, 54.8%, 62.9% and more like bronchus distortion and air bronchogram, 72.6% and 17.7% respectively (Zhou et al., 2020).

During the evolution of health issues globally, CT modality plays a key role in early COVID-19 recognition and screening (Rehman et al., 2021d). The study of COVID-19-infected patients from three different hospitals in China was analyzed, and reported GGO, crazy-paving pattern, infection fringe spread, healthy CT and morphology opacities of 57%, 19%, 33%, 14% and 33%, respectively. Lymphadenopathy, pleural declarations, lung nodules, and cavitation of lung were not found. The CT scans were recorded in the supine position of infected patients. The patients were described as confirmed COVID-19 patients according to the oropharyngeal swab, endotracheal aspirate, bronchoalveolar lavage, nasopharyngeal swab samples and lab evaluations (Chung et al., 2020).

The results reported consecutive CT image assessment of coronavirus-infected patients and showed low development in the virus in non-pregnant patients. The clinical manifestation evolutions reported CT as an important commodity in SARS-CoV-2 detection. It is widely used for any lung infection, in either mild or severe cases. The report evidenced US screening as a most-used commodity in pregnant women for SARS-CoV-2 detection, due to its safety and lack of radioactive nature (Buonsenso et al., 2020).

Individuals with lung cancer (LC) were exceptionally affected in the coronavirus outbreak, due to fast evolution and death rate. The COVID-19 influence and LC and progression due to lack of diagnosis and preventive procedures were examined in South Korea. The comparison between new and previous three-year LC identified patients' history from three different hospitals. The study reported LC detected in 612 patients for three years. A 16% reduction in patients in terms of lung treatment in hospital during the COVID-19 outbreak compared with the preceding year was reported. isolation wards were created for the treatment of infected patients. Several detections and preemptive procedures were executed, LC diagnosis suspension was reduced, and patients were protected from coronavirus infection from hospital acquaintance. Exceptional growth was recorded in advanced non-small-cell LC (NSCLC) stage percentage during these three years and the diagnostic percentage was similar (Park et al., 2020).

The experiments showed worsened lung function in patients with pragmatic deficient breathing assistance during the pandemic's early days. There is no strong medical or scientific evidence for patient self-inflicted lung injury (P-SILI). The lack of sympathetic investigations reported that P-SILI affects lungs due to local different stress and high respiration. ARDS and remote respirational abnormalities are complex during SARS-CoV-2 lung injury. Hypoxemia is occasionally the main source of lung injury. However, COVID-19 infection may spread into more organs than the lungs, such as the kidney and liver. The study was distinctly limited due to the lack of detection of respiratory infections during SARS-CoV-2 lung injury, and P-SILI examination processes have not yet proved that P-SILI is a cause of lung infection evolution (Cruces et al., 2020).

CXR is usually used for the screening and detection of lung infections, due to several preventive problems associated with CT, such as CT accessibility locally or globally, and CT room sanitization inadequacies. CXR may reduce the threat of cross-infection arising from the CT distillation's subsequently infected cases that disorder the accessibility of the radiological facility reported by the American College of Radiology. CT and CXR both detect lung infection in any doubt that is created in the diagnosis of some patients. Distinctive redundancy of symptoms, i.e. reticular, hazy, irregular, GGO, and others, were reported in infected and non-infected patients on CXR (Jacobi et al., 2020).

Lung infections occurred by oral viruses through the inferior respiratory area. Cough, fever, deprived oral sterility, abnormal respiration, and mechanical ventilation

(MV) cause lung diseases. Oral microbiota-initiated anaerobes are facilitated by SARS-CoV-2 indications and lung hypoxia. Lung infection may worsen due to the COVID-19 and microbiota relationship through variations in T-cell retorts, cytokines, and host situations. Precautions for controlling lung infections in SARS-CoV-2 infected individuals are essential due to the COVID-19 and oral microbiome relationship (Bao et al., 2020; Khan et al., 2021).

Current examinations present the efficacy of LUS for SARS-CoV-2 detection but there is a lack of LUS-based data concerning infected children. In hospitals in Rome, ten infected children were reported sequentially. US was used for infection detection from lungs. The results reported 10%, 60%, 70% and 10% of white lung regions, pleural abnormalities, vertical portions, and subpleural consolidations, respectively; pleural effusions were not created in any of the patients. The results showed monotonous LUS usage in the screening of coronavirus- or non-coronavirus-infected children, due to its radiation-free, safety and spread qualities (Musolino et al., 2020).

MV usage is reported in ARDS infection while the excessive death rate is 39%. When MV is used inadequately, the death rate in ARDS individuals will grow. The inflated local lung stress triggers ventilation-induced lung injury (VILI), according to current research. VILI and local active strain reduction will occur when the ventilation approach keeps the lung's extravascular space constant and similar. A time-controlled adaptive ventilation (TCAV) technique has been proposed that reduces active alveolar stress through breath regulation conferring automatic lung features. The result showed a normal and healthy lung decreasing local strain and having better lung safety. ARDS infections were reported in over 3 million patients yearly worldwide, including 10% intensive care unit (ICU) patients. The death rate is 75k, while 2 million are identified as having ARDS infection annually in the United States (Nieman et al., 2020) (Figure 1.1).

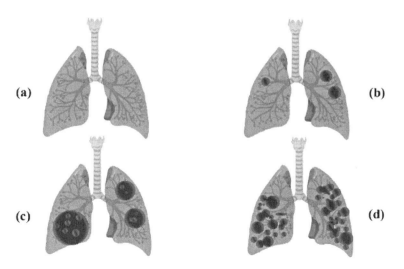

FIGURE 1.1 General structure of healthy and severe-stage lungs: (a) healthy lungs (b) early-stage infected lungs (c) advanced-stage lungs (d) severe-stage lungs.

According to state-of-the-art analysis reports, pulmonary fibrosis occurs in infected COVID-19 patients due to old age, severity of disease, long stays in ICU and on ventilation, smoking and alcohol consumption. Lung prevalence and protection are recommended for other infections, due to lack of pulmonary fibrosis procedures. Some patients' features may lead to augmented danger, such as lung damage, death, or pulmonary fibrosis (Ojo et al., 2020).

1.3 DIAGNOSTICS

A novel SARS-CoV-2, named COVID-19, pandemic emerged in late December 2019 worldwide (Rodriguez-Morales et al., 2020). Research studies of influenza, tuberculosis, and pneumonia for decades have shown that lung infections worldwide could be fatal diseases. Clinical symptoms such as cough, fever, tiredness, myalgia and dyspnea are common in COVID-19 patients (Fu et al., 2020).

A study of eight patients reported the examination of bronchoalveolar lavage samples from COVID-19-infected lungs through microbial composition and occupied by oral and upper respiratory tract bacteria like traditional pneumonia. The analysis recognized that acute respiratory distress syndrome might be predicted through microbial signs and antiviral immunity improvement by gut microbiota (Ahlawat & Sharma, 2020). The cause of lung infection is typically an external offense, damaged tissue, or immune syndrome. COVID-19 caused a pandemic since December 2019 when the early infected cases were reported from China and its basics and origin are still unknown (Minucci et al. 2020).

A study was made of a COVID-19-infected patient (age 72 years) with cough, fever, diabetes, and hypertension history. The reports presented COVID-19 through early examination of the throat and pharyngeal swab samples. After the necessary therapies, using imaging modalities such as CT to examine and analyze samples obtained from various parts of the lungs, the patient died and the study has several limitations; even with antidote therapies, the shortness of breath was sustained. As it was not possible to acquire larger tissue samples, the aim was to solve the sample issues using autopsies or thoracoscopic lung biopsies. In two cases COVID-19 was detected during adenocarcinoma operations (lung lobectomies). The patients did not have any fever or pneumonia symptoms before or at the time of the operation, which signifies that the early COVID-19 procedures varied, due to investigations reporting lung edema, multinucleated giant cells, irregular seditious cellular penetration, and proteinaceous exudate in patients (Tian et al., 2020).

The innovative automated segmentation pipeline technique was proposed for issues of limited datasets (overfitting) by considering them as modified datasets. This method uses numerous preprocessing image techniques before training. A neural network-based 3D U-Net model was executed for supplementary overfitting reduction. The model validated on 20 COVID-19-infected CT scans resulted in 0.956 and 0.761 dice similarity coefficients for lungs and COVID-19, respectively. This was accurate and robust in limited datasets deprived of overfitting coronavirus and lung (Müller et al., 2020).

ML and DL techniques such as convolutional neural networks (CNNs) are mainly used to automatically diagnose medical imaging (Iqbal et al., 2018, 2019; Amin et al., 2019a, 2019b, 2019c, 2019d). There are two methodologies, segmentation and classification, used along with these techniques. The purpose of segmentation is to extract features (regions of interest or ROIs) from each pixel of an image and label it while classification is used to label the whole image and categorize it into a distinct group (Sadad et al., 20121a, 2021b; Ejaz et al., 2018a, 2018b). DL techniques such as U-Net, VB-Net, and other U-Net modifiers are used for segmentation while DenseNet, ResNet, and Inception-v3 are used for classification, and have accomplished better performance than humans (Shi et al., 2020; Javed et al., 2019, 2020; Javed et al., 2019a, 2019b, 2020a, 2020b).

Percentage of infection (POI) has been used for inclusive prediction of infection registration and imaging, whereas better and accurate data extraction through various lung parts and infection segmentation is required for analysis (Shi et al., 2020; Müller et al., 2020). COVID-19 infection has also been diagnosed in some lung transplant individuals (Michel et al., 2020). The study has limited value, due to the following issues: data recorded for analysis was only from China, more data is essential such as more regions, countries, or worldwide for in-depth analysis of SARS-CoV-2. The data and clinical reports about COVID-19 were limited at the time of analysis (Rodriguez-Morales et al., 2020; Mughal et al., 2018a; Rahim et al., 2017a; Ramzan et al., 2020a).

A prognostic model was proposed using LUS imaging and medical features for COVID-19. The study analyzed 100 patients, 31 of whom were confirmed COVID-19-infected patients through reverse transcription polymerase chain reaction (RT-PCR). COVID-19-infected examination is autonomously related to fast organs injury serial wise, consolidation and lower site hard pleura. The statistical analysis reported 0.82, 97%, 62%, 54% and 98% of receiver operating characteristic (ROC), sensitivity, specificity, PPV, and NPV. This study is limited by the following limitations: the proposed technique is not more effective, due to the misclassification of infected patients recorded in some sensitivity tests; and LUS examination distinguishes other pulmonary and cardiac disease but cannot distinguish the pneumonia type and influenza, whereas it differentiates it in COVID-19 variations (Wang et al., 2020). The research aimed to examine more lung regions to improve the model's performance (Figure 1.2).

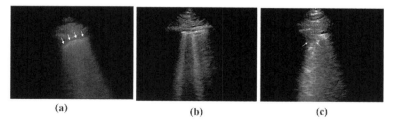

(a)   (b)   (c)

FIGURE 1.2 LUS results in COVID-19 detection shows (a) pleural line abnormalities; (b) confluent B-lines; (c) subpleural consolidation.

Youssef et al. (2020) reported that RT-PCR is effective during immediate SARS-CoV-2 disease diagnosis while CT and X-ray are better in COVID-19 analysis having some limitations. LUS only detects features deeply and images next to pleura. It cannot detect any other infections like tumors. LUS depends on skilled sonographers with sophisticated analytical accuracy. It also depends on the patient. If the patient is overweight, the examination will be complex because of the extent of soft tissues. The performance is not as effective as CT scans. LUS is recommended for COVID-19 patients, due to its visualization of pneumonia signs, which distinguishes lung infection in COVID-19 (Fu et al., 2020).

A VB-Net neural network was implemented for infected regions, automatic segmentation from COVID-19 patients' CT images based on DL. The study evaluated 249 infected patients for the training model; 300 images were used for testing. The model's analysis reported 91.6%, 10.0% dice similarity coefficients for both automatic and manual segmentation and 0.3% POI and accuracy. This research study is limited, due to a lack of evaluated data. It was obtained from one hospital, not from large hospitals or more regions. This system was implemented only for infection evaluation, not specifically for any severe infection like pneumonia. The study aimed to extend it to multiple regions, centers and for the severity of pneumonia infections (Shan et al., 2020).

There is a lack of research studies based on analysis of autopsies of COVID-19 patients after death because of severe infectiousness and unavailability of accurate treatment conduction methods and procedures. During diagnostic procedures, clinical manifestations reported that COVID-19 obviously leads to infections or injury to many tissues and organs, such as the lungs. It occurs in and infects animals and humans, spreading through respiratory, hepatic, bowel, neurologic and renal systems (Wan et al., 2020).

1.4 DL METHODOLOGY

CAD systems have been developed for misclassification and false-negative rates reduction. These assist doctors in automatically detecting and diagnosing infections explicitly and in early stages from CT images (Rehman, 2020, 2021; Saba et al., 2018a, 2020b; Yousaf et al., 2019a, 2019b). Deep neural networks (DNNs) based on DL have recently increased researchers' extensive attention to medical imaging and deep feature understanding. Several DL models' implementation and development for big data reported significant effectiveness in performance (Khan et al., 2019a, 2019e, 2020a, 2020b).

DL techniques are based primarily on two categories; one is supervised learning (consisting of neural networks (NNs), CNNs, recurrent neural networks (RNNs) and others) while the second is unsupervised learning (consisting of auto-encoders (AEs), stacked auto-encoders (SAEs), deep belief networks (DBNs), and others). CNNs based on DL techniques result in high-performance accuracy and improved medical imaging performance. Numerous state-of-the-art models have been developed based on CNN, such as AlexNet (Khan et al., 2019b, 2019c, 2019d, 2020), ResNet (He et al., 2020; Khan et al., 2020b), DenseNet (Huang et al., 2017), and Inception (Khan et al., 2020c, 2020d).

DL techniques include the following basic steps, followed during the whole process: preprocessing, segmentation, feature extraction, and selection and classification (Rehman et al., 2018a, 2020a, 2021b) (Figure 1.3).

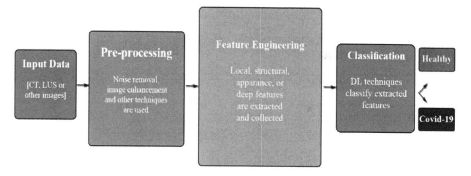

FIGURE 1.3 General CAD architecture for COVID-19 detection.

1.4.1 DATASETS

Datasets play a vital role in state-of-the-art experiments and results comparisons (Rad et al., 2013, 2016). The dataset section aimed to collect and provide an overview of the different publicly available datasets that might be useful for future research during COVID-19 or any other severe disease (Rehman et al., 2021a, 2021b, 2021c; Saba et al., 2018b, 2019c, 2020a).

The COVIDx dataset was created by collecting chest radiography image data from five publicly available databases: (1) COVID-19 Image Data Collection; (2) COVID-19 Chest X-ray Dataset Initiative; (3) ActualMed COVID-19 Chest X-ray Dataset Initiative; (4) RSNA Pneumonia Detection Challenge dataset; and (5) COVID-19 radiography database. The whole dataset contains CXR images and cases of 13,970 and 13,870, respectively (Wang et al., 2020). The COVID-19 X-ray images dataset is collected from three different datasets that contain 381 images of COVID-19, pneumonia and normal patients divided equally. The collection of 20 COVID-19 CT scans labeled by and verified by two radiologists as left and right lung is named as COVID-19 CT Lung and Infection Segmentation Dataset (Shuja et al., 2020). The dataset was prepared by images labeling from 64 videos simultaneously. It consists of 1,103 LUS patient images comprising COVID-19, bacterial pneumonia, and healthy images of 654, 277, and 172 images, respectively (Born et al., 2020; Mughal et al., 2017, 2018b).

There is a dataset collecting Twitter tweets regarding COVID-19 from different countries and languages since January 22, 2020, simultaneously through API and hashtags (virus, coronavirus and others) when infected patients were reported as <600 globally (Lopez et al., 2020). The dataset called NAIST COVID (Multilingual COVID-19 Twitter and Weibo Dataset) contains multilanguage Twitter and Weibo of more than 20 million microblogs between January 20 and March 24, 2020. It comprises related microblog IDs, timestamps and keywords (Gao et al., 2020). The Weibo-COV dataset was found to consist of more than 40 million tweets from December 1, 2019 to April 30, 2020, containing tweet variations, retweeted data, and more.

The dataset CoAID (COVID-19 Healthcare Misinformation Dataset) consists of 183,564 user-related engagements from 516 social platform posts and 1,896 news

TABLE 1.1
Datasets and Their Access Links

Dataset	Access Links
COVID19_Tweets	https://github.com/lopezbec/COVID19_Tweets_Dataset
LUNA16	https://luna16.grand-challenge.org/
COVID-19	https://github.com/sociocom/covid19_dataset
LIDC/IDRI	https://wiki.cancerimagingarchive.net/display/Public/LIDC-IDRI
covid19 Pocus Ultrasound	https://github. com/jannisborn/covid19_pocus_ultrasound
COVID-19-InstaPostIDs	https://github.com/kooshazarei/COVID-19-InstaPostIDs
ChestX-ray14	https://academictorrents.com/details/557481faacd824c83fbf57dcf7b6da9383b3235a
CoAID	https://github.com/cuilimeng/CoAID
COVID-Net	https://github.com/lindawangg/COVID-Net
convid19-X-rays	https://www.Kaggle.com/andrewmvd/convid19-X-rays
COVID-19 lung and infection segmentation	https://zenodo.org/record/3757476#.XxKOkigzbIU
Weibo public-opinion	https://github.com/nghuyong/weibo-public-opinion-datasets
Evacuees from Wuhan	https://figshare.com/articles/Evacuees_from_Wuhan/11859207/1
COVID-Q	https://github.com/JerryWei03/COVID-Q
CXR pneumonia	https://www.kaggle.com/paultimothymooney/chest-xray-pneumonia
Lung1	https://wiki.cancerimagingarchive.net/display/Public/NSCLC-Radiomics

and labels. The data was collected after the evaluation of a fact-based strategy for fake and true news confirmation of articles, posts and websites as well as users' social signals (Cui & Lee, 2020). The dataset based on Instagram collected English-language data of over posts, likes, and comments of 5.3k, 329k, and 18.5k, respectively of 2.5 publisher distribution from January 5 to March 30, 2020 (covid19, virus, corona, and other) hashtags. It is categorized into post content, features, comments, and publisher data (Zarei et al., 2020). The dataset based on online questions was collected from 13 sources, including Google, Yahoo and Quora, of 1,690 questions about COVID-19. It is labeled into 207 classes and 15 groups (Wei et al., 2020). The city of Wuhan, China, was evacuated of 2,666 foreigners after isolation and their health was monitored for any infection; 12 COVID-19-infected individuals were reported. The data was named as evacuee dataset (Zhou, 2020) (Table 1.1).

1.4.2 Preprocessing

Preprocessing is an important stage of DL techniques for medical image examinations. The process comprises noise removal (e.g. of images, undesirable and unnecessary parts) and improves image quality. It also used for labeling and tagging the desired classes (Marie-Sainte et al., 2019a, 2019b; Mittal et al., 2020; Liaqat et al., 2020; Majid et al., 2020; Mashood Nasir et al., 2020; Perveen et al., 2020).

1.4.3 SEGMENTATION

The segmentation process is the most substantial, critical, and complex image-processing task, due to the results being dependent on these features. After noise contrast removal and other enhancements, it divides images into parts for analysis (Nazir et al., 2019; Qureshi et al., 2020; Rehman et al., 2018b, 2018c). Several techniques are implemented, such as edge, region, thresholding, and clustering, based on segmentation. Numerous techniques are used for image segmentation, such as artificial neural networks (ANNs), Gaussian mixture model, and Gaussian kernel fuzzy c-means (FCM) (Ragab et al., 2019; Sadad et al., 2018; Ullah et al., 2019; Yousaf et al., 2019). No inclusive segmentation technique was reported that validates effectiveness in image types and for a specific medical imaging modality.

1.4.4 FEATURE EXTRACTION AND SELECTION

In this step, the features are extracted from medical images for infections of disease diagnosis. This section can categorize many features, such as local, texture, or statistical features from patients' images (Saba et al., 2019a, 2019b). Several techniques are used, such as local binary pattern (LBP), center symmetric local binary pattern (CS-LBP), and scale invariant feature transform (SIFT), for feature extraction (Iqbal et al., 2017; Lung et al., 2014; Rahim et al., 2017b; Ramzan et al., 2020b; Rehman et al., 2018, 2021a).

1.4.5 CLASSIFICATION

The final step of DL techniques is classification. This helps the researcher build an effective model for diagnosis based on the feature vector created during the feature extraction and selection processes (Saba, 2017, 2018, 2019, 2020). The classification section is basically used to classify patients' images through labels, either healthy or infected. Several techniques are used for classification, such as CNN, ResNet, AlexNet, and DensNet (Fahad et al., 2018; Husham et al., 2016; Hussain et al., 2020; Khan et al., 2017; Iftikhar et al., 2017; Al-Ameen et al., 2015; Jamal et al., 2017; Khan et al., 2020).

1.5 ANALYSIS AND FINDINGS

COVID-19 is a newly emerged pandemic that originated in December 2019. The mortality rate is very high globally and a serious threat for the health community. There were no specific treatment procedures and guidelines available. The chest CT modality was mainly used for infection detection and diagnosis to assist doctors in controlling the virus. Analysis of lung infection and COVID-19 detection and diagnosis based on different DL techniques through different performance measures, such as area under the curve (AUC) and ROC, were presented.

The SARS-CoV-2 causes ARDS, due to considerable lung inflammation risk. ARDS is considered a result of classified COVID-19 lung injury. The analysis is still not accurate that severe hypoxemia is associated with an insignificant reduction in lung acquiescence and recommended supplementary procedures for hypoxia contribution to parenchymal damage procedures. The analysis of COVID-19 patients' lungs' reports presented that monocyte polymorphonuclear neutrophils and other leukocytes conscription follow-on due to proinflammatory intermediaries and paracrine activation autocrine cytokine stimulation (Polidoro et al., 2020). In COVID-19, preference is given to LUS in diagnosis over CT and X-ray, on account of past epidemics, affordable cost, the low virus spread threat, and easy usage (Tsai et al., 2014). It helps doctors distinguish pneumonia and dyspnea and is recommended where there is a lack of molecular examinations (Convissar et al., 2020).

In treatment procedures, patients requiring oxygen assessment as a result of severe infection are almost 15%–20%, while others have a common infection (Wang et al., 2020). Proinflammatory cytokines are released due to lung epithelial cells and alveolar macrophages' initiation in some lung-infected patients. The DL-based algorithm's performance reported 94.4%, 11% and 5% of AUC, false positives and negatives, respectively, for National Lung Screening Trial (NLST) images. The algorithm Mask-R-CNN and cancer ROI detection technique were applied on the Lung Image Database Consortium (LIDC), NLST and LUng Nodule Analysis (LUNA) datasets. 3D U-Net and CapNets examined the effectiveness of lung nodule data. The technique resulted in accuracy, sensitivity, specificity, and AUC of 84.5%, 92.9%, 0.84, and 70%. The specificity is low because of benign samples' extravagance. The performance is limited, due to less data.

Fever frequency rate (FR) was low in children (92.8%, 95%CI 89.4–96.2%) but high in adults (43.9%, 95% CI 28.2%–59.6%) (Rodriguez-Morales et al., 2020). The clinical manifestations examined and evaluated more, having 11,950 COVID-19-infected patients from 24 studies reporting fever, cough and fatigue of 77%, 60% and 38%, respectively. The study is limited, due to lack of lab result variations, full data access, patients with various sickness stages, and poor treatment procedures (Ebrahimi et al., 2020). Chest CT procedure was reported to be effective and fast during COVID-19 diagnosing. The COVIDx dataset, consisting of CXR images of COVID-19 patients, was created and evaluated through a proposed COVID-Net deep CNN, resulting in 91% sensitivity and 98.9% high PPV. Low false positive COVID-19 detection was reported, with misclassification of one normal patient CT image as COVID-19 (Wang et al., 2020).

Lack of throat swab samples and detection procedure led to 30%–60% sensitivity being recorded for RT-PCR. Negative RT-PCR resulted, due to positive CT outcomes. There was 97% CT sensitivity and 90% of patients were diagnosed and analyzed for COVID-19 having different CT results (Wang et al., 2020). Positive likelihood ratio, negative likelihood ratio, specificity, and sensitivity of 1.17%, 0.48%, 25% and 88% respectively were reported for CT (Siordia, 2020). A total of 64 patients were studied, including 26 56-year-old male patients at baseline chest radiography. 91% (58), 69% (44), and 59% (38) patients were reported infected for COVID-19 and abnormal respectively by diagnosing through RT-PCR. 69% sensitivity was recorded for the proposed method, despite its having some limitations: the diagnostic and treatment procedures were not followed accordingly. The effectiveness during the process probably changed due to the interruption of CT and RT-PCR diagnosis. A lack of radiologists also affects specificity and sensitivity (Wong et al., 2020). COVID-19-associated irregular results and variables 19% leucopenia, 13% lymphocytes, high body temperature and 31% creatine kinase MB (high level) and 17% procalcitonin were

TABLE 1.2

Performance Measures Based on DL Techniques

Author	Techniques	Database	AUC	Accuracy	Sensitivity	Specificity	F1 Score	PPV	NPV
Wang et al. (2020)	prognostic model	Ultrasound images	N/A	N/A	97%	62%	N/A	54%	98%
He et al. (2020)	(DECAPS)	CT images	96.10%	87.60%	N/A	N/A	N/A	N/A	N/A

expressively reduced (Qiu et al., 2020). Performance measures, such as accuracy, ROC, AUC, and the sensitivity of different research studies, are presented in Table 1.2.

1.6 CONCLUSIONS AND FUTURE CHALLENGES

A novel coronavirus named COVID-19 by WHO proved the biggest health threat due to its fast-spreading nature globally and particularly for health practitioners. Symptoms such as fever, cough and tiredness were most commonly reported initially for SARS-CoV-2. Individuals who have prior contact or travel history to coronavirus epidemic areas must follow clinical procedures, guidelines, and protection through lab tests. An effective vaccine development is on its way to control virus spread and death rate globally. Many effective DL techniques have been developed for detection and diagnosis to predict COVID-19 with better efficiency. However, this chapter reviewed and analyzed lung infection due to COVID-19 based on DL techniques, including an overview of COVID-19 diagnosis and public datasets. This study aimed for the future to resolve the complicated association of lung microbiome, immunity, and the COVID-19 virus and their difficulties. There is still no medication available for COVID-19 treatment, but research on chloroquine and remdesivir medications is being carried out in the United States and China. However, discussion of COVID-19 vaccine evolution is beyond the scope of the present study.

In conclusion, coronavirus affected humanity globally and left the world to face many future challenges. Many researchers studied and analyzed future challenges and possible solutions for COVID-19; there were GDP, economic, and unemployment crises due to lockdown globally. The growth of air contamination and the control of medical waste of infected patients require policies and strategies to be implemented rapidly. Medical nurses need protection while assisting patients; there were new challenges during lockdown for communication between doctors and orthopedic patients, and for the monitoring of arthroscopy, surgery or fractures.

REFERENCES

Abbas, A., Saba, T., Rehman, A., Mehmood, Z., Javaid, N., Tahir, M., Khan, N.U., Ahmed, K.T., & Shah, R. (2019b) Plasmodium species aware based quantification of malaria, parasitemia in light microscopy thin blood smear, *Microscopy Research and Technique*, 82(7), 1198–1214. doi:10.1002/jemt.23269.

Abbas, N., Saba, T., Mehmood, Z., Rehman, A., Islam, N., & Ahmed, K. T. (2019a). An automated nuclei segmentation of leukocytes from microscopic digital images. *Pakistan Journal of Pharmaceutical Science*, 32(5), 2123–2138.

Abbas, N. Saba, T. Mohamad, D. Rehman, A. Almazyad, A.S., & Al-Ghamdi, J.S. (2018a). Machine aided malaria parasitemia detection in Giemsa-stained thin blood smears, *Neural Computing and Applications*, 29(3), 803–818, doi:10.1007/s00521-016-2474-6.

Abbas, N., Saba, T., Rehman, A., Mehmood, Z., Kolivand, H., Uddin, M. & Anjum, A. (2018b). Plasmodium life cycle stage classification-based quantification of malaria parasitaemia in thin blood smears. *Microscopy Research and Technique.* doi:10.1002/jemt.23170

Adeel, A., Khan, M. A., Akram, T., Sharif, A., Yasmin, M., Saba, T., & Javed, K. (2020). Entropy-controlled deep features selection framework for grape leaf diseases recognition. *Expert Systems.* doi:10.1111/exsy.12569.

Afza, F., Khan, M. A., Sharif, M., & Rehman, A. (2019). Microscopic skin laceration segmentation and classification: A framework of statistical normal distribution and optimal feature selection. *Microscopy Research and Technique*, 82(9), 1471–1488.

Afza, F., Khan, M. A., Sharif, M., Saba, T., Rehman, A., & Javed, M. Y. (2020). *Skin Lesion Classification: An Optimized Framework of Optimal Color Features Selection.* In *2020 2nd International Conference on Computer and Information Sciences (ICCIS)* (pp. 1–6). IEEE.

Ahlawat, S., & Sharma, K. K. (2020). Immunological co-ordination between gut and lungs in SARS-CoV-2 infection. *Virus Research*, 286, 198103.

Al-Ameen, Z. Sulong, G. Rehman, A., Al-Dhelaan, A. Saba, T., & Al-Rodhaan, M. (2015) An innovative technique for contrast enhancement of computed tomography images using normalized gamma-corrected contrast-limited adaptive histogram equalization. *EURASIP Journal on Advances in Signal Processing*, 32. doi:10.1186/s13634-015-0214-1.

Amin, J., Sharif, M., Rehman, A., Raza, M., & Mufti, M. R. (2018). Diabetic retinopathy detection and classification using hybrid feature set. *Microscopy Research and Technique*, 81(9), 990–996.

Amin, J., Sharif, M., Raza, M., Saba, T., & Anjum, M. A. (2019a). Brain tumor detection using statistical and machine learning method. *Computer Methods and Programs in Biomedicine*, 177, 69–79.

Amin, J., Sharif, M., Raza, M., Saba, T., & Rehman, A. (2019b). *Brain Tumor Classification: Feature Fusion.* In *2019 International Conference on Computer and Information Sciences (ICCIS)* (pp. 1–6). IEEE.

Amin, J., Sharif, M., Raza, M., Saba, T., Sial, R., & Shad, S. A. (2019c). Brain tumor detection: A long short-term memory (LSTM)-based learning model. *Neural Computing and Applications*, 1–9.

Amin, J., Sharif, M., Rehman, A., Raza, M., & Mufti, M. R. (2019d). Diabetic retinopathy detection and classification using hybrid feature set. *Microscopy Research and Technique*, 81(9), 990–996.

Amin, J., Sharif, M., Yasmin, M. T Saba, & Raza, M. (2019e) "Use of machine intelligence to conduct analysis of human brain data for detection of abnormalities in its cognitive functions". *Multimed Tools Applications*, 79(15), 10955–10973. doi:10.1007/s11042-019-7324-y

Bao, L., Zhang, C., Dong, J., Zhao, L., Li, Y., & Sun, J. (2020). Oral microbiome and SARS-CoV-2: Beware of lung co-infection. *Frontiers in Microbiology*, 11, 1840.

Born, J., Brändle, G., Cossio, M., Disdier, M., Goulet, J., Roulin, J., & Wiedemann, N. 2020. POCOVID-Net: Automatic detection of COVID-19 from a new lung ultrasound imaging dataset (POCUS). arXiv preprint arXiv:2004.12084.

Buonsenso, D., Raffaelli, F., Tamburrini, E., Biasucci, D. G., Salvi, S., Smargiassi, A..... & Moro, F. (2020). Clinical role of lung ultrasound for the diagnosis and monitoring of COVID-19 pneumonia in pregnant women. *Ultrasound in Obstetrics & Gynecology*, 56(1), 106–109.

Chung, M., Bernheim, A., Mei, X., Zhang, N., Huang, M., Zeng, X.,.... & Jacobi, A. (2020). CT imaging features of 2019 novel coronavirus (2019-nCoV). *Radiology*, 295(1), 202–207.

Convissar, D., Gibson, L. E., Berra, L., Bittner, E. A., & Chang, M. G. (2020). Application of lung ultrasound during the coronavirus disease 2019 pandemic: A narrative review. *Anesthesia and Analgesia*, 131(2), 1.

Cruces, P., Retamal, J., Hurtado, D. E., Erranz, B., Iturrieta, P., González, C., & Díaz, F. (2020). A physiological approach to understand the role of respiratory effort in the progression of lung injury in SARS-CoV-2 infection. *Critical Care*, 24(1), 1–10.

Cui, L., & Lee, D. 2020. CoAID: COVID-19 healthcare misinformation dataset. arXiv preprint arXiv:2006.00885.

Ebrahimi, M., Saki Malehi, A., & Rahim, F. 2020. COVID-19 infection in medical staffs versus patients: A systematic review and meta-analysis of laboratory findings, comorbidities, and clinical outcome. Fakher, COVID-19 infection in medical staffs versus patients: A systematic review and meta-analysis of laboratory findings, Comorbidities, and clinical outcome (April 20, 2020).

Ejaz, K., Rahim, D. M. S. M., Rehman, D. A., & Ejaz, E. F. (2018b). An image-based multimedia database and efficient detection though features. *VFAST Transactions on Software Engineering*, 14(1), 6–15.

Ejaz, K., Rahim, M. S. M., Bajwa, U. I., Chaudhry, H., Rehman, A., & Ejaz, F. (2021) *Hybrid Segmentation Method with Confidence Region Detection For Tumor Identification*, IEEE Access. doi:10.1109/ACCESS.2020.3016627

Ejaz, K., Rahim, M. S. M., Bajwa, U. I., Rana, N., & Rehman, A. (2019). *An unsupervised learning with feature approach for brain tumor segmentation using magnetic resonance imaging*. In *Proceedings of the 2019 9th International Conference on Bioscience, Biochemistry and Bioinformatics*, New York, (pp. 1–7).

Ejaz, K., Rahim, M. S. M., Rehman, A., Chaudhry, H., Saba, T., & Ejaz, A. (2018a). Segmentation method for pathological brain tumor and accurate detection using MRI. *International Journal of Advanced Computer Science and Applications*, 9(8), 394–401.

Fahad, H. M., Khan, M.U.G., Saba, T, Rehman, A, & Iqbal, S. (2018) Microscopic abnormality classification of cardiac murmurs using ANFIS and HMM. *Microscopy Research and Technique*, 81(5), 449–457. doi:10.1002/jemt.22998.

Fu, L., Wang, B., Yuan, T., Chen, X., Ao, Y., Fitzpatrick, T., … & Luo, G. (2020). Clinical characteristics of coronavirus disease 2019 (COVID-19) in China: A systematic review and meta-analysis. *Journal of Infection*, 80(6), 656–665.

Gao, Z., Yada, S., Wakamiya, S., & Aramaki, E., 2020. Naist covid: Multilingual COVID-19 twitter and weibo dataset. arXiv preprint arXiv:2004.08145.

He, K., Zhang, X., Ren, S., & Sun, J., 2020. *Deep residual learning for image recognition*. In *Proceedings of the IEEE Conference on Computer Vision and Pattern Recognition* (pp. 770–778).

Huang, G., Liu, Z., Van Der Maaten, L., & Weinberger, K.Q., 2017. *Densely connected convolutional networks*. In *Proceedings of the IEEE Conference on Computer Vision and Pattern Recognition* (pp. 4700–4708).

Husham, A., Alkawaz, M. H., Saba, T., Rehman, A., & Alghamdi, J.S. (2016) Automated nuclei segmentation of malignant using level sets, *Microscopy Research and Technique*, 79(10), 993–997, doi. 10.1002/jemt.22733.

Hussain, N., Khan, M. A., Sharif, M., Khan, S. A., Albesher, A. A., Saba, T., & Armaghan, A. (2020). A deep neural network and classical features-based scheme for objects recognition: An application for machine inspection. *Multimedia Tools and Applications*. doi:10.1007/s11042-020-08852-3.

Iftikhar, S. Fatima, K. Rehman, A. Almazyad, A.S., & Saba, T. (2017). An evolution based hybrid approach for heart diseases classification and associated risk factors identification. *Biomedical Research* 28 (8), 3451–3455

Iqbal, S. Ghani, M.U. Saba, T., & Rehman, A. (2018). Brain tumor segmentation in multispectral MRI using convolutional neural networks (CNN). *Microscopy Research and Technique*, 81(4), 419–427. doi:10.1002/jemt.22994.

Iqbal, S., Khan, M. U. G., Saba, T. Mehmood, Z. Javaid, N., Rehman, A., & Abbasi, R. (2019) Deep learning model integrating features and novel classifiers fusion for brain tumor segmentation. *Microscopy Research and Technique*, 82(8), 1302–1315. doi:10.1002/jemt.23281

Iqbal, S., Khan, M. U. G., Saba, T., & Rehman, A. (2017). Computer assisted brain tumor type discrimination using magnetic resonance imaging features. *Biomedical Engineering Letters*, 8(1), 5–28. doi:10.1007/s13534-017-0050-3.

Jacobi, A., Chung, M., Bernheim, A., & Eber, C. (2020). Portable chest X-ray in coronavirus disease-19 (COVID-19): A pictorial review. *Clinical Imaging*, 64, 35–42.

Jamal, A., Alkawaz, M. H., Rehman, A., Saba, T. (2017) Retinal imaging analysis based on vessel detection. *Microscopy Research and Technique*, 80(17),799–811. doi:10.1002/jemt.

Javed, R., Rahim, M. S. M., & Saba, T. (2019a) An improved framework by mapping salient features for skin lesion detection and classification using the optimized hybrid features, *International Journal of Advanced Trends in Computer Science and Engineering*, 8(1), 95–101.

Javed, R., Rahim, M. S. M., Saba, T., & Rashid, M. (2019b). Region-based active contour JSEG fusion technique for skin lesion segmentation from dermoscopic images. *Biomedical Research*, 30(6), 1–10.

Javed, R., Rahim, M. S. M., Saba, T., & Rehman, A. (2020a) A comparative study of features selection for skin lesion detection from dermoscopic images. *Network Modeling Analysis in Health Informatics and Bioinformatics*, 9 (1), 4.

Javed, R., Saba, T., Shafry, M., & Rahim, M. (2020b). *An intelligent saliency segmentation technique and classification of low contrast skin lesion dermoscopic images based on histogram decision*. In *2019 12th International Conference on Developments in eSystems Engineering (DeSE)*, New York, (pp. 164–169).

Khan, M. A., Akram, T., Sharif, M., Javed, K., Raza, M., & Saba, T. (2020a). An automated system for cucumber leaf diseased spot detection and classification using improved saliency method and deep features selection. *Multimedia Tools and Applications*, 1–30.

Khan, M. A., Akram, T. Sharif, M., Saba, T., Javed, K., Lali, I. U., Tanik, U.J., & Rehman, A. (2019d). Construction of saliency map and hybrid set of features for efficient segmentation and classification of skin lesion, *Microscopy Research and Technique*, 82(5), 741–763, doi:10.1002/jemt.23220

Khan, M. A., Ashraf, I., Alhaisoni, M., Damaševičius, R., Scherer, R., Rehman, A., & Bukhari, S. A. C. (2020d). Multimodal brain tumor classification using deep learning and Robust feature selection: A machine learning application for radiologists. *Diagnostics*, 10, 565.

Khan, M. A., Javed, M. Y., Sharif, M., Saba, T., & Rehman, A. (2019a). *Multi-model deep neural network based features extraction and optimal selection approach for skin lesion classification*. In *2019 International Conference on Computer and Information Sciences (ICCIS)* (pp. 1–7). IEEE.

Khan, M. A. Kadry, S., Zhang, Y. D., Akram, T., Sharif, M., & Saba, T. (2021) Prediction of COVID-19 - Pneumonia based on selected deep features and one class Kernel extreme learning machine. *Computers & Electrical Engineering*, 90, 106960.

Khan, M. A., Lali, I. U. Rehman, A. Ishaq, M. Sharif, M. Saba, T., Zahoor, S., & Akram, T. (2019c) Brain tumor detection and classification: A framework of marker-based watershed algorithm and multilevel priority features selection, *Microscopy Research and Technique*, 82(6), 909–922. doi:10.1002/jemt.23238

Khan, M. A., Sharif, M. Akram, T., Raza, M., Saba, T., & Rehman, A. (2020b). Hand-crafted and deep convolutional neural network features fusion and selection strategy: An application to intelligent human action recognition. *Applied Soft Computing*, 87, 105986

Khan, M. A., Sharif, M. I., Raza, M., Anjum, A., Saba, T., & Shad, S. A. (2019b). Skin lesion segmentation and classification: A unified framework of deep neural network features fusion and selection. *Expert Systems*, e12497.

Khan, M. W., Sharif, M., Yasmin, M., & Saba, T. (2017). CDR based glaucoma detection using fundus images: A review. *International Journal of Applied Pattern Recognition*, 4(3), 261–306.

Khan, M. Z., Jabeen, S., Khan, M. U. G., Saba, T., Rehmat, A., Rehman, A., & Tariq, U. (2020c). A realistic image generation of face from text description using the fully trained generative adversarial networks. *IEEE Access*. doi:10.1109/ACCESS.2020.3015656

Khan, S. A., Nazir, M., Khan, M. A., Saba, T., Javed, K., Rehman, A., ... & Awais, M. (2019e). Lungs nodule detection framework from computed tomography images using support vector machine. *Microscopy Research and Technique*, 82(8), 1256–1266.

Liaqat, A., Khan, M. A., Sharif, M., Mittal, M., Saba, T., Manic, K. S., & Al Attar, F. N. H. (2020). Gastric tract infections detection and classification from wireless capsule endoscopy using computer vision techniques: A review. *Current Medical Imaging*, 16(10), 1229–1242.

Lomoro, P., Verde, F., Zerboni, F., Simonetti, I., Borghi, C., Fachinetti, C., ... & Martegani, A. (2020). COVID-19 pneumonia manifestations at the admission on chest ultrasound, radiographs, and CT: Single-center study and comprehensive radiologic literature review. *European Journal of Radiology Open*, 7, 100231.

Lopez, C.E., Vasu, M., & Gallemore, C., 2020. Understanding the perception of COVID-19 policies by mining a multilanguage Twitter dataset. arXiv preprint arXiv:2003.10359.

Ludvigsson, J. F. (2020). Systematic review of COVID-19 in children shows milder cases and a better prognosis than adults. *Acta Paediatrica*, 109(6), 1088–1095.

Lung, J. W. J., Salam, M. S. H., Rehman, A., Rahim, M. S. M., & Saba, T. (2014) Fuzzy phoneme classification using multi-speaker vocal tract length normalization, *IETE Technical Review*, 31 (2), 128–136. doi:10.1080/02564602.2014.892669.

Majid, A., Khan, M. A., Yasmin, M., Rehman, A., Yousafzai, A., & Tariq, U. (2020). Classification of stomach infections: A paradigm of convolutional neural network along with classical features fusion and selection. *Microscopy Research and Technique*, 83(5), 562–576.

Marie-Sainte, S. L. Aburahmah, L., Almohaini, R., & Saba, T. (2019a). Current techniques for diabetes prediction: Review and case study. *Applied Sciences*, 9(21), 4604.

Marie-Sainte, S. L., Saba, T., Alsaleh, D., Alotaibi, A., & Bin, M. (2019b). An improved strategy for predicting diagnosis, survivability, and recurrence of breast cancer. *Journal of Computational and Theoretical Nanoscience*, 16(9), 3705–3711.

Mashood Nasir, I., Attique Khan, M., Alhaisoni, M., Saba, T., Rehman, A., & Iqbal, T. (2020). A hybrid deep learning architecture for the classification of superhero fashion products: An application for medical-tech classification. *Computer Modeling in Engineering & Sciences*, 124(3), 1017–1033.

Mayor-Ibarguren, A., & Robles-Marhuenda, Á. (2020). A hypothesis for the possible role of zinc in the immunological pathways related to COVID-19 infection. *Frontiers in Immunology*, 11, 1736.

Michel, S., Witt, C., Gottlieb, J., & Aigner, C. (2020). Impact of COVID-19 on lung transplant activity in Germany—A cross-sectional survey. *The Thoracic and Cardiovascular Surgeon*, 69(01), 092–094.

Minucci, S. B., Heise, R. L., & Reynolds, A. M. (2020). Review of mathematical modeling of the inflammatory response in lung infections and injuries. *Frontiers in Applied Mathematics and Statistics*, 6, 36.

Mittal, A., Kumar, D., Mittal, M., Saba, T., Abunadi, I., Rehman, A., & Roy, S. (2020). Detecting pneumonia using convolutions and dynamic capsule routing for chest X-ray images. *Sensors*, 20(4), 1068.

Mughal, B., Muhammad, N., Sharif, M., Rehman, A., & Saba, T. (2018a). Removal of pectoral muscle based on topographic map and shape-shifting silhouette. *BMC cancer*, 18(1), 1–14.

Mughal, B. Muhammad, N. Sharif, M. Saba, T. Rehman, A. (2017) Extraction of breast border and removal of pectoral muscle in wavelet domain, *Biomedical Research*, 28 (11), 5041–5043.

Mughal, B., Sharif, M., Muhammad, N., & Saba, T. (2018b). A novel classification scheme to decline the mortality rate among women due to breast tumor. *Microscopy Research and Technique*, 81(2), 171–180.

Müller, D., Rey, I. S., & Kramer, F. (2020). Automated chest CT image segmentation of COVID-19 lung infection based on 3D U-Net. arXiv preprint arXiv:2007.04774.

Musolino, A. M., Supino, M. C., Buonsenso, D., Ferro, V., Valentini, P., Magistrelli, A., & Campana, A. (2020). Lung ultrasound in children with COVID-19: Preliminary findings. *Ultrasound in Medicine & Biology*, 46(8), 2094–2098.

Nazir, M., Khan, M. A., Saba, T., & Rehman, A. (2019). *Brain Tumor Detection from MRI images using Multi-level Wavelets*. In *2019, IEEE International Conference on Computer and Information Sciences (ICCIS)* (pp. 1–5).

Nieman, G. F., Gatto, L. A., Andrews, P., Satalin, J., Camporota, L., Daxon, B., … & Aiash, H. (2020). Prevention and treatment of acute lung injury with time-controlled adaptive ventilation: Physiologically informed modification of airway pressure release ventilation. *Annals of Intensive Care*, 10(1), 1–16.

Ojo, A. S., Balogun, S. A., Williams, O. T., & Ojo, O. S. (2020). Pulmonary fibrosis in COVID-19 survivors: Predictive factors and risk reduction strategies. *Pulmonary Medicine*, 2020, 6175964. doi:10.1155/2020/6175964.

Park, J. Y., Lee, Y. J., Kim, T., Lee, C. Y., Kim, H. I., Kim, J. H., & Jang, S. H. (2020). Collateral effects of the coronavirus disease 2019 pandemic on lung cancer diagnosis in Korea. *BMC Cancer*, 20(1), 1–8.

Perveen, S., Shahbaz, M., Saba, T., Keshavjee, K., Rehman, A., & Guergachi, A. (2020). Handling irregularly sampled longitudinal data and prognostic modeling of diabetes using machine learning technique. *IEEE Access*, 8, 21875–21885.

Polidoro, R. B., Hagan, R. S., de Santis Santiago, R., & Schmidt, N. W. (2020). Overview: Systemic inflammatory response derived from lung injury caused by SARS-CoV-2 infection explains severe outcomes in COVID-19. *Frontiers in Immunology*, 11, 1626.

Qiu, H., Wu, J., Hong, L., Luo, Y., Song, Q., & Chen, D. 2020. Clinical and epidemiological features of 36 children with coronavirus disease 2019 (COVID-19) in Zhejiang, China: An observational cohort study. *The Lancet Infectious Diseases*, 20(6), 689–696.

Qureshi, I., Khan, M. A., Sharif, M., Saba, T., & Ma, J. (2020) Detection of glaucoma based on cup-to-disc ratio using fundus images. *International Journal of Intelligent Systems Technologies and Applications*, 19(1), 1–16. doi:10.1504/IJISTA.2020.105172

Rad, A. E., Rahim, M. S. M., Rehman, A. Altameem, A., & Saba, T. (2013) Evaluation of current dental radiographs segmentation approaches in computer-aided applications *IETE Technical Review*, 30(3), 210–222

Rad, A. E., Rahim, M. S. M., Rehman, A., & Saba, T. (2016) Digital dental X-ray database for caries screening. *3D Research*, 7(2), 1–5 doi:10.1007/s13319-016-0096-5

Ragab, D. A., Sharkas, M., Marshall, S., & Ren, J. 2019. Breast cancer detection using deep convolutional neural networks and support vector machines. *Peer Journal*, 7, e 6201.

Rahim, M. S. M., Norouzi, A. Rehman, A., & Saba, T. (2017a) 3D bones segmentation based on CT images visualization, *Biomedical Research*, 28(8), 3641–3644

Rahim, M. S. M., Razzali, N., Sunar, M. S., Abdullah, A. A., & Rehman, A. (2012a) Curve interpolation model for visualizing disjointed neural elements, *Neural Regeneration Research*, 7(21), 1637–1644. doi:10.3969/j.issn.1673-5374.2012.21.006

Rahim, M. S. M., Rehman, A. Kurniawan, F., & Saba, T. (2017b) Ear biometrics for human classification based on region features mining, *Biomedical Research*, 28 (10), 4660–4664

Ramzan, F., Khan, M. U. G., Iqbal, S., Saba, T., & Rehman, A. (2020b). Volumetric segmentation of brain regions from MRI scans using 3D convolutional neural networks. *IEEE Access*, 8, 103697–103709.

Ramzan, F., Khan, M. U. G., Rehmat, A., Iqbal, S., Saba, T., Rehman, A., & Mehmood, Z. (2020a). A deep learning approach for automated diagnosis and multi-class classification of Alzheimer's disease stages using resting-state fMRI and residual neural networks. *Journal of Medical Systems*, 44(2), 37.

Rehman, A. (2020). *Ulcer Recognition based on 6-Layers Deep Convolutional Neural Network.* In *Proceedings of the 2020 9th International Conference on Software and Information Engineering (ICSIE)* (pp. 97–101). Cairo Egypt.

Rehman, A. (2021). Light microscopic iris classification using ensemble multi-class support vector machine. *Microscopic Research & Technique.* doi:10.1002/jemt.23659

Rehman, A., Abbas N., Saba, T., Mahmood, T., & Kolivand, H. (2018c). Rouleaux red blood cells splitting in microscopic thin blood smear images via local maxima, circles drawing, and mapping with original RBCs. *Microscopic Research and Technique*, 81(7), 737–744. doi:10.1002/jemt.23030

Rehman, A., Abbas, N., Saba, T. Mehmood, Z. Mahmood, T., & Ahmed, K.T. (2018b) Microscopic malaria parasitemia diagnosis and grading on benchmark datasets, *Microscopic Research and Technique*, 81(9), 1042–1058. doi:10.1002/jemt.23071

Rehman, A., Abbas, N., Saba, T., Rahman, S. I. U., Mehmood, Z., & Kolivand, K. (2018a). Classification of acute lymphoblastic leukemia using deep learning. *Microscopy Research & Technique*, 81(11), 1310–1317. doi:10.1002/jemt.23139

Rehman, A., Khan, M. A. Mehmood, Z. Saba, T. Sardaraz, M., & Rashid, M. (2020). Microscopic melanoma detection and classification: A framework of pixel-based fusion and multilevel features reduction. *Microscopy Research and Technique*, 83(4), 410–423. doi:10.1002/jemt.23429

Rehman, A., Khan, M. A., Saba, T., Mehmood, Z., Tariq, U., & Ayesha, N. (2021a). Microscopic brain tumor detection and classification using 3D CNN and feature selection architecture. *Microscopy Research and Technique*, 84(1), 133–149. doi:10.1002/jemt.23597.

Rehman, A., Saba, T., Ayesha, N., & Tariq, U (2021c). Deep learning-based COVID-19 Detection using CT and X-ray images: Current analytics and comparisons. *IEEE IT Professional.* 23(3), 63–68. doi:10.1109/MITP.2020.3036820

Rehman, A., Sadad, T. Saba, T., Hussain, A., & Tariq, U. (2021b). Real-time diagnosis system of COVID-19 using X-Ray images and deep learning. *IEEE IT Professional.* doi:10.1109/MITP.2020.3042379

Rodriguez-Morales, A. J., Cardona-Ospina, J. A., Gutiérrez-Ocampo, E., Villamizar-Peña, R., Holguin-Rivera, Y., Escalera-Antezana, J. P., ... & Paniz-Mondolfi, A. (2020). Clinical, laboratory and imaging features of COVID-19: A systematic review and meta-analysis. *Travel Medicine and Infectious Disease*, 34, 101623.

Saba, T. (2019). Automated lung nodule detection and classification based on multiple classifiers voting. *Microscopy Research and Technique*, 82(9), 1601–1609.

Saba, T. (2020). Recent advancement in cancer detection using machine learning: Systematic survey of decades, comparisons and challenges. *Journal of Infection and Public Health*, 13(9), 1274–1289.

Saba, T. (2021) Computer vision for microscopic skin cancer diagnosis using handcrafted and non-handcrafted features. *Microscopy Research and Technique*. doi:10.1002/jemt.23686,2021

Saba, T., Bokhari, S. T. F., Sharif, M., Yasmin, M., & Raza, M. (2018b). Fundus image classification methods for the detection of glaucoma: A review. *Microscopy Research and Technique*, 81(10), 1105–1121.

Saba, T., Haseeb, K., Ahmed, I., & Rehman, A. (2020b). Secure and energy-efficient framework using Internet of Medical Things for e-healthcare. *Journal of Infection and Public Health*, 13(10), 1567–1575.

Saba, T., Khan, M. A., Rehman, A., & Marie-Sainte, S. L. (2019a). Region extraction and classification of skin cancer: A heterogeneous framework of deep CNN features fusion and reduction. *Journal of Medical Systems*, 43(9), 289.

Saba, T., Khan, S. U., Islam, N., Abbas, N., Rehman, A., Javaid, N., & Anjum, A., (2019b). Cloud-based decision support system for the detection and classification of malignant cells in breast cancer using breast cytology images. *Microscopy Research and Technique*, 82(6), 775–785.

Saba, T., Mohamed, A. S., El-Affendi, M. Amin, J., & Sharif, M. (2020a). Brain tumor detection using fusion of hand crafted and deep learning features. *Cognitive Systems Research*, 59, 221–230

Saba, T., Rehman, A., & AlGhamdi, J. S. (2017). Weather forecasting based on hybrid neural model. *Applied Water Science*, 7(7), 3869–3874.

Saba, T., Rehman, A. Mehmood, Z., Kolivand, H., & Sharif, M. (2018a). Image enhancement and segmentation techniques for detection of knee joint diseases: A survey. *Current Medical Imaging Reviews*, 14(5), 704–715. doi:10.2174/1573405613666170912164546

Saba, T., Sameh, A., Khan, F., Shad, S. A., & Sharif, M. (2019c). Lung nodule detection based on ensemble of hand crafted and deep features. *Journal of Medical Systems*, 43(12), 332.

Sadad, T., Khan, A. R., Hussain, A., Tariq, U., Fati, S. M., Bahaj, S. A., & Munir, A. (2021b). Internet of medical things embedding deep learning with data augmentation for mammogram density classification. *Microscopy Research and Technique*, doi. 10.1002/jemt.23773

Sadad, T., Munir, A., Saba, T., & Hussain, A. (2018) Fuzzy C-means and region growing based classification of tumor from mammograms using hybrid texture feature. *Journal of Computational Science*, 29, 34–45

Sadad, T., Rehman, A., Hussain, A., Abbasi, A. A., & Khan, M. Q. (2021c). A review on multi-organs cancer detection using advanced machine learning techniques. *Current Medical Imaging*, 17, 1–00.

Sadad, T., Rehman, A., Munir, A., Saba, T., Tariq, U., Ayesha, N., & Abbasi, R. (2021a). Brain tumor detection and multi-classification using advanced deep learning techniques. *Microscopy Research and Technique*. 84(6), 1296–1308. doi:10.1002/jemt.23688.

Shan, F., Gao, Y., Wang, J., Shi, W., Shi, N., Han, M., ... & Shi, Y. (2020). Lung infection quantification of COVID-19 in CT images with deep learning. arXiv preprint arXiv:2003.04655.

Shi, F., Wang, J., Shi, J., Wu, Z., Wang, Q., Tang, Z., … & Shen, D. (2020). Review of artificial intelligence techniques in imaging data acquisition, segmentation and diagnosis for COVID-19. *IEEE Reviews in Biomedical Engineering.* doi:10.1109/RBME.2020. 2987975.

Shuja, J., Alanazi, E., Alasmary, W., & Alashaikh, A., 2020. COVID-19 datasets: A survey and future challenges. medRxiv.

Singhal, T. (2020). A review of coronavirus disease-2019 (COVID-19). *The Indian Journal of Pediatrics*, 1–6.

Siordia Jr, J. A. (2020). Epidemiology and clinical features of COVID-19: A review of current literature. *Journal of Clinical Virology*, 127, 104357.

Tian, S., Hu, W., Niu, L., Liu, H., Xu, H., & Xiao, S. Y. (2020). Pulmonary pathology of early phase 2019 novel coronavirus (COVID-19) pneumonia in two patients with lung cancer. *Journal of Thoracic Oncology*, 15(5), 700–704.

Tsai, N. W., Ngai, C. W., Mok, K. L., & Tsung, J. W. (2014). Lung ultrasound imaging in avian influenza A (H7N9) respiratory failure. *Critical Ultrasound Journal*, 6(1), 6.

Ullah, H., Saba, T., Islam, N., Abbas, N., Rehman, A., Mehmood, Z., & Anjum, A. (2019). An ensemble classification of exudates in color fundus images using an evolutionary algorithm based optimal features selection. *Microscopy Research and Technique*, 82(4), 361–372.

Umakanthan, S., Sahu, P., Ranade, A. V., Bukelo, M. M., Rao, J. S., Abrahao-Machado, L. F., … & Dhananjaya, K. V. (2020). Origin, transmission, diagnosis and management of coronavirus disease 2019 (COVID-19). *Postgraduate Medical Journal*, 96(1142), 753–758.

Wan, Y., Shang, J., Graham, R., Baric, R. S., & Li, F. (2020). Receptor recognition by the novel coronavirus from Wuhan: An analysis based on decade-long structural studies of SARS coronavirus. *Journal of Virology*, 94(7), e00127-20. doi:10.1128/JVI.00127-20.

Wang, D., Hu, B., Hu, C., Zhu, F., Liu, X., Zhang, J., … & Zhao, Y. (2020a). Clinical characteristics of 138 hospitalized patients with 2019 novel coronavirus–infected pneumonia in Wuhan, China. *JAMA*, 323(11), 1061–1069.

Wang, H., Wei, R., Rao, G., Zhu, J., & Song, B. (2020c). Characteristic CT findings distinguishing 2019 novel coronavirus disease (COVID-19) from influenza pneumonia. *European Radiology*, 30(9), 4910–4917.

Wang, L. S., Wang, Y. R., Ye, D. W., & Liu, Q. Q. (2020b). A review of the 2019 Novel Coronavirus (COVID-19) based on current evidence. *International Journal of Antimicrobial Agents*, 55(6), 105948.

Wei, J., Huang, C., Vosoughi, S., & Wei, J. 2020. What are people asking about COVID-19? A question classification dataset. arXiv preprint arXiv:2005.12522.

Wong, H. Y. F., Lam, H. Y. S., Fong, A. H. T., Leung, S. T., Chin, T. W. Y., Lo, C. S. Y., Lui, M. M. S., Lee, J. C. Y., Chiu, K. W. H., Chung, T., & Lee, E. Y. P., 2020. Frequency and distribution of chest radiographic findings in COVID-19 positive patients. *Radiology*, 201160.

Yousaf, K., Mehmood, Z., Awan, I. A., Saba, T., Alharbey, R., Qadah, T., & Alrige, M. A. (2019a). A comprehensive study of mobile-health based assistive technology for the healthcare of dementia and Alzheimer's disease (AD). *Health Care Management Science*, 1–23.

Yousaf, K. Mehmood, Z. Saba, T. Rehman, A. Munshi, A. M. Alharbey, R., & Rashid, M. (2019b). Mobile-health applications for the efficient delivery of health care facility to people with dementia (PwD) and support to their carers: A survey. *BioMed Research International*, 2019, 1–26

Youssef, A., Cavalera, M., Azzarone, C., Serra, C., Brunelli, E., Casadio, P., & Pilu, G. (2020). The use of lung ultrasound during the COVID-19 pandemic: A narrative review with specific focus on its role in pregnancy. *Journal of Population Therapeutics and Clinical Pharmacology*, 27(SP1), e64–e75.

Zarei, K., Farahbakhsh, R., Crespi, N., & Tyson, G., 2020. A first instagram dataset on COVID-19. arXiv preprint arXiv:2004.12226.

Zhou, C., (2020). Evaluating new evidence in the early dynamics of the novel coronavirus COVID-19 outbreak in Wuhan, China with real time domestic traffic and potential asymptomatic transmissions. medRxiv.

Zhou, S., Wang, Y., Zhu, T., & Xia, L. (2020). CT features of coronavirus disease 2019 (COVID-19) pneumonia in 62 patients in Wuhan, China. *American Journal of Roentgenology*, 214(6), 1287–1294.

2 Artificial Intelligence in Coronavirus Detection
Recent Findings and Future Perspectives

Syed Ale Hassan, Sahar Gull, Shahzad Akbar, Israr Hanif, Sajid Iqbal, and Muhammad Waqas Aziz

CONTENTS

2.1 INTRODUCTION

World history has faced several disease outbreaks that have threatened the survival of humanity. The World Health Organization (WHO), national authorities, and various other clinicians worldwide have fought pandemics. The COVID-19 pandemic has posed such a challenge for humanity since the first coronavirus case was detected on December 8, 2019, in Wuhan, China, and is continuing to spread worldwide (Khan et al., 2021; Khan et al., 2019a, b). On January 30, 2020, WHO declared a coronavirus pandemic and announced that the outbreak constituted a public health emergency of international concern (WHO, 2020a). The new virus (SARS-CoV-2) has now spread worldwide, infecting 84.8 million individuals and causing 1.84 million deaths as of January 3, 2021 (WHO, 2021b). Besides, its second wave has now been active, and clinicians and scientists claim that this attack will be more lethal than the first wave (WHO, 2020c).

In an early outbreak of COVID-19, cases were doubled in a week and were transported to other parts of China during celebrations of Chinese New Year. Later, on January 20, 2020, nearly 140 cases were reported in a single day and it was reported that around 6,174 people had developed symptoms (diagnosed later). The WHO declared its indication in human transmission and warned the world of its pandemic potential (WHO, 2021a). On January 30, 2020, almost 7,818 cases were confirmed in about 19 countries, thus stated as a public health emergency to international alarm and as a pandemic on March 11, 2020, as the number of cases worldwide, leading Japan, South Korea, Iran, and Italy, overtook China. Italy's first confirmed case was reported on March 13, 2020, Italy overtook China with the majority of deaths, and WHO declared Europe the active center (Italian COVID-19 news, 2020). By March 26, 2020, the United States had overtaken Italy and China with the highest number of verified cases. As the number of cases was declining, many countries and scientists considered this was due to changes in the weather conditions.

Moreover, on June 29, 2020, WHO notified that the deadly virus was still increasing its speed with the reopening of countries' economies. Later, a WHO leaders' meeting was called in October 2020 declared that infection had been spread as one of ten people in the world as translated as 780 million people been infected, while 37 million have been confirmed (WHO, 2020b). Besides, in early November, an outbreak of a unique variant of the deadly virus has been transmitted and is assumed to have a 60% faster transmission rate. However, its state of toxicity is yet to be confirmed. On December 14, 2020, Public Health England reported a novel variant of COVID-19 and renamed as VCO-202012/01 and has depicted spike protein changes that could be more infectious (Novel Variant of COVID-19, 2020). Worldwide the reported number of patients has been approximately 80.7 million since December 28, 2020, and more than 1.76 million have died, while 45.6 million have recovered. These confirm the laboratory tests; however, the reported cases would be eight times the reported cases per the disease prevention and control axis cases.

2.1.1 Vaccination

It has been reported that candidate vaccines are under trial; more or less, five names are under consideration around the world—namely, Moderna (mRNA-1273), Pfizer/BioNTech/Fosun Pharma, Merck/Sharpe & Dohme, Johnson &Johnson/Janssen, AstraZeneca (O'Callaghan et al., 2020). Studies are under consideration in terms of their efficacy and safety results. These candidates have not been authorized by WHO EUL/PQ, though Pfizer is expected to get an assessment certification by January 2021. The Pfizer candidate vaccine trial results were released on November 9, 2020, and the company proclaimed its effectiveness of around 90% (COVID-19 Vaccine, 2020). Later, Novavax also entered a Food and Drug Administration (FDA) application for the vaccine. WHO coordinates with its partners for the safe and efficacious use of the developed vaccine and formulates critical steps to facilitate equitable access to the billions of people worldwide (WHO Vaccine Coordination, 2020).

The WHO researchers, practitioners, and scientists about medical industries are searching for innovative techniques and methods to trace, monitor, and screen infected patients at various levels to contain the virus. The techniques may also be capable of finding the best clinical trials, tracing contacts of infections, controlling its spread, and developing an effective treatment for curing patients. Early and fast detection, diagnosis, and screening are critically significant to prevent the spread of the pandemic, and develop a cost-effective treatment to save human lives. Artificial intelligence (AI) is one such technology that can assist practitioners, clinicians, and scientists in detecting and tracking viruses for their spread, predicting mortality risk, identifying high-risk patients, and controlling infection in real time. AI can help effectively deal with population screening, notifications, medical help, predicting potential hotspots, infection control suggestions, planning, treatment, and reported outcomes of infected patients (Haleem et al., 2020; Bai et al., 2020).

The major AI applications for the novel virus COVID-19 are classified as: (i) Detection and tracing: irregular symptoms can be monitored through AI and alert patients and healthcare officials for quick decision making for its diagnosis and management. It provides information for infected cases through (a) radio imaging technology akin to CT, (b) magnetic resonance imaging (MRI), (c) X-ray, (d) clinical blood sample. (ii) Monitoring and treatment: AI can develop automated monitoring, prediction, and treatment of the pandemic spread. (iii) Tracing: AI can analyze the infection level and categorize hot spots and clusters and trace individuals for future disease and its future recurrence. (iv) Prediction: AI technique can forecast the virus's nature, its associated risks, and pattern for its spread. (v) Drugs and vaccine: drug delivery design and development can significantly accelerate the process with AI. (vi) Disease prevention: the influx of virus, probable sites for infection can be proactively identified using AI algorithms, thus offering predictive and preventive healthcare management. Thus, AI can significantly reduce this pandemic with treatment consistency, prevention strategies, drug and vaccine development, and decision making based on designed algorithms (Abbas et al., 2018, 2019a, 2019b; Mughal et al., 2017, 2018a, 2018b).

This chapter provides a detailed discussion on the potential contributions of AI in managing the challenging situation against COVID-19, in addition to the current limitations on its contributions. Moreover, this article aims to draw attention to the research community to advance knowledge in utilizing AI and ML techniques that serve as an input to conveniently and instantly respond to COVID-19 research, policy, and medical analysis.

2.1.2 Datasets

At present, in many countries, medical capabilities are minimal (Mittal et al., 2020; Mashood Nasir et al., 2020; Lung et al., 2014). Due to the lockdowns in different countries, the flow of knowledge and exchange between doctors and scientists is restricted (Rad et al., 2016). Therefore, more extensive datasets with COVID-19 cases' image data are either unavailable or publicly accessible. A typical dataset contains a range of records, and these COVID-19 datasets include the repositories of X-ray and CT images. Several researchers typically access and utilize these publicly accessible datasets, easily retrieved using their specific links. A portion of the all-out images is normally downloaded from these datasets to train new ML models and algorithms and achieve testing goals in various experimental studies. In the following list, we briefly describe the publicly available datasets along with their accessibility.

2.1.2.1   Kaggle

This COVID-19 imaging database is an open-source X-ray and CT imaging database accessible at the website of Kaggle (Kaggle COVID-19 Dataset, 2020). In collaboration with medical doctors of higher education institutions from Qatar, Doha, Bangladesh, Malaysia, and Pakistan, a database has been developed for chest X-ray (CXR) pictures of COVID-19-positive patients besides images of pneumonia caused by a virus. In their latest database, a total of 1,341 regular images with 1,345 pictures of viral pneumonia and 219 COVID-19 healthy images have been released for the research community for study and analysis (Rehman et al., 2018a, 2018b, 2018c).

2.1.2.2   Github Repository

Github (Github COVID-19 Dataset, 2020) contains the publicly available dataset for CXR images of COVID-19-positive patients. The latest release of images has been referred to in identifying the medical analysis and diagnosis of treatable disease, specifically through image-based deep learning (DL). At GitHub, various images are available for both training and testing purposes to detect COVID-19 in various ML ventures. This chapter aims to briefly develop the basis for recognition strategies of COVID-19 and the usage of AI techniques to effectively deal with the pandemic.

2.1.3   COVID-19 DETECTION THROUGH IMAGING TECHNOLOGIES

The real-time reverse transcription-polymerase chain reaction (RT-PCR) is now a general strategy for diagnosing patients with COVID-19. CXR and chest CT imaging have a significant role in the early diagnosis and treatment of this deadly virus.

2.1.3.1   CT Imaging

This imaging technique is also referred to as computerized axial tomography (CAT) scanning. It comprises diagnostic imaging procedures about X-rays in developing cross-sectional images ('slices') of the body. These cross-sections are restructured with coefficients' measurements from X-ray beam attenuation within the particular case's object volume (Husham et al., 2016; Hussain et al., 2020). CT imaging is primarily based on the principle of computing attenuation coefficient and tissue density passed through an X-ray beam for measurement (Ucar & Korkmaz, 2020). Chest CT images (a) standard (b) infected are depicted in Figure 2.1 (Perumal et al., 2020).

It is an intricate test about pathophysiology that sheds light on multiple evolutions and disease detection stages. The quick diagnosis of CoV-2, a pattern of infection, features like glass opacities in lungs, round opacity, enhanced intro-infiltrate vessel, and critical illness (Harmon et al., 2020). It appears to be an evolving and valuable investigative tool for the clinical management of COVID-19 testing relevant to the

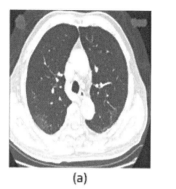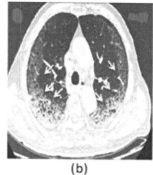

(a)   (b)

FIGURE 2.1   (a) shows normal chest CT scan and (b) shows infected chest CT scan.

lungs. In recent consensus, chest CT observing deteriorating respiratory function for COVID-19 patients depicts high pretest probability. Since the rise in new and suspected cases of novel coronavirus, AI techniques for characterization and detection may be improved with CT imaging. The CT with AI algorithms has allowed precise and swift process with an additional benefit of automation and reproducibility of biomarkers for quantification and sorting of CoV-2 disease. The developed AI offers robust, reliable, and almost 90% accurate testing of the disease and maintains high specificity for normal humans.

2.1.3.2   X-Ray Imaging

To create images inside the lungs, a very minute dose of ionizing radiation is used in CXR. It measures the walls within the heart, lungs, and chest (Iftikhar et al., 2017; Amin et al., 2019a; Rad et al., 2013; Rahim et al., 2017a, b). It also measures and diagnoses continual cough, breathing problems, chest pain, and fever or injury (Javed et al., 2019, 2020a, 2020b). It also helps to identify a variety of lung disorders, such as pneumonia, emphysema, and cancer, which allows taking care, diagnosing and controlling its further spread (Ramzan et al., 2020a, b; Ramzan et al., 2020a, b). Due to its quick and easy measurement of CXRs, this technique is instrumental in emergency diagnosis and treatment (Loey et al., 2020). In Figure 2.2a, the standard indications of a healthy chest are depicted. In contrast, in Figure 2.2b, right infrahilar airspace opacities are shown as chest radiograph of COVID-19-infected patients (Ilyas et al., 2020).

The tracing and detection of CoV-2 through X-ray imaging offer investigations for clinical management. Several approaches have been demonstrated by employing AI techniques in utilizing X-ray images to detect COVID-19 by bone suppression. The authors (Rahimzadeh & Attar, 2020) have classified CXRs of COVID-19 patients based on concatenated convolutional neural network (CNN) Xception and ResNet50V2 models. Additionally, the researchers (Ucar & Korkmaz, 2020) have detected COVID-19 patients by deploying deep architecture on available X-ray images. The researchers (Apostolopoulos & Mpesiana, 2020) employed a CNN and transfer learning strategy to extract essential CXRs for detecting COVID-19 cases. The authors (Bandyopadhyay & Dutta, 2020) proposed a novel AI model that has

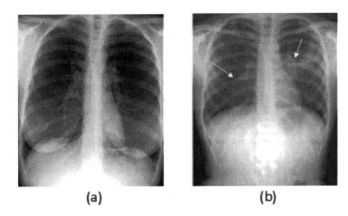

(a)   **(b)**

FIGURE 2.2   (a) shows normal CXR scan and (b) shows infected CXR scan.

classified the confirmed, negative, recovered and death cases of COVID-19. The researchers Khan et al. (2020a, 2020b) and (Sethy & Behera, 2020) presented a DL network to detect infected patients from their CXRs.

2.1.3.3   MRI

The MRI imaging strategy generates images of the cross-sectional body for diagnosis and analysis purposes (Rehman et al., 2020, 2021; Perveen et al., 2020). It does not emit radiation, unlike CT scans. However, it releases magnetic fields that spread over the spot, and high-resolution images of soft tissues and bones are taken through the sophisticated computer (Saba et al., 2018). It is strictly prohibited during MRI scan to have a metal clip, implants, pacemakers, or other metal objects within the body (Khan et al., 2020c, d; Ullah et al., 2019; Nazir et al., 2019). The ionizing radiation in CT develops concerns in COVID-19 patients who need dynamic observation. Besides, CT's unavailability during pandemics and lung lesions may be challenging with X-ray imaging; an alternate MRI diagnostic method is emerging. MRI offers the severity of lung damage in patients, detects viral pneumonia symptoms with less scan time, and provides detailed information (Saba et al., 2020, Saba, 2020; Vasilev et al., 2020).

2.2   LITERATURE REVIEW

2.2.1   Overview of ML

Preliminary recognition of some disease is crucial for early healing to save more lives (Xie et al., 2020; Abbas et al., 2019c; Adeel et al., 2020). The rapid process of detection and screening facilitates avoiding the spread of pandemic diseases like COVID-19. Implementing a skilled framework for health supports the novel arrangement of classification management and screening of COVID-19 carriers through extra cost-effectiveness compared to the conventional approach (Afza et al., 2020; Al-Ameen et al., 2015; Amin et al., 2019b). ML is used to maximize the diagnostic and screening process for recognized patients using radio imaging technology similar to X-ray, CT, and data of clinical blood samples. Experts have conceptualized four essential medical features, a mixture of demographic, laboratory, and clinical information via the use of CD3, GHS percentage, age of the patient and full amount protein, using Support Vector Machine (SVM) as the most important characteristic for the model of classification (Saba et al., 2019a, b, Sun et al., 2020). To explain the above point, Table 2.1 shows select knowledge on the detection of COVID-19 via ML (Figure 2.3).

TABLE 2.1
COVID-19 Detection through Conventional ML Classifiers

Author	Year	Technique	Dataset	Results
Tuncer et al.	(2020)	ResExLBP, Iterative ReliefF (IRF), the SVM classifier	321 CXR images collected from Kaggle and Github	Accuracy of IRF 100%,ResExLBP 99.69% with SVM classifier
Hassanien et al.	(2020)	SVM	40 CXR images of a public dataset	Specificity was 99.7%, accuracy 97.48%, and sensitivity of 95.76%
Kausani et al.	(2020)	VGGNet, DenseNet, ResNet50, InceptionV3	117 CXR images, 20 CT images from Kaggle repository	DenseNet121 accuracy 99% ResNet50 accuracy 98%
Sharma	(2020)	ResNet architecture	2,200 Lungs CT scans images of a public dataset	Accuracy of 91%
Wang et al.	(2020a)	Xception and SVM model	1,102 CXR images of a public dataset	Accuracy was 99.33% sensitivity 99.27%, specificity 99.38%, and AUC 99.32%
Ohata et al.	(2020)	MobileNet architecture through SVM classifiers	388 CXR images of a public dataset	SVM accuracy of 98.5% and the F1-score 98.5%. MLP accuracy of 95.6% and the F1-score 95.6%.
Abbas et al.	(2020)	DeTraC deep CNN	1,764 CXR images of a public dataset	Accuracy 95.12 %, specificity 91.87, and the sensitivity 97.91%
Al-Timmy et al.	(2020)	ResNet-50 model	2,186 five-class images of the public dataset	Accuracy was 91.6% ± 2.6% with five-class classification
Jokandan et al.	(2020)	VGG16 and SVM classifier	100 CXR, 746 CT scan images of a public dataset	Accuracy was 98.6 ± 2.1 with the VGG16 model
Barstugan et al.	(2020)	GLSZM and SVM classifier	618 CT scan Images of a public dataset	Accuracy was 99.68% with 10-fold cross-validation
Alimadadi et al.	(2020)	CRISPR method	Open Research Dataset CORD-19	Improved Accuracy
Mustafiz and Mohsin	(2020)	Decision Support System	At least 50 images of each class (CT scan, CXR) of a public dataset	Precision was 96.8%

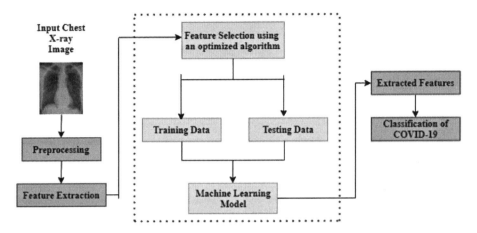

FIGURE 2.3   ML model for classification of COVID-19.

2.2.1.1   COVID-19 Detection Through Conventional ML Classifiers

The researchers (Tuncer et al., 2020) proposed an automated method for detecting coronavirus consisting of a step before processing level, feature extraction, as well as feature variety included with the use of X-ray images. For feature extraction and iterative ReliefF (IRF) used for feature collection, the proposed ResExLBP technique was used. For categorizing patients with coronavirus, leave-one-out cross-validation (LOOCV) and holdout validation were used for testing and training, and the classifiers decision tree (DT), SVM, K-nearest neighbors (KNN), linear discriminant (LD), and subspace discriminant (SD) were utilized. The 321 CXR image samples were used from the Kaggle and Github dataset that is publicly available. The result showed that the proposed IRF, ResExLBP methods have 100% and 99.69% classification accuracy using an SVM classifier through the 10-fold cross-validation (CV) and LOOCV on the public dataset.

The researchers (Hassanien et al., 2020) proposed an automated method for detecting coronavirus patients based on ML algorithms. The proposed approach included an SVM and multilevel thresholding algorithm for the detection and classification, using CXR images for coronavirus patients. The ML algorithms demonstrated the highest efficiency and computation complexity. The dataset of 40 X-ray images was the same size with 512 * 512 pixels. The results showed that the planned technique achieved specificity of 99.7%, accuracy of 97.48%, and sensitivity of 95.76% of the lungs classification on the public dataset.

The authors (Kausani et al., 2020) planned a technique based upon the ML algorithms for the automated classification and detection of coronavirus patients. The VGGNet, DenseNet, ResNet50, InceptionV3, and other models were used to obtain accurate features. The features were extracted through multiple ML classifiers to classify the cases of COVID-19. This method is better for unseen data rather than the task-specific data. The 117 CXR images and 20 CT image samples were collected from the publicly available Kaggle dataset of COVID-19. The result showed that the DenseNet121 model attained the highest accuracy of 99%. The ResNet50 model trained through Light GBM achieved an accuracy of 98% for the classification of coronavirus patients on the public dataset.

The researchers (Sharma, 2020) proposed an automated method based on ML CNN architecture to identify the coronavirus using CT scan images. The developed method used Microsoft Azure software based on ML algorithms for the testing and training phase. ResNet architecture was used in the training phase. The screening/test for real-time RT-PCR for COVID-19 was different from the other viral pneumonia. A total of 2,200 lung CT scan trials was taken from the publicly available dataset from Italy's hospitals of COVID-19 patients. The result showed that the proposed method attained an accuracy of 91% on the publicly available dataset.

The authors (Wang et al., 2020a) planned an automated technique consisting of ML and DL models to identify COVID-19 with the aid of CXR images. The proposed methodology used a mixture of ML and DL models (transfer learning) with a 96.75% accuracy shown by the Xception model. In-depth features were used for the classification through the Xception model and SVM classifier to enhance accuracy. Two datasets were used, which contained 1,102 CXR images, to detect and categorize COVID-19.The result showed that the accuracy was 99.33%, 2.58% enhanced, sensitivity 99.27%, specificity 99.38%, and AUC 99.32% with Xception and SVM model.

The authors (Ohata et al., 2020) developed a method that consisted of the transfer learning CNN model for the automatic detection of coronavirus cases through CXR images. For categorizing COVID-19 patients, the distinct CNN architectures skilled on ImageNet and KNN, SVM, MLP, RF, naive Bayes classifiers were used. Two publicly available datasets were used for COVID-19 patients, 388 CXR images across 194 COVID-19 patients. The outcome showed that the DenseNet201 architecture by MLP got an accuracy of 95.6% and the F1-score 95.6% in one of the datasets to identify COVID-19 cases. The MobileNet architecture through the SVM classifier attained an accuracy of 98.5% and the F1-score 98.5% for identifying COVID-19 cases on the public dataset.

The researchers (Abbas et al., 2020) proposed a technique that consisted of the DeTraC deep model of CNN for the classification of coronavirus patients with CXR images. By examining the class boundaries that used a class decomposition process, DeTraC can contract with the CXR dataset abnormalities. The CXR samples were collected from two publicly available datasets from numerous hospitals containing 1,764 CXR images across 949 coronavirus patients. The result showed that the DeTraC method attained an accuracy of 95.12%, specificity of 91.87, and sensitivity of 97.91% on the publicly available dataset.

The researchers (Al-Timemy et al., 2020) built a method based on an ML classifier and extracted deep features through a CXR to detect COVID-19 and tuberculosis (TB). The publicly available 2,186 five-class dataset of CXR images was used. This study also compared the accuracy of DL networks for extracting in-depth features with ML classifiers. The ResNet-50 model was used for the classification of five classes that attained the best performance. The result showed that the accuracy was 91.6% ± 2.6% attained with the five-class classification. The accuracy was 98.6% ± 1.4% and 99.9% ± 0.5% with three-class and two-class classification to identify COVID-19 on the public dataset.

The authors (Jokandan et al., 2020) developed a method based on CNN for the detection of coronavirus. The DL models were used to train the four CNN architectures (DenseNet121, VGG16, ResNet50, and InceptionResNetV2) to detect COVID-19 cases. These models were used to extract in-depth features with the exercise of CXR images and CT scans. Different ML classifiers and statistical modeling techniques were used for the classification of the COVID-19 cases. A publicly available dataset was used for coronavirus cases containing 100 CXR images across 25 COVID-19-positive patients and 746 CT scans across 349 COVID-19-positive cases. The linear SVM and neural network performed better than the other ones. The finding showed that the linear SVM classifier obtained the best accuracy of 98.6 ± 2.1 with the VGG16 model on the public dataset.

The authors (Barstugan et al., 2020) gave an automated scheme consisted of ML algorithms to detect coronavirus patients. To enhance the classification efficiency, the feature extraction technique was applied. For feature extraction, the algorithms GLCM, LDP, GLRLM, GLSZM, and DWT were used. Through the use of CT images, the SVM ML algorithm classified the extracted features. For categorizing coronavirus patients, a publicly available dataset of 618 CT scan images was used. The results demonstrated that the planned method got an accuracy that was 99.68% with 10-CV and the GLSZM model on a public dataset.

The researchers (Alimadadi et al., 2020) used advanced ML techniques in the taxonomic classification of COVID-19 cases. The open-source (Open Research) dataset was used that was a weekly update of COVID-19 patients. This study required real-time data for the classification. The large data of COVID-19 patients were evaluated and implemented into ML algorithms to increase the model's speed and accuracy. The advanced ML algorithms were used to classify COVID-19 genomes, COVID-19 detection based on CRISPR, and predict the continued existence of patients of COVID-19 on the open-source dataset.

The authors (Mustafiz and Mohsin, 2020) developed an automated ML system to categorize COVID-19 cases through XCR and CT images. For the detection and classification of the COVID-19 cases, ML algorithms were used. The decision support system that was based on ML was used for the classification. A publicly available dataset of at least 50 images of each class of CXR and CT images was used to detect COVID-19 patients. The result showed that the trained network attained 96.8% precision for the web application, which has smartphone-based real-time inference on a public dataset.

Table 2.1 shows the comparative study used and gained effectiveness in the segmentation of ML methods. ML can be used to segment a COVID-19 scan, which requires a lot of research endeavor and is not yet widely operational.

2.2.2 Overview of DL

The DL algorithm depicts considerable performance improvements for detecting and diagnosing diseases in medical imaging compared to ML methods (Ejaz et al., 2018; Fahad et al., 2018; Iqbal et al., 2017, 2018, 2019; Khan et al., 2017, 2019c, 2019d, 2019e, 2020e). DL has processed such radiologic thoracic images as lung reconstruction and segmentation (Gaal et al., 2020; Souza et al., 2019), pneumonia (Rajpurkar et al., 2017), and tuberculosis (Lakhani &

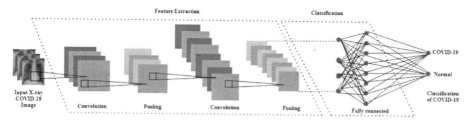

FIGURE 2.4 DL model for classification of COVID-19.

Sundaram, 2017). The scientific society has paid attention to conducting CXR images, contributing to clinically evaluating COVID-19 patients (Figure 2.4).

2.2.2.1 COVID-19 Detection Through DL Classifiers

Nour et al. (2020) proposed an automated technique for categorizing coronavirus based on the DL CNN architecture through CXR images. Scratch's demonstrated CNN model was trained rather than pre-trained CNN and this model was designed for five convolutional layers. The CNN model was applied for the feature extraction. The extracted features were utilized to nourish the ML algorithms: SVM, KNN, and DT. A public database of 2,905 CXR images was utilized for the experiments, and it was divided into 2 pieces (training, testing) with 70% and 30% rates. The results demonstrated that the SVM classifier has the highest accuracy of 98.97%, F-score 96.72%, sensitivity 89.39%, specificity 99.75%. In this study, the proposed CNN model detected positive COVID-19 cases with high sensitivity of 94.61%, accuracy 97.14%, F-score 95.75%, and specificity 98.29% on the public datasets.

Jain et al. (2020) made a CNN model based on DL for categorizing and detecting COVID-19 infected patients. In this study, the researchers occupied the posteroanterior (PA) view of CXR, as well as data augmentation techniques, applied for the patients of COVID-19. The CNN models (Inception V3, Xception, and ResNeXt model) were compared based on their performance. The Kaggle repository was used to collect the 6,432 CXR image samples. The results confirmed that, instead of the other models for detecting COVID-19 affected patients on the publicly available dataset, the Xception model got the maximum accuracy of 97.97%.

Al-Bawi et al. (2020) proposed an enhanced VGG network consisting of deep CNN used for the automated recognition of coronavirus cases. The enhanced VGG deep neural network was built through the convolutional COVID (CC) block. For the identification and classification of coronavirus patients CXR images on a publicly available dataset were used, and the VGG screening model was proposed. The test dataset included 1,828 images of X-ray of which patients of COVID-19 were 310 images of X-ray, viral pneumonia was 864 images, and healthy people were 654 images. The result demonstrated that the enhanced VGG network got the highest performance. The enhanced VGG model's accuracy was 98.52% with the two classes, and the accuracy was 95.34% with the three classes on the public dataset.

Khan & Aslam (2020) gave a technique consisting of the DL models to recognize COVID-19 patients. The four DL models (ResNet50, VGG-16, VGG-19, and DenseNet121) were used for COVID-19 patients' detection through X-ray images.

The proposed model consisted of the preprocessing phase, transfer learning, and data argumentation techniques. The CXR image samples were taken from the four open-source datasets, which contained 1,683 X-ray images and were divided into two binary classes. The results illustrated that the planned VGG-16 and VGG-19 models attained the sensitivity of 100%, specificity of 99.3% and accuracy was 99.38% on open-source datasets.

Horry et al. (2020) proposed an automated method consisting of the deep CNN model. The VGG19 model was utilized to categorize and detect COVID-19 cases and develop a preprocessing image phase for testing the DL models. The proposed approach reduced the noise from the images and detected the COVID-19 cases through specific features. Ultrasound images, X-Ray images, and CT scans compose the dataset of coronavirus that is available publicly. The VGG19 model achieved the best performance through ultrasound. The result showed that the proposed approach got 86% precision with X-ray images, 84% with CT scans, and 100% with ultra-sound images on the public dataset.

Ali Narin and Pamuk (2020) developed a method based on five CNN models for the automated recognition of coronavirus cases through CXR images attained from four classes (regular, bacterial, viral pneumonia, COVID-19 patients). Three available public datasets were used in which 341 CXR images were selected from the GitHub repository, the Chest X-ray8 dataset used 2,800 ordinary people's X-ray images, and 2,772 bacterial and 1,493 viral pneumonia CXR images were selected from the Kaggle repository with 5-fold CV. The outcomes showed that the best performance was the ResNet50 model compared to the four models with three binary classified datasets. The ResNet50 model accuracy for dataset-1 was 96.1%, for dataset-2, the accuracy was 99.5%, and for dataset-3, the accuracy was 99.7%.

Ozturk et al. (2020) developed a method for the automated recognition and categorization of coronavirus patients based on deep neural networks. The DarkNet model was used for the classification of COVID-19 patients from CXR images. The proposed model provided the best performance for binary class and multi-class classification. In the planned technique, 17 convolutional layers were implemented, and on each layer, different filters were introduced. The proposed model's (DarkCovidNet) performance was evaluated through expert radiologists and tested by a larger dataset. The result showed that the proposed model attained an accuracy of 98.08% for the binary classification and 87.02% accuracy for the multi-class classification on the publicly available dataset.

Islam et al. (2020) developed a DL technique based on CNN and long short-term memory (LSTM) for the automated detection of COVID-19 patients. The CNN was used for the feature extraction, and the LSTM was used to detect coronavirus cases. The images of CXR samples were taken from the publicly available dataset, which contained 4,575 X-ray images and 1,525 images of COVID-19 patients. The proposed technique improved from the other models using CNN and LSTM and showed the best performance. The results showed that the proposed technique had accuracy of 99.4%, specificity of 99.2%, sensitivity of 99.3%, AUC of 99.9%, and the F1-score of 98.9% on the public dataset.

Lokwani et al. (2020) proposed an automated method for coronavirus detection based on the CNN architecture with the use of CT scans. The proposed model was

2D segmentation through U-Net architecture, given the region of infection's output to recognize COVID-19 patients. This model originated a logic that converted the slice-level prediction to the scan level, which helped lessen the false positives. The CT scan samples were collected from the open-source available dataset, which contained 3,285 CT scan slices for training, 597 slices for validation, and 1,330 slices for testing. The result showed that the proposed model attained the sensitivity of 0.964 (95% CI: 0.88–1), specificity of 0.884 (95% CI: 0.82–0.94), and F1-score of 0.794 (95% CI: 0.68-0.89) on the open-source dataset.

Halgurd S. Maghdid et al. (2020) developed a method that was based on DL and transfer learning for the automated recognition of coronavirus cases. The CNN and AlexNet modified models were used to detect COVID-19 by utilizing images of X-rays and CT scan datasets. This study plans to generate a dataset of various sources for X-ray images and CT scans. The result showed that the AlexNet model achieved accuracy f up to 98%, sensitivity 100%, specificity 96% with the X-ray, and accuracy 82%, sensitivity 72%, and specificity 100% with the CT scans. The modified CNN model's accuracy was 94.1%, sensitivity 90%, specificity 100% with CT scans, and accuracy 94%, sensitivity 100%, and specificity 88% with X-ray, on the public dataset.

Wang et al. (2020b) planned a process that consisted of the DL-CNN architecture for the recognition of coronavirus cases. For the detection of coronavirus from CXR images, the proposed COVID-Net was used. Through the automated analysis approach in such an initiative, the suggested method generates predictions not just to provide deeper insights into essential cases of COVID-19. The available COVIDx largest dataset of coronavirus patients was used, containing 13,975 CXR images across 13,870 coronavirus patients. The result showed that the proposed COVID-Net attained an accuracy of 93.3% and the sensitivity of coronavirus patients was 91.0% on the open-source COVIDx dataset.

Panwar et al. (2020) planned a technique based on the DL and grad-CAM color visualization technique for the automated detection of coronavirus patients with CXRs and chest CT scans. A relationship between COVID-19 and viral pneumonia patients was established using the proposed approach, which discovered the outline between pneumonia and COVID-19 images. The three CXRs (COVID, pneumonia, normal), and CT scan dataset consisted of 1,252 images for COVID-19 patient diagnosis. The COVID-19 positive patients were examined by the proposed method ≤2 seconds faster than the RT-PCR studies. The result showed that the planned technique achieved an overall accuracy of 95%, precision 0.95, recall 0.94, F1 score 0.95 on the publicly available dataset.

Ragb et al. (2020) developed an approach based on CNN for the detection of coronavirus. The DL algorithms were used to train the CNN models (ResNet18, DenseNet 121, and ResNet50) to detect coronavirus through CXR images on the ImageNet dataset. The neural networks (NN) were stuck in parallel, and the final classification decision was dependent on the voting criteria. The outcomes of NN were showed a single vote. Numerous experiments were conducted on the open-source COVID-CT-CXR images dataset of coronavirus patients, containing 288 CXR images across 263 COVID-19 patients. The outcomes showed that the proposed approach's 99.7% accuracy was obtained using k-fold CV and bagging classifier. The sensitivity was 100% of the coronavirus patients on the open-source dataset.

Narayan Das et al. (2020) built a method consisting of transfer learning to detect coronavirus patients through the CXR images with the Xception model. Three classes of publicly accessible datasets containing 70% of training data, 10% of validation data, as well as 20% of testing data were used. The 150 CXR image samples were contained from the Kaggle repository. The result showed that the proposed method in training analysis achieved accuracy of 0.995246, F-measure 0.986271, sensitivity 0.991236, specificity 0.994615, Kappa statistics 0.980833 and in testing analysis, the proposed model achieved accuracy of 0.974068, F-measure 0.969697, sensitivity 0.970921, specificity 0.972973, and Kappa statistics 0.971924 on the publicly available dataset.

Manapure et al. (2020) developed an approach that was based on the DL model. The DL-based method was developed for the automatic segmentation, identification, and quantification of coronavirus infection patients through X-ray images. The open-source Kaggle dataset of 50 CXR images was used for the identification of coronavirus cases. In this study, a system was built, exploiting two-dimensional and three-dimensional DL algorithms, changing and adapting the existing AI models. The outcomes showed that the proposed method attained an accuracy of ~90%–92%, sensitivity 100%, and specificity 80% for automatic detection of coronavirus cases on the open-source dataset of CXR images.

Xu et al. (2020) developed a screening model built on the DL to detect coronavirus cases through CT scans. A publicly available dataset of COVID-19 patients containing a total of 618 samples of CT scan was used, including 219 samples across 110 COVID-19 patients, 175 samples from healthy people, and 224 samples from IAVP. The DL models were used for the early screening of the coronavirus cases in this study. The CT scan samples were collected from three different COVID-19 hospitals in China. The result showed that the planned method attained an overall accuracy of 86.7% through CT scan samples on the benchmark dataset.

Gupta et al. (2020) developed a method that was based on DL for detection of coronavirus patients. The DL algorithms were used to pre-train models (NASNet, ResNet101, MobileNet, etc.) to detect coronavirus through CXR images. The publicly available dataset was used for coronavirus patients, which contained 2,905 CXR images across 219 coronavirus patients. The outcomes showed that the developed method attained 99.08% accuracy on the 3 class and 99.53% on the 2 class. The proposed model attained a recall of 99%, F1 score of 99%, and precision of 99% on ternary classification, on the binary class, the recall 99%, precision 100%, and on the COVID-19 class attained 100% precision and a 99% recall on the public dataset.

Hemdan et al. (2020) built an approach based on the DL to detect coronavirus cases. The proposed method COVIDX-Net consists of seven different CNN architectures of DL, such as VGG19, Google MobileNet, and DenseNet, to detect coronavirus through CXR images. The 80%–20% of data were used for the training and testing phase. For classification, a dataset that was publicly available of 50 CXR images was used. The outcomes illustrated that the worst categorization of the InceptionV3 model through F1-scores for ordinary people was 0.67 and 0.00 for COVID-19 patients. The VGG19, DenseNet model attained the best performance.

The F1-scores of VGG19, DenseNet model were 0.89 for ordinary people and 0.91 for COVID-19 patients on the public dataset.

Acar et al. (2020) proposed a method based on DL for the detection of coronavirus. The proposed method was segmentation, GAN, and data augmentation techniques to enhance the models' performance. The DL algorithms were used to train the CNN architectures (VGG16, Xception, ResNet-50, DenseNet-121, DenseNet-169, InceptionV3, ResNet50V2, InceptionResNetV2, and VGG19) to detect the coronavirus with the use of 2,900 CT scan images which consist of 1,232 coronavirus patients and 1,668 healthy CT scan images. The result showed accuracy of 99% attained for the lung segmentation. The DenseNet169 model attained the 99.8% highest accuracy, InceptionResNetV2 accuracy was 99.65%, Xception and InceptionV3 models achieved an accuracy of 99.60%. The highest accuracy was 99.8%, precision 99.8%, F1-score 99.8%, recall 99.8%, through DL models on a public dataset.

Table 2.2 shows the comparative research in the detection of DL methods employed and obtained efficiently. It has been found that CNN is assumed as the most popular DL method in the field of imaging that constitutes many staked convolutional layers. The prime advantage of this method is that it offers end-to-end learning by detaching a handcrafted engine. Therefore, its deep structure comprises layers: convolutional, maximum or average pooling, non-linear, batch normalization, fully connected, and softmax. These layers offer unique features to boost translational invariance, strengthen the network, and extract parameters for classification.

TABLE 2.2
COVID-19 Detection through DL Classifiers

Author	Year	Technique	Dataset	Results
Nour et al.	(2020)	CNN,KNN,DT,SVM classifier	2,905 X-ray images of public dataset	Sensitivity 94.61%, accuracy 97.14%, F-score 95.75%, specificity 98.29% with CNN
Jain et al.	(2020)	Inception V3, Xception, and ResNeXt model	6,432 CXR images	Accuracy 97.97% with Xception
Gilani et al.	(2020)	Xception model	746 CT scans images of public dataset	Accuracy was 0.85, precision 0.857, and recall 0.854
Lansana et al.	(2020)	VGG-16, Inception-V2, and DT	360 CXR images of public dataset	Accuracy 91%, precision 100%, recall 94%, F1-score 97% with VGG-16
Al-Bawi et al.	(2020)	VGG model	1,828 CXR images of a public dataset	98.52% accuracy with the two classes, 95.34% accuracy with the three classes
Khan and Aslam et al.	(2020)	VGG-16, VGG-19 model	1,683 CXR images of a public dataset	Sensitivity 100%, specificity 99.3%, and accuracy was 99.38%

(Continued)

TABLE 2.2
(Continued)

Author	Year	Technique	Dataset	Results
Horry et al.	(2020)	VGG19 model	1,531 CXR images, CT scan, ultrasound of a public dataset	Precision up to 86% with X-ray images, 84% with CT scans, and 100% with Ultrasound images
Ali Narin and Pamuk	(2020)	ResNet50	341 CXR images from GitHub repository, ChestX-ray8 dataset, Kaggle repository	ResNet50 model accuracy for dataset-1 was 96.1%, for dataset-2 was 99.5%, and for dataset-3 was 99.7%
Ozturk et al.	(2020)	DarkNet	CXR images of a public dataset	Accuracy was 98.08% for binary classification and accuracy 87.02% for multi-class classification
Islam et al.	(2020)	CNN and LSTM	6,100 CXR images of a public dataset	Accuracy 99.4%, the specificity of 99.2%, the sensitivity of 99.3%, AUC of 99.9%, and the F1-score of 98.9%
Lokwani et al.	(2020)	CNN,U-Net Architecture	5,212 CT scan images of open-source dataset	Specificity was 0.884 (95% CI: 0.82–0.94), sensitivity 0.964 (95% CI: 0.88–1), and F1-score 0.794 (95% CI: 0.68–0.89)
Maghdid et al.	(2020)	CNN, AlexNet models	170 CXR images and356 CT scan images of public dataset	AlexNet accuracy was up to 98% with X-ray, and accuracy 82%with CT scans. CNN model accuracy 94.1%with CT scans, and accuracy 94% with chest X-Ray
Wang et al.	(2020b)	CNN COVID-Net	13,975 CXR images of open-source COVIDx dataset	Accuracy of 93.3%, sensitivity 91.0%
Panwar et al.	(2020)	DL and grad-CAM color visualization technique	1,252 CT scan images, CXR images of a public dataset	Accuracy of 95%, Precision 0.95, Recall 0.94, F1 Score 0.95
Ragab et al.	(2020)	k-fold cross-validation and bagging classifier	288 CXR images (COVID-CT-CXR) open-source dataset	Accuracy of 99.7% and the sensitivity was 100%
Narayan Das et al.	(2020)	Xception	150 chest X-ray images from Kaggle repository	Accuracy was 0.974068, F-measure 0.969697, Sensitivity 0.970921, Specificity 0.972973, Kappa statistics 0.971924

(Continued)

TABLE 2.2
(Continued)

Author	Year	Technique	Dataset	Results
Manapure et al.	(2020)	CNN	50 CXR images from Kaggle repository	Accuracy of ~90%–92%, sensitivity 100%, and specificity 80%
Xu et al.	(2020)	DL-CNN	618 CT scan images of the benchmark dataset	Accuracy was 86.7%
Gupta et al.	(2020)	ResNet101, NASNet, Xception, MobileNet, InceptionV3	2,905 CXR images of a public dataset	Recall 99%, precision 100% on the binary class and 100% precision and 99% recall on COVID-19 class
Hemdan et al.	(2020)	VGG19, Google MobileNet, DenseNet	50 CXR images of a public dataset	F1-scores of VGG19 and DenseNet model was 0.89 for normal people and 0.91 for COVID-19 patients
Wang et al.	(2020b)	Inception transfer learning model	1,065 CT scan images of a public dataset	Accuracy was 79.3%, specificity 0.83, sensitivity 0.67 with the testing dataset
Acar et al.	(2020)	VGG16, Xception, ResNet-50, DenseNet-121, DenseNet-169, InceptionV3, ResNet50V2, InceptionResNetV2, VGG19	2,900 CT scan images of public dataset	DenseNet169 model attained the 99.8% highest accuracy, InceptionResNetV2 accuracy 99.65%, Xception, InceptionV3 models accuracy 99.60%

2.3 DISCUSSION

To protect against infections of COVID-19 and save lives, quick detection with precise computation time is assumed to be a practical approach. It is also helpful to constrain the widening and mutation of the disease through a considerable AI model. According to the UN Global Pulse research, they have reviewed the number of intelligent technique applications for detecting COVID-19. Their research has proposed that imaging methods, including CT plus X-ray scans, offer a quick and exact recognition of COVID-19 infections based on AI models. Moreover, in today's advanced world, the researchers have now proposed using a digital phone with digital apps to instantly inspect CT images and detect the deadly virus instantly. The scientists, researchers, and medical professionals are compelled to contest for containing the pandemic, seeking a ubiquitous alternative approach for quick prediction and screening processes, contact tracing, and associated vaccine or drug development and delivery with robust and dependable operation. AI and ML are effective, potential methods exercised by many healthcare suppliers to achieve such goals (Akbar et al.,

2017, 2018a, 2018b, 2019; Jamal et al., 2017). The breakthroughs from recent studies that address the struggles and challenges of the researchers in assisting medical experts from multiple perspectives are addressed by this study. This chapter has also considered implications based on AI/ML-based model designs for researchers, medical experts, and policymakers. This is due to identifying a few errors encountered in the existing circumstances while handling and managing the alarming situation. The majority of the study employed profound learning breakthroughs. Nevertheless, according to the rapid increase of the pandemic situation, the requirement of an enhanced model and a screening of high-end performance with diverse mammographic, clinical, and demographic information is mandatory now. We can use AI in the future to contain the pandemic by scanning public places and to emphasize social distancing and lockdown measures.

2.4   CONCLUSION

The AI and ML-dependent techniques proposed in writing to detect COVID-19 show propitious results like ResNet50V2 with 99.6% accuracy, Sgdm-SqueezeNet 96.70%, CNN-LSTM with 99.20% of accuracy, Bayes-SqueezeNet with 98.30, and InceptionV3 with 98%. All the planned techniques used binary categorization methods, while pneumonia has a lot of other reasons too. It is troublesome to struggle with COVID-19 due to its inexplicable presence and foreign biological origin. To achieve the goal of preventing further public health outbreaks, including MERS-CoV and SARS-CoV, we have learned the lesson to undertake prudent measures. The chances of spreading this pandemic would be reduced with enough awareness that hygiene, breathable masks, social distancing, isolation, and quarantine are obligatory. One of the many potential therapies for COVID-19 is convalescent plasma. The future provocations of AI for the disclosure of COVID-19 are education and examining the diverse DL architectures with massive databases and all genera of CXR images damaged via diverse kinds of viral pneumonia patients.

Furthermore, ML and AI can extensively augment and advance treatment strategies for prediction, screening, and medication more significantly, developing vaccines or drugs for the pandemic devoid of human intervention in medical preparation. However, most models are not predisposed to show their real-world maneuvers and are assumed to be sufficient in handling the pandemic.

CONFLICT OF INTEREST

The authors of this work have no conflict of interest.

ACKNOWLEDGMENTS

This study is supported by Artificial Intelligence Research & Innovation Group (AIRIG), Riphah International University, Faisalabad Campus, Pakistan.

REFERENCES

Abbas, A., Abdelsamea, M. M., & Gaber, M. M. (2020). Classification of COVID-19 in chest X-ray images using DeTraC deep convolutional neural network. arXiv(less.IV).

Abbas, A., Saba, T., Rehman, A., Mehmood, Z., Javaid, N., Tahir, M., Khan, N. U., Ahmed, K. T., & Shah, R. (2019b). Plasmodium species aware based quantification of malaria, parasitemia in light microscopy thin blood smear, *Microscopy Research and Technique*, 82(7), 1198–1214. doi:10.1002/jemt.23269.

Abbas, N., Saba, T., Mehmood, Z., Rehman, A., Islam, N., & Ahmed, K. T. (2019a). An automated nuclei segmentation of leukocytes from microscopic digital images. *Pakistan Journal of Pharmaceutical Sciences*, 32(5), 2123–2138.

Abbas, N. Saba, T. Mohamad, D. Rehman, A. Almazyad, A. S., & Al-Ghamdi, J. S. (2018) Machine aided malaria parasitemia detection in Giemsa-stained thin blood smears. *Neural Computing and Applications*, 29(3), 803–818, doi:10.1007/s00521-016-2474-6.

Abbas, N., Saba, T., Rehman, A., Mehmood, Z., Kolivand, H., Uddin, M., & Anjum, A. (2019c). Plasmodium life cycle stage classification-based quantification of malaria parasitaemia in thin blood smears. *Microscopy Research and Technique*, 82(3), 283–295. doi:10.1002/jemt.23170.

Acar, E., Engin, Ş., & İhsan, Y. 2020. Improving effectiveness of different deep learning-based models for detecting COVID-19 from computed tomography (CT) images. *Neural Computing and Applications*. medRxiv.

Adeel, A., Khan, M. A., Akram, T., Sharif, A., Yasmin, M., Saba, T., & Javed, K. (2020). Entropy-controlled deep features selection framework for grape leaf diseases recognition. *Expert Systems*, 1–17. doi:10.1111/exsy.12569.

Afza, F., Khan, M. A., Sharif, M., Saba, T., Rehman, A., & Javed, M. Y. (2020). *Skin Lesion Classification: An Optimized Framework of Optimal Color Features Selection*. In *2020 2nd International Conference on Computer and Information Sciences (ICCIS)* (pp. 1–6). IEEE.

Akbar, S., Akram, M. U., Sharif, M., Tariq, A., & Khan, S. A. 2018b. Decision support system for detection of hypertensive retinopathy using arteriovenous ratio. *Artificial Intelligence in Medicine*, 90, 15–24.

Akbar, S., Akram, M. U., Sharif, M., Tariq, A., & Ullah Yasin, U., 2017. Decision support system for detection of papilledema through fundus retinal images. *Journal of Medical Systems*, 41(4), 66.

Akbar, S., Akram, M. U., Sharif, M., Tariq, A., & Ullah Yasin, U. 2018a. Arteriovenous ratio and papilledema based hybrid decision support system for detection and grading of hypertensive retinopathy. *Computer Methods and Programs in Biomedicine*, 154, 123–141.

Akbar, S., Sharif, M., Akram, M. U., Saba, T., Mahmood, T., & Kolivand, M. 2019. Automated techniques for blood vessels segmentation through fundus retinal images: A review. *Microscopy Research and Technique*, 82(2), 153–170.

Al-Ameen, Z. Sulong, G. Rehman, A., Al-Dhelaan, A. Saba, T., & Al-Rodhaan, M. (2015) An innovative technique for contrast enhancement of computed tomography images using normalized gamma-corrected contrast-limited adaptive histogram equalization. *EURASIP Journal on Advances in Signal Processing*, 32, 1–12, doi:10.1186/s13634-015-0214-1

Al-Bawi, A., Al-Kaabi, K., Jeryo, M., & Al-Fatlawi, A. 2020. Block: An effective use of deep learning for automatic diagnosis of COVID-19 using X-ray images. *Research on Biomedical Engineering*, 1–10.

Ali Narin, C. K., & Pamuk, Z. (2020). Automatic detection of coronavirus disease (COVID-19) using X-ray images and deep convolutional neural networks. *Electrical Engineering and Systems Science*. doi:10.1007/s10044-021-00984-y.

Alimadadi, A., Aryal, S., Manandhar, I., Munroe, P. B., Joe, B., & Cheng, X. (2020). Artificial intelligence and machine learning to fight COVID-19. *Physiol Genomics*, 52, 200–202.

Al-Timemy, A. H., Khushaba, R. N., Mosa, Z. M., & Escudero, J. (2020). An efficient mixture of deep and machine learning models for COVID-19 and tuberculosis detection using X-ray images in resource-limited settings. ArXiv, abs/2007.08223.

Amin, J., Sharif, M., Raza, M., Saba, T., & Rehman, A. (2019a). *Brain Tumor Classification: Feature Fusion*. In *2019 International Conference on Computer and Information Sciences (ICCIS)* (pp. 1–6). IEEE.

Amin, J., Sharif, M., Yasmin, M. Saba, T., & Raza, M. (2019b). Use of machine intelligence to conduct analysis of human brain data for detection of abnormalities in its cognitive functions. *Multimed Tools and Applications*, 79 (15), 10955–10973. doi:10.1007/s11042-019-7324-y

Apostolopoulos, I. D., & Mpesiana, T. A. 2020. Covid-19: Automatic detection from X-ray images utilizing transfer learning with convolutional neural networks. *Physical and Engineering Sciences in Medicine*, 43(2), 635–640.

Bai, H. X., Hsieh, B., Xiong, Z., Halsey, K., Choi, J. W., Tran, T. M. L., Pan, I., Shi, L.-B., Wang, D.-C., & Mei, J. 2020. Performance of radiologists in differentiating COVID-19 from non-COVID-19 viral pneumonia on chest CT. *Radiology*, 296(2), E46–E54.

Bandyopadhyay, S. K., & Dutta, S. 2020. Machine learning approach for confirmation of covid-19 cases: Positive, negative, death and release. medRxiv.

Barstugan, M., Ozkaya, U., & Ozturk, S. (2020). Coronavirus (COVID-19) classification using CT images by machine learning methods. arXiv preprint arXiv:2003.09424

COVID-19 vaccine overview. 2020. [Online] Available: https://www.pfizer.com/news/press-release/press-release-detail/pfizer-and-biontech-announce-vaccine-candidate-against (Accessed:10 November 2020).

Ejaz, K., Rahim, M. S. M., Rehman, A., Chaudhry, H., Saba, T., & Ejaz, A. (2018). Segmentation method for pathological brain tumor and accurate detection using MRI. *International Journal of Advanced Computer Science and Applications*, 9(8), 394–401.

Fahad, H. M., Khan, M. U. G., Saba, T., Rehman, A., & Iqbal, S. (2018). Microscopic abnormality classification of cardiac murmurs using ANFIS and HMM. *Microscopy Research and Technique*, 81(5), 449–457. doi:10.1002/jemt.22998.

Gaal, G., Maga, B., & Lukacs, A. 2020. Attention u-net-based adversarial architectures for chest x-ray lung segmentation. arXiv preprint arXiv:2003.10304.

Gilani, P., Shalbaf, A., & Vafaeezadeh, M. 2020. Automated detection of COVID-19 using ensemble of transfer learning with deep convolutional neural network based on CT scans. *International Journal of Computer Assisted Radiology and Surgery*, doi:10.1007/s11548-020-02286-w.

Github COVID-19 imaging dataset 2020. [Online] Available: https://www.kaggle.com/imdevskp/corona-virus-report/download (Accessed:12 November 2020)

Gupta, A., Anjum, G. S., & Kataria, R. 2020. InstaCovNet-19: A deep learning classification model for the detection of COVID-19 patients using Chest X-ray. *Applied Soft Computing*, 99, 106859.

Haleem, A., Javaid, M., & Vaishya, R. 2020. Effects of COVID 19 pandemic in daily life. *Current Medicine Research and Practice*, 10(2), 78–79.

Harmon, S. A., Sanford, T. H., Xu, S., Turkbey, E. B., Roth, H., Xu, Z., Yang, D., Myronenko, A., Anderson, V., & Amalou, A. 2020. Artificial intelligence for the detection of COVID-19 pneumonia on chest CT using multinational datasets. *Nature Communications*, 11, 1–7.

Hassanien, A. E., Mahdy, L. N., Ezzat, K. A., Elmousalami, H. H., & Ella, H. A. 2020. Automatic x-ray covid-19 lung image classification system based on multilevel thresholding and support vector machine. medRxiv.

Hemdan, E. E.-D., Shouman, M. A., & Karar, M. E. (2020). COVIDX-net: A framework of deep learning classifiers to diagnose COVID-19 in X-ray images. ArXiv, abs/2003.11055.

Horry, M. J., Chakraborty, S., Paul, M., Ulhaq, A., Pradhan, B., Saha, M., & Shukla, N. 2020. COVID-19 detection through transfer learning using multimodal imaging data. *IEEE Access*, 8, 149808–149824.

Husham, A., Alkawaz, M. H., Saba, T., Rehman, A., & Alghamdi, J. S. (2016). Automated nuclei segmentation of malignant using level sets, *Microscopy Research and Technique*, 79(10), 993–997. doi:10.1002/jemt.22733.

Hussain, N., Khan, M. A., Sharif, M., Khan, S. A., Albesher, A. A., Saba, T., & Armaghan, A. (2020). A deep neural network and classical features based scheme for objects recognition: An application for machine inspection. *Multimedia Tools and Applications*. doi:10.1007/s11042-020-08852-3.

Iftikhar, S. Fatima, K. Rehman, A. Almazyad, A. S., & Saba, T. (2017). An evolution-based hybrid approach for heart diseases classification and associated risk factors identification. *Biomedical Research*, 28 (8), 3451–3455

Ilyas, M., Rehman, H., & Nait-Ali, A. 2020. Detection of covid-19 from chest X-ray images using artificial intelligence: An early review. arXiv preprint arXiv:2004.05436.

Iqbal, S. Ghani, M. U. Saba, T., & Rehman, A. (2018). Brain tumor segmentation in multi-spectral MRI using convolutional neural networks (CNN). *Microscopy Research and Technique*, 81(4), 419–427. doi:10.1002/jemt.22994.

Iqbal, S., Khan, M. U. G., Saba, T. Mehmood, Z. Javaid, N., Rehman, A., & Abbasi, R. (2019). Deep learning model integrating features and novel classifiers fusion for brain tumor segmentation. *Microscopy Research and Technique*, 82(8), 1302–1315. doi:10.1002/jemt.23281

Iqbal, S., Khan, M. U. G., Saba, T., & Rehman, A. (2017). Computer-assisted brain tumor type discrimination using magnetic resonance imaging features. *Biomedical Engineering Letters*, 8(1), 5–28. doi:10.1007/s13534-017-0050-3.

Islam, M. Z., Islam, M. M., & Asraf, A. 2020. A combined deep CNN-LSTM network for the detection of novel coronavirus (COVID-19) using X-ray images. *Informatics in Medicine Unlocked*, 20, 100412.

Italian's COVID-19 News. (2020). [Online] Available: https://www.cidrap.umn.edu/news-perspective/2020/03/italian-covid-19-deaths-pass-chinas-total-cases-surge-europe (Accessed:12 September 2020)

Jain, R., Gupta, M., Taneja, S., & Hemanth, D. J. 2020. Deep learning-based detection and analysis of COVID-19 on chest X-ray images. *Applied Intelligence*, 51(3), 1690–1700.

Jamal, A., Hazim Alkawaz, M., Rehman, A., & Saba, T. (2017). Retinal imaging analysis based on vessel detection. *Microscopy Research and Technique*, 80(17), 799–811. doi:10.1002/jemt.

Javed, R., Rahim, M. S. M., Saba, T., & Rashid, M. (2019). Region-based active contour JSEG fusion technique for skin lesion segmentation from dermoscopic images. *Biomedical Research*, 30(6), 1–10.

Javed, R., Rahim, M. S. M., Saba, T., & Rehman, A. (2020a) A comparative study of features selection for skin lesion detection from dermoscopic images. *Network Modeling Analysis in Health Informatics and Bioinformatics*, 9 (1), 4.

Javed, R., Saba, T., Shafry, M., & Rahim, M. (2020b). *An Intelligent Saliency Segmentation Technique and Classification of Low Contrast Skin Lesion Dermoscopic Images Based on Histogram Decision*. In *2019 12th International Conference on Developments in eSystems Engineering (DeSE)* (pp. 164–169).

Jokandan, A. S., Asgharnezhad, H., Jokandan, S. S., Khosravi, A., Kebria, P. M., Nahavandi, D., Nahavandi, S., & Srinivasan, D. (2020). An uncertainty-aware transfer learning-based framework for covid-19 diagnosis. arXiv: [eess.IV].

Kaggle COVID-19 imaging dataset. (2020) [Online] Available: https://www.kaggle.com/imdevskp/corona-virus-report/download (Accessed:12 November 2020)

Kausani, S. H., Kausani, P. H., Wesolowski, M. J., Schneider, K. A., & Deters, R. 2020. Automatic detection of coronavirus disease (COVID-19) in X-ray and CT images: A machine learning-based approach. arXiv preprint arXiv:2004.10641.

Khan, A. I., Shah, J. L., & Bhat, M. M. 2020a. Coronet: A deep neural network for detection and diagnosis of COVID-19 from chest X-ray images. *Computer Methods and Programs in Biomedicine*, 196, 105581.

Khan, I. U., & Aslam, N. 2020. A deep-learning-based framework for automated diagnosis of COVID-19 using X-ray images. *Information*, 11(9), 419.

Khan, M. A., Akram, T., Sharif, M., Javed, K., Raza, M., & Saba, T. (2020c). An automated system for cucumber leaf diseased spot detection and classification using improved saliency method and deep features selection. *Multimedia Tools and Applications*, 79, 1–30. doi:10.1007/s11042-020-08726-8.

Khan, M. A.; Akram, T. Sharif, M., Saba, T., Javed, K., Lali, I. U., Tanik, U. J., & Rehman, A. (2019d). Construction of saliency map and hybrid set of features for efficient segmentation and classification of skin lesion. *Microscopy Research and Technique*, 82(5), 741–763, doi:10.1002/jemt.23220

Khan, M. A., Ashraf, I., Alhaisoni, M., Damaševičius, R., Scherer, R., Rehman, A., & Bukhari, S. A. C. (2020e). Multimodal brain tumor classification using deep learning and Robust feature selection: A machine learning application for radiologists. *Diagnostics*, 10, 565.

Khan, M. A., Javed, M. Y., Sharif, M., Saba, T., & Rehman, A. (2019a). *Multi-model deep neural network-based features extraction and optimal selection approach for skin lesion classification*. In *2019 International Conference on Computer and Information Sciences (ICCIS)* (pp. 1–7). IEEE.

Khan, M. A. Kadry, S., Zhang, Y. D., Akram, T., Sharif, M., Rehman, A., & Saba, T. (2021) Prediction of COVID-19 - Pneumonia based on selected deep features and one class Kernel extreme learning machine, computers & electrical engineering, 90, 106960.

Khan, M. A., Lali, I. U. Rehman, A. Ishaq, M. Sharif, M. Saba, T., Zahoor, S., & Akram, T. (2019c). Brain tumor detection and classification: A framework of marker-based watershed algorithm and multilevel priority features selection, *Microscopy Research and Technique*, 82(6), 909–922, doi:10.1002/jemt.23238

Khan, M. A., Sharif, M. Akram, T., Raza, M., Saba, T., & Rehman, A. (2020d). Hand-crafted and deep convolutional neural network features fusion and selection strategy: An application to intelligent human action recognition. *Applied Soft Computing* 87, 105986

Khan, M. A., Sharif, M. I., Raza, M., Anjum, A., Saba, T., & Shad, S. A. (2019b). Skin lesion segmentation and classification: A unified framework of deep neural network features fusion and selection. *Expert Systems*, e12497.

Khan, M. W., Sharif, M., Yasmin, M., & Saba, T. (2017). CDR based glaucoma detection using fundus images: A review. *International Journal of Applied Pattern Recognition*, 4(3), 261–306.

Khan, M. Z., Jabeen, S., Khan, M. U. G., Saba, T., Rehmat, A., Rehman, A., & Tariq, U. (2020b). A realistic image generation of face from text description using the fully trained generative adversarial networks. *IEEE Access*. doi:10.1109/ACCESS.2020.3015656.

Khan, S. A., Nazir, M., Khan, M. A., Saba, T., Javed, K., Rehman, A., … & Awais, M. (2019e). Lungs nodule detection framework from computed tomography images using support vector machine. *Microscopy Research and Technique*, 82(8), 1256–1266.

Lakhani, P., & Sundaram, B. 2017. Deep learning at chest radiography: Automated classification of pulmonary tuberculosis by using convolutional neural networks. *Radiology*, 284, 574–582.

Lansana, D., Kumar, R., Bhattacharjee, A., Hemanth, D. J., Gupta, D., Khanna, A., & Castillo, O. 2020. Early diagnosis of COVID-19-affected patients based on X-ray and computed tomography images using deep learning algorithm. *Soft Computing*, 1–9. doi:10.1007/s00500-020-05275-y.

Loey, M., Smarandache, F., & Khalifa, N. E. M. 2020. Within the lack of chest COVID-19 X-ray dataset: A novel detection model based on GAN and deep transfer learning. *Symmetry*, 12, 651.

Lokwani, R., Gaikwad, A., Kulkarni, V., Pant, A., & Kharat, A. 2020. Automated detection of COVID-19 from CT scans using convolutional neural networks. arXiv: Image and Video Processing (less.IV).

Lung, J. W. J., Salam, M. S. H., Rehman, A., Rahim, M. S. M., & Saba, T. (2014). Fuzzy phoneme classification using multi-speaker vocal tract length normalization. *IETE Technical Review*, 31 (2), 128–136. doi:10.1080/02564602.2014.892669.

Maghdid, H. S., Asaad, A. T., Ghafoor, K. Z., Sadiq, A. S., & Khan, M. K. (2020). Diagnosing COVID-19 pneumonia from X-ray and CT images using deep learning and transfer learning algorithms. arXiv:2004.00038 [eess.IV].

Manapure, P., Chaudhari, N., & Likhar, K. 2020. Detecting COVID-19 in X-ray images with Keras, tensor flow, and deep learning. *Artificial & Computational Intelligence*. https://acors.org/ijacoi/VOL1_ISSUE3_09.pdf

Mashood Nasir, I., Attique Khan, M., Alhaisoni, M., Saba, T., Rehman, A., & Iqbal, T. (2020). A hybrid deep learning architecture for the classification of superhero fashion products: an application for Medical-Tech classification. *Computer Modeling in Engineering & Sciences*, 124(3), 1017–1033.

Mittal, A., Kumar, D., Mittal, M., Saba, T., Abunadi, I., Rehman, A., & Roy, S. (2020). Detecting pneumonia using convolutions and dynamic capsule routing for chest X-ray images. *Sensors*, 20(4), 1068.

Mughal, B., Muhammad, N., Sharif, M., Rehman, A., & Saba, T. (2018a). Removal of pectoral muscle based on topographic map and shape-shifting silhouette. *BMC Cancer*, 18(1), 1–14.

Mughal, B. Muhammad, N. Sharif, M. Saba, T., & Rehman, A. (2017) Extraction of breast border and removal of pectoral muscle in wavelet domain, *Biomedical Research*, 28(11), 5041–5043.

Mughal, B., Sharif, M., Muhammad, N., & Saba, T. (2018b). A novel classification scheme to decline the mortality rate among women due to breast tumor. *Microscopy Research and Technique*, 81(2), 171–180.

Mustafiz, R., & Mohsin, K. (2020). Assessing automated machine learning service to detect COVID-19 from X-Ray and CT images: A real-time smartphone application case study. arXiv preprint arXiv:2010.02715

Narayan Das, N., Kumar, N., Kaur, M., Kumar, V., & Singh, D. 2020. Automated deep transfer learning-based approach for detection of COVID-19 infection in chest X-rays. *Ing Rech Biomed.* 10.1016/j.irbm.2020.07.001.

Nazir, M., Khan, M. A., Saba, T., & Rehman, A. (2019). *Brain tumor detection from MRI images using multilevel wavelets.* In *2019, IEEE International Conference on Computer and Information Sciences (ICCIS)* (pp. 1–5).

Nour, M., Comert, Z., & Polat, K. 2020. A novel medical diagnosis model for COVID-19 infection detection based on deep features and bayesian optimization. *Applied Soft Computing*, 97, 106580.

Novel Variant of COVID-19. (2020). Reported in England [Online] Available: https://www.gov.uk/government/news/phe-investigating-a-novel-variant-of-covid-19 (Accessed:22 December 2020)

O'Callaghan, K. P., Blatz, A. M., & Offit, P. A. 2020. Developing a SARS-CoV-2 vaccine at warp speed. *JAMA*, 324(5), 437–438. doi:10.1001/jama.2020.12190.

Ohata, E. F., Bezerra, G. M., Chagas, J. V. S. D., Lira Neto, A. V., Albuquerque, A. B., Albuquerque, V. H. C. D., & Reboucas Filho, P. P. 2020. Automatic detection of COVID-19 infection using chest X-ray images through transfer learning. *IEEE/CAA Journal of Automatica Sinica*, 8(1), 1–10.

Ozturk, T., Talo, M., Yildirim, E. A., Baloglu, U. B., Yildirim, O., & Rajendra, A. U. 2020. Automated detection of COVID-19 cases using deep neural networks with X-ray images. *Computers in Biology and Medicine*, 121, 103792.

Panwar, H., Gupta, P. K., Siddiqui, M. K., Morales-Menendez, R., Bhardwaj, P., & Singh, V. 2020. A deep learning and grad-CAM based color visualization approach for fast detection of COVID-19 cases using chest X-ray and CT-Scan images. *Chaos Solitons Fractals*, 140, 110190.

Perumal, V., Narayanan, V., & Rajasekar, S. J. S. 2020. Detection of COVID-19 using CXR and CT images using Transfer Learning and Haralick features. *Applied Intelligence*, 1–18.

Perveen, S., Shahbaz, M., Saba, T., Keshavjee, K., Rehman, A., & Guergachi, A. (2020). Handling irregularly sampled longitudinal data and prognostic modeling of diabetes using machine learning technique. *IEEE Access*, 8, 21875–21885.

Rad, A. E., Rahim, M. S. M., Rehman, A. Altameem, A., & Saba, T. (2013) Evaluation of current dental radiographs segmentation approaches in computer-aided applications *IETE Technical Review*, 30(3), 210–222

Rad, A. E., Rahim, M. S. M., Rehman, A., & Saba, T. (2016) Digital dental X-ray database for caries screening. *3D Research*, 7(2), 1–5. doi:10.1007/s13319-016-0096-5

Ragb, H. K., Dover, I. T., Ali, R. (2020). Fused deep convolutional neural network for precision diagnosis of COVID-19 using chest X-ray images. arXiv (less.IV).

Rahim, M. S. M., Norouzi, A. Rehman, A., & Saba, T. (2017a). 3D bones segmentation based on CT images visualization, *Biomedical Research*, 28(8), 3641–3644

Rahim, M. S. M., Rehman, A., & Kurniawan, F., Saba, T. (2017b). Ear biometrics for human classification based on region features mining. *Biomedical Research*, 28 (10), 4660–4664

Rahimzadeh, M., & Attar, A. 2020. A new modified deep convolutional neural network for detecting COVID-19 from X-ray images. arXiv preprint arXiv:2004.08052

Rajpurkar, P., Irvin, J., Zhu, K., Yang, B., Mehta, H., Duan, T., Ding, D., Bagul, A., Langlotz, C., & Shpanskaya, K. 2017. Chenet: Radiologist-level pneumonia detection on chest x-rays with deep learning. arXiv preprint arXiv:1711.05225.

Ramzan, F., Khan, M. U. G., Iqbal, S., Saba, T., & Rehman, A. (2020b). Volumetric segmentation of brain regions from MRI scans using 3D convolutional neural networks. *IEEE Access*, 8, 103697–103709.

Ramzan, F., Khan, M. U. G., Rehmat, A., Iqbal, S., Saba, T., Rehman, A., & Mehmood, Z. (2020a). A deep learning approach for automated diagnosis and multi-class classification of alzheimer's disease stages using resting-state fMRI and residual neural networks. *Journal of Medical Systems*, 44(2), 37.

Rehman, A., Abbas N., Saba, T., Mahmood, T., & Kolivand, H. (2018c). Rouleaux red blood cells splitting in microscopic thin blood smear images via local maxima, circles drawing, and mapping with original RBCs, *Microscopic Research and Technique*, 81(7), 737–744. doi:10.1002/jemt.23030

Rehman, A. Abbas, N. Saba, T. Mehmood, Z. Mahmood, T., & Ahmed, K. T. (2018b). Microscopic malaria parasitemia diagnosis and grading on benchmark datasets. *Microscopic Research and Technique*, 81(9), 1042–1058. doi:10.1002/jemt.23071

Rehman, A., Abbas, N., Saba, T., Rahman, S. I. U., Mehmood, Z., & Kolivand, H. (2018a). Classification of acute lymphoblastic leukemia using deep learning, *Microscopy Research & Technique*, 81(11), 1310–1317. doi:10.1002/jemt.23139

Rehman, A., Khan, M. A., Mehmood, Z., Saba, T., Sardaraz, M., & Rashid, M. (2020) Microscopic melanoma detection and classification: A framework of pixel-based fusion and multilevel features reduction, *Microscopy Research and Technique*, 83(4), 410–423, doi:10.1002/jemt.23429

Rehman, A., Khan, M. A., Saba, T., Mehmood, Z., Tariq, U., & Ayesha, N. (2021). Microscopic brain tumor detection and classification using 3D CNN and feature selection architecture. *Microscopy Research and Technique*, 84(1), 133–149. doi:10.1002/jemt.23597.

Saba, T. (2020). Recent advancement in cancer detection using machine learning: Systematic survey of decades, comparisons and challenges. *Journal of Infection and Public Health*, 13(9), 1274–1289.

Saba, T., Haseeb, K., Ahmed, I., & Rehman, A. (2020). Secure and energy-efficient framework using Internet of Medical Things for e-healthcare. *Journal of Infection and Public Health*, 13 (10), 1567–1575.

Saba, T., Khan, M. A., Rehman, A., & Marie-Sainte, S. L. (2019a). Region extraction and classification of skin cancer: A heterogeneous framework of deep CNN features fusion and reduction. *Journal of Medical Systems*, 43(9), 289.

Saba, T., Khan, S. U., Islam, N., Abbas, N., Rehman, A., Javaid, N., & Anjum, A. (2019b). Cloud-based decision support system for the detection and classification of malignant cells in breast cancer using breast cytology images. *Microscopy Research and Technique*, 82(6), 775–785.

Saba, T., Rehman, A. Mehmood, Z., Kolivand, H., & Sharif, M. (2018) Image enhancement and segmentation techniques for detection of knee joint diseases: a survey, *Current Medical Imaging Reviews*, 14(5), 704–715. doi:10.2174/1573405613666170912164546

Sethy, P. K.,& Behera, S. K. 2020. Detection of coronavirus disease (covid-19) based on deep features. *Preprints*, 2020030300, 2020.

Sharma, S. 2020. Drawing insights from COVID-19-infected patients using CT scan images and machine learning techniques: A study on 200 patients. *Environmental Science and Pollution Research*, 27, 37155–37163.

Souza, J. C., Diniz, J. O. B., Ferreira, J. L., Da Silva, G. L. F., Silva, A. C., & De Paiva, A. C. 2019. An automatic method for lung segmentation and reconstruction in chest X-ray using deep neural networks. *Computer Methods and Programs in Biomedicine*, 177, 285–296.

Sun, L., Liu, G., Song, F., Shi, N., Liu, F., Li, S., Li, P., Zhang, W., Jiang, X., & Zhang, Y. 2020. A combination of four clinical indicators predicts the severe/critical symptom of patients infected COVID-19. *Journal of Clinical Virology*, 128, 104431.

Tuncer, T., Dogan, S., & Ozyurt, F. 2020. An automated residual exemplar local binary pattern and iterative ReliefF based corona detection method using lung X-ray image. *Chemometrics and Intelligent Laboratory Systems*, 203, 104054.

Ucar, F., & Korkmaz, D. 2020. COVIDiagnosis-net: Deep Bayes-SqueezeNet based diagnostic of the coronavirus disease 2019 (COVID-19) from X-Ray images. *Medical Hypotheses*, 140, 109761.

Ullah, H., Saba, T., Islam, N., Abbas, N., Rehman, A., Mehmood, Z., & Anjum, A. (2019). An ensemble classification of exudates in color fundus images using an evolutionary algorithm based optimal features selection. *Microscopy Research and Technique*, 82(4), 361–372.

Vasilev, Y., Sergunova, K., Bazhin, A., Mazri, A., Vasileva, Y., Suleumanov, E., Semenov, D., Kudryavtsev, N., Panina, O., & Khoruzhaya, A. 2020. MRI of the lungs in patients with COVID-19: Clinical case. medRxiv.

Wang, D., Mao, J., Zhou, G., Xu, L., & Liu, Y. 2020a. An efficient mixture of deep and machine learning models for COVID-19 diagnosis in chest X-ray images. *PLoS One*, 15, e0242535.

Wang, L., Lin, Z. Q., & Wong, A. 2020b. COVID-Net: A tailored deep convolutional neural network design for detection of COVID-19 cases from chest X-ray images. *Scientific Reports*, 10, 19549.

Wang, S., Kang, B., Ma, J., Zeng, X., Xiao, M., Guo, J., Cai, M., Yang, J., Li, Y., Meng, X., & Xu, B. 2020c. A deep learning algorithm using CT images to screen for Corona Virus 1 Disease (COVID-19). medRxiv

WHO. (2020a). *Declares coronavirus outbreak a global health emergency* [Online]. Available at: https://www.who.int/news/item/27-04-2020-who-timeline---covid-19 (Accessed:15 October 2020)

WHO. 2020b *Director General's Remarks on COVID-19* [Online]. Available at: https://www.who.int/director-general/speeches/detail/who-director-general-s-opening-remarks-at-executive-board-meeting (Accessed:15 October 2020)

WHO. 2020c. *Coordination on vaccine of COVID-19* [Online]. Available at: https://www.who.int/emergencies/diseases/novel-coronavirus-2019/global-research-on-novel-coronavirus-2019-ncov/accelerating-a-safe-and-effective-covid-19-vaccine (Accessed:27 December 2020)

WHO. 2021a *COVID-19 advice for public* [Online]. Available at: https://www.who.int/news/item/27-04-2020-who-timeline---covid-19 (Accessed:02 January 2021)

WHO. 2021b *Current statistics of COVID-19* [Online]. Available at: https://www.who.int/emergencies/diseases/novel-coronavirus-2019/advice-for-public (Accessed:03 January 2021).

WHO Vaccine Coordination. 2020. *Public statement for collaboration on COVID-19 vaccine development* [Online]. Available at: https://www.who.int/news/item/13-04-2020-public-statement-for-collaboration-on-covid-19-vaccine-development

Xie, X., Zhong, Z., Zhao, W., Zheng, C., Wang, F., & Liu, J. 2020. Chest CT for typical 2019-nCoV pneumonia: Relationship to negative RT-PCR testing. *Radiology*, 200343.

Xu, X., Jiang, X., Ma, C., Du, P., Li, X., Lv, S., Yu, L., Ni, Q., Chen, Y., Su, J., Lang, G., Li, Y., Zhao, H., Liu, J., Xu, K., Ruan, L., Sheng, J., Qiu, Y., Wu, W., Liang, T., & Li, L. 2020. A deep learning system to screen novel coronavirus disease 2019 pneumonia. *Engineering (Beijing)*, 6, 1122–1129.

3 Solutions of Differential Equations for Prediction of COVID-19 Cases by Homotopy Perturbation Method

Nahid Fatima and Monika Dhariwal

CONTENTS

3.1 INTRODUCTION

On January 9, 2020, WHO said in a statement that, according to the preliminary determination of Chinese researchers, there was a new kind of coronavirus. By March 20, 2020, more than 6,000 deaths had occurred due to coronavirus (COVID-19) worldwide. There have been cases of COVID-19 from around 180 countries around the world, including India. Coronavirus is a large virus family that causes common colds and serious toxins, such as severe acidity in the case of SARSOM and M/s. SARS-CoV-2. Coronavirus is similar to the virus that caused SARS. Many coronaviruses are zoonotic, that is, they reach animals via humans. Like other viruses, the SARS-CoV-2 virus is round in shape, with an outer side like an onion. WHO has declared COVID-19 as a pandemic. Coronavirus symptoms appear within 2 to 14 days of exposure and include fever, cough, runny nose, and shortness of breath.

3.2 HOW DOES THIS DISEASE SPREAD?

The coronavirus (COVID-19) is believed to spread basically from close contact from one person to another. It may also spread from a person who does not show any symptoms of the virus. We are still in the learning phase on understanding how this virus will behave in the future, and the intensity of illness in the case of an individual. Individuals who are physically close (within 6 feet) to an individual with COVID-19 or have coordinate contact with that individual are at most risk of contracting this virus. Respiratory droplets are produced when a COVID-infected person coughs or sneezes. Contaminations happen basically through the introduction of respiratory beads when an individual is in near contact with somebody who has COVID-19. Respiratory beads cause disease when they are breathed in or kept on mucous layers, such as those that line the interior of the nose and mouth (Batista, 2020).

Infectious diseases like tuberculosis, measles, and chickenpox are spread through this kind of transmission, commonly known as **airborne transmission**. Such contamination can occur as a result of poorly ventilated space and rooms. Available data indicate that it is much more common for the virus that causes COVID-19 to spread through close contact with a person who has COVID-19 than through airborne transmission.

Respiratory droplets can also land on surfaces and objects. It is possible that a person could get COVID-19 by touching a surface or object that has the virus on it and then touching their own mouth, nose, or eyes. Spread from touching surfaces is not thought to be a common way that COVID-19 spreads.

3.3 EDUCATION SYSTEM AFFECTED BY COVID-19

The closure of schools, colleges, and educational institutions as a result of coronavirus has affected the education of 157 crore students in 191 countries of the world, which is 91.3% of the total number of students enrolled at various levels (Aristovnik et al., 2020). This information has been revealed in a study by the UN Educational, Scientific and Cultural Organization (UNESCO). It revealed that the major impact of the closure of schools is affecting students and particularly girls of the disadvantaged sections.

Experts say that the virus will definitely change the face of the field of education. The epidemic will change the way of reading and teaching and its widespread impact will be seen in every field and at every level of education. Looking from the perspective of India, it shows that children will be deterred from going to another country or even studying in other states within the country because the situation will not be the same as before.

The spread of COVID-19 has influenced instructive frameworks around the world, driving to the near-total closures of schools and colleges. Nearly 1.07 billion students are affected by the closure of schools due to this pandemic. One of the UNICEF reports suggests that nearly 53 countries worldwide are following nationwide closure. This has impacted almost 61% of the world's population of students enrolled.

Most top-level institutions across the globe, such as Cambridge International Examination and the International Baccalaureate, have cancelled or postponed their courses/examinations. Additionally, placements, SATs, and ACTs have been cancelled or scheduled for online. The effect has been more serious for deprived children and their families, causing hindered learning, compromised sustenance, and child-care issues, and resulting in financial expenditure to families who may not be able to work. UNESCO has proposed to use online/distance learning programs in response to the closure of institutions so that remote students can be reached and this would limit the disturbance of education.

3.3.1 HOMOTOPY PERTURBATION METHOD (HPM)

Ji-Huan He first proposed the method of homotopy perturbation (He, 1999, 2009). This method has proved tremendously effective in solving various mathematical problems (He, 2005a, 2005b) so that it has been widely used by researchers and scientists across the globe.

Let us presume:

$$B(u) - g(s) = 0, s \in \xi \tag{3.1}$$

with starting condition

$$C\left(u, \frac{\partial u}{\partial n} = 0, s \in \lambda\right) \tag{3.2}$$

point B is a general operative, $g(s)$ is an accepted analytic action, C is an initial operative, also λ is the boundary of the domain ξ. Operative B is divided into two functions, K and M, point K being a linear and M being a nonlinear operative. Equation (3.1) could be, so,

$$K(u) + M(u) - g(s) = 0 \tag{3.3}$$

proving the homotopy formula, our created homotopy $v(s, p)$: $\xi \times [0, 1] \rightarrow S,$.

$$(H(v,p) = (1-p)\left[K(v) - K(u_0)\right] + p\left[B(v) - g(s)\right] = 0 \tag{3.4}$$

Or

$$H(v,p) = K(v) - K(u_0) + pK(u_0) + p\left[M(v) - g(s)\right] = 0 \tag{3.5}$$

point $p \in [0, 1]$ is called homotopy limitation and u_0 is an original proximate as the explanation about (3.1), and that satisfy the initial condition. certainly, about (3.4) or (3.5), we get

$$H(v, 0) = K(v) - K(u_0) = 0 \tag{3.6}$$

$$H(v, 1) = B(v) - g(s) = 0 \tag{3.7}$$

Along with altering action like p from zero to unity is just that of $H(v, p)$ from $K(v) - K(u_0)$ to $B(v) - g(s)$.

In topology, here is called deformities $K(v) - K(u_0)$ and $B(v) - g(s)$ are called homotopy.

We suppose that the explanation as (3.4) or (3.5) we write as a series in p as pursue [4, 5]:

$$V = v_0 + pv_1 + p^2v_2 + p^3v_3 + \ldots \qquad (3.8)$$

Putting p =1 result in the proximate explanation as (3.1)

$$U = \lim_{p \to 1} v = v_0 + v_1 + v_2 + v_3 + \ldots \qquad (3.9)$$

The series (3.9) is convergence for most cases [4, 6, 7, 8].

3.4 SIR MODEL

The SIR model was first introduced by W. O. Kermach and A. G McKendrick in 1927 and many other scientists. Brauer (2017) and Hai et al. (2020) used the SIR model as the best model deployed for an infectious disease. This model divides the population into three groups, namely:

S (t) is the susceptible people at the time

I (t) is the infected people at the time

R (t) is the recovered people at the time.

This method has been used successfully many times before for spreading diseases such as yellow fever, plague, influenza, and avian influenza (Pell et al., 2019). Therefore, we have tried to solve the differential equations of COVID-19 using this method. This method is very helpful in giving a mathematical model to COVID-19 (Dhariwal & Fatima, 2020).

Now we have solved the differential equations of COVID-19, with the help of the SIR model. In these equations, we took the data of September 29, 2020 people of India who were caught by the COVID-19 epidemic.

Total confirmed cases on September 29, 2020 in India was 6,223,519.

Deaths	97,527
Recovered	5,184,723
Active cases	940,384
Susceptible persons	74,196,729

Therefore, we take the

$$s(0) = 74.196729$$

$$i(0) = 6.223519$$

$$r(0) = 97527 + 5184723 = 5.282250$$

$$g = \frac{\text{active cases in India on September 29, 2020, for COVID} - 19}{\text{susceptible people of india on September 29, 2020, for COVID} - 19}$$

$$g = \frac{940384}{74.196729} = 0.0012674$$

$$f = \frac{1}{14} = 0.0714$$

$$\frac{ds}{dt} = -gsi$$

$$\frac{di}{dt} = gsi - fi$$

$$\frac{dr}{dt} = fi$$

Where t is the independent variable s, i, r are dependent variables, i.e.

- **s** represents the susceptible person at the time t
- **i** represents the infected person at the time t
- **r** represents the recovered person at the time t
- **g** is transmission coefficient
- f is recovery,

Now we will solve these equations with the help of the HPM method. By the homotopy technique (3.1–3.3) we get

$$(1-p)\frac{dS}{dt} + p\left(\frac{dS}{dt} + si\right)$$

$$\frac{dS}{dt} = p(-0.0012674si) \tag{3.10}$$

$$(1-p)\frac{di}{dt} + p\left(\frac{di}{dt} - 0.0012674si + 0.0714i\right)$$

$$\frac{di}{dt} = p(0.0012674si - 0.0714i) \tag{3.11}$$

$$\frac{dr}{dt} = p0.0714i \tag{3.12}$$

$$s = s_0 + p^1 s_1 + p^2 s_2 + \dots$$

$$i = i_0 + p^1 i_1 + p^2 i_2 + \dots$$

$$r = r_0 + p^1 r_1 + p^2 r_2 + \dots$$

Putting value s, i, r in Equations (3.4–3.6) we get

$$\frac{ds}{dt} = p\left(-0.0012674\left\{s_0 + p^1 s_1 + p^2 s_2 + \ldots\right\}\left\{i_0 + p^1 i_1 + p^2 i_2 + \ldots\right\}\right) \tag{3.13}$$

$$\frac{di}{dt} = p\left(\begin{array}{l}0.0012674\left[\left\{s_0 + p^1 s_1 + p^2 s_2 + \ldots\right\}\left\{i_0 + p^1 i_1 + p^2 i_2 + \ldots\right\}\right]\\-0.0714\left\{i_0 + p^1 i_1 + p^2 i_2 + \ldots\right\}\end{array}\right) \tag{3.14}$$

$$\frac{dr}{dt} = p\,0.0714\left\{i_0 + p^1 i_1 + p^2 i_2 + \ldots\right\} \tag{3.15}$$

Both sides comparing the coefficient of p in Equations (3.7–3.9) we get

$$s_0 = 74.196729$$
$$i_0 = 6.223519$$
$$r_0 = 5.282250$$

$$\frac{ds_1}{dt} = -0.0012674\,(74.196729)\,(6.223519)$$

$$s_1 = -0.585240t$$

$$\frac{di_1}{dt} = 0.0012674\,(74.196729)\,(6.223519) - 0.0714\,(6.223519)$$

$$i_1 = 0.140881t$$

$$\frac{dr_1}{dt} = 0.0714\left\{6.223519\right\}$$

$$r_1 = 0.444359t$$

So, by the HPM we get the solution of Equations (3.1–3.3) (Table 3.1)

$$s(t) = 74.196729 - 0.585240t + \ldots$$

$$i(t) = 6.223519 + 0.140881t + \ldots$$

$$r(t) = 5.282250 + 0.444359t + \ldots$$

With the help of the above table, we will plot these curves (Figures 3.1–3.4):

- **S(t)** reflects the number of susceptible people at the time, **I**(t) shows infected people at the time.
- **R(t)** shows recovered people at the time.

In all the above tables we are plotting dates on the x-axis and number of people on the y-axis. We can see the recovered rate is increasing and it is at the maximum on October 14.

TABLE 3.1

SIR Methods Simulation and Results of HPM from September 29 to October 14, 2020.

S.No.	Date	S	I	R	S+I+R
1	29/09/2020	74.196729	6.223519	5.282250	85.702498
2	30/09/2020	73.611489	6.364399	5.726609	85.702498
3	1/10/2020	73.026249	6.505281	6.170968	85.702498
4	2/10/2020	72.441009	6.646162	6.615327	85.702498
5	3/10/2020	71.855769	6.787043	7.059686	85.702498
6	4/10/2020	71.270529	6.927924	7.504045	85.702498
7	5/10/2020	70.685289	7.068805	7.948404	85.702498
8	6/10/2020	70.100049	7.209686	8.392763	85.702498
9	7/10/2020	69.514809	7.350567	8.837122	85.702498
10	8/10/2020	68.929569	7.491448	9.281481	85.702498
11	9/10/2020	68.344329	7.632329	9.72584	85.702498
12	10/10/2020	67.759089	7.77321	10.17019	85.702498
13	11/10/2020	67.173849	7.914091	10.614558	85.702498
14	12/10/2020	66.588609	8.054972	11.058917	85.702498
15	13 / 10 /2020	66.003369	8.195853	11.503276	85.702498
16	14 / 10 /2020	65.418129	8.336734	11.947635	85.702498

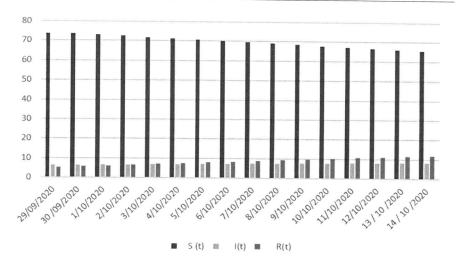

FIGURE 3.1 Number of susceptible (S), infected (I), and recovered (R) COVID-19 cases to October 14, 2020.

We have observed that on October 14 S is 65.418129, I is 8.336734 and R is 11.947635 while the total confirmed cases on September 29, 2020, in India was 6,223,519, after solving the differential equations of COVID-19, which is made with the help of the SIR model.

We have obtained the solution of the three-dimensional mathematical model of Novel Coronavirus differential equation by using HPM and we have found that the HPM method is a very helpful and useful method to solve the Novel Coronavirus

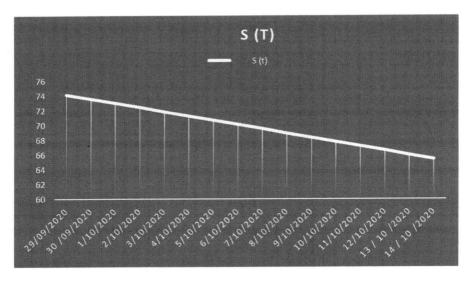

FIGURE 3.2   Number of susceptible (S) to COVID-19.

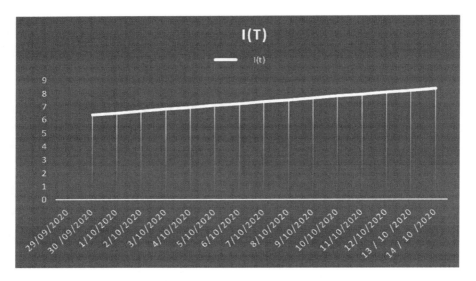

FIGURE 3.3   Number of infected with COVID-19.

equation. In all the models for prediction of number of corona cases we can find the solutions using HPM is not very difficult; hence this method can be easily used to solve the Novel Coronavirus equation (Sonnino 2020; Yu et al., 2020).

Now we will solve the differential equations of COVID-19, with the help of the SIR model. In these equations, we took the data of October 29, 2020, people of India who were caught by COVID-19 epidemics.

Total confirmed cases on October 29, 2020 in India were 8,088,049.

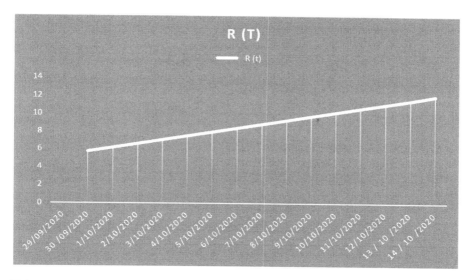

FIGURE 3.4   Number of recovered from COVID-19.

Deaths	121,132
Recovered	7,371,906
Active cases	8,088,049
Susceptible persons	**107,728,088**

So, we take

$$s(0) = 10.7728088$$

$$i(0) = 8.088049$$

$$r(0) = 7371906 + 12132 = 7.384038$$

$$g = \frac{\text{active cases of India on October 29, 2020, for COVID} - 19}{\text{susceptible people of India on Cctober 29, 2020, for COVID} - 19}$$

$$g = \frac{8088049}{10.7728088} = 0.000750783$$

$$f = \frac{1}{14} = 0.0714$$

$$\frac{ds}{dt} = -gsi$$

$$\frac{di}{dt} = gsi - fi$$

$$\frac{dr}{dt} = fi$$

Where t is the independent variable and s, i, r are dependent variables, i.e.

- **s** denotes the susceptible person at the time t
- **i** denotes the infected person at the time t
- **r** denotes the recovered person at the time t
- **g** is transmission coefficient
- **f** is recovery.

Now we will solve these equations with the help of the HPM method.
By the homotopy theory (3.1–3.3) we get

$$(1-p)\frac{dS}{dt} + p\left(\frac{dS}{dt} + si\right)$$

$$\frac{dS}{dt} = p(-0.00750783si) \qquad (3.16)$$

$$(1-p)\frac{di}{dt} + p\left(\frac{di}{dt} - 0.00750783si + 0.0714i\right)$$

$$\frac{di}{dt} = p(0.00750783si - 0.0714i) \qquad (3.17)$$

$$\frac{dr}{dt} = p0.0714i \qquad (3.18)$$

$$s = s_0 + p^1 s_1 + p^2 s_2 + \dots$$

$$i = i_0 + p^1 i_1 + p^2 i_2 + \dots$$

$$r = r_0 + p^1 r_1 + p^2 r_2 + \dots$$

Putting value s, i, r in Equations (3.4–3.6) we get

$$\frac{ds}{dt} = p\left(-0.000750783\{s_0 + p^1 s_1 + p^2 s_2 + \dots\}\{i_0 + p^1 i_1 + p^2 i_2 + \dots\}\right) \qquad (3.19)$$

$$\frac{di}{dt} = p\left(\begin{array}{l} 0.00750783\left[\{s_0 + p^1 s_1 + p^2 s_2 + \dots\}\{i_0 + p^1 i_1 + p^2 i_2 + \dots\}\right] \\ -0.0714\{i_0 + p^1 i_1 + p^2 i_2 + \dots\} \end{array}\right) \qquad (3.20)$$

$$\frac{dr}{dt} = p0.0714\{i_0 + p^1 i_1 + p^2 i_2 + \dots\} \qquad (3.21)$$

Both sides comparing the coefficient of p in the above equation we get

$$s_0 = 10.7728088$$
$$i_0 = 8.088049$$
$$r_0 = 7.384038$$
$$\frac{ds_1}{dt} = -0.00750783(10.7728088)(8.088049)$$
$$s_1 = -0.654164t$$
$$\frac{di_1}{dt} = 0.000750783(10.7728088)(8.88049) - 0.0714(8.088049)$$
$$i_1 = 0.071857t$$
$$\frac{dr_1}{dt} = 0.0714\{8.088049\}$$
$$r_1 = 0.577486t$$

So, by the HPM we get the following solution of the above equations

$$s(t) = 10.7728088 - 0.654164t + \ldots$$
$$i(t) = 8.088049 + 0.071857t + \ldots$$
$$r(t) = 7.384038 + 0.577486t + \ldots$$

We introduce (Figures 3.5–3.8) Table 3.2:

We will repeat the above procedure of COVID-19 data from November 29, 2020, creating differential equations with the help of the SIR model and simplifying these equations with the help of HPM.

Total confirmed cases on November 29, 2020, in India was 9,432,075.

Deaths	137,177
Recovered	846,313
Active cases	446,411
Susceptible persons	**140,379,976**

So, we will choose,

$$s(0) = 14.0379976$$

$$i(0) = 9.432075$$

$$r(0) = 8846313 + 137177 = 8.983490$$

$$g = \frac{\text{active cases in India on November 29, 2020, for COVID}-19}{\text{susceptible people of India on November 29, 2020, for COVID}-19}$$

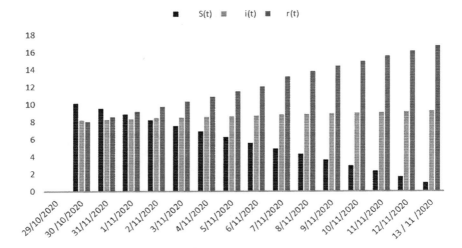

FIGURE 3.5 shows the number of susceptible (S), infected (I), and recovered (R) COVID-19 cases in India. Susceptible is shown in blue, recovered in green, and infected in red. The number reflects the data from October 29 to November 13, 2020. We can clearly see an increasing trend in the total recovered cases across these dates.

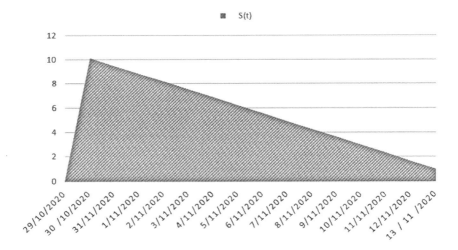

FIGURE 3.6   Number of susceptible (S) COVID-19 cases to November 13, 2020.

$$g = \frac{446411}{14.0379976} = 0.00031800$$

$$f = \frac{1}{14} = 0.0714$$

$$\frac{ds}{dt} = -gsi$$

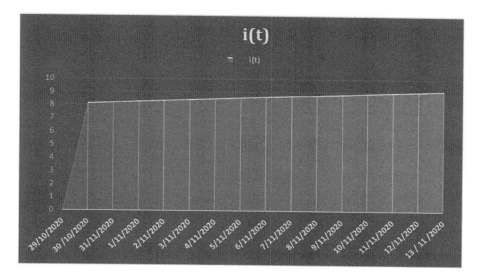

FIGURE 3.7 Number of infected COVID-19 cases in India to November 13, 2020.

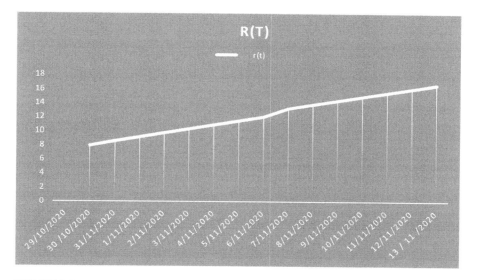

FIGURE 3.8 Number of recovered COVID-19 cases to November 13, 2020.

$$\frac{di}{dt} = gsi - fi$$

$$\frac{dr}{dt} = fi$$

Where t is the independent variable and s, i, r are dependent variables, i.e.

TABLE 3.2

SIR Methods Simulation and Results of HPM to November 13, 2020

S.No.	Date	S	I	R	S+I+R
1	29/10/2020	10.772808	8.088049	7.384038	26.2448958
2	30 /10/2020	10.1211688	8.159906	7.961524	26.2425988
3	31/11/2020	9.4695288	8.231763	8.53901	26.2403018
4	1/11/2020	8.8178888	8.30362	9.116496	26.2380048
5	2/11/2020	8.1662488	8.375477	9.693982	26.2357078
6	3/11/2020	7.5146088	8.447334	10.271468	26.2334108
7	4/11/2020	6.8629688	8.519191	10.848954	26.2311138
8	5/11/2020	6.2113288	8.591048	11.42644	26.2288168
9	6/11/2020	5.5596888	8.662905	12.003926	26.2265198
10	7/11/2020	4.9080488	8.734762	13.158898	26.8017088
11	8/11/2020	4.2564088	8.806619	13.736384	26.7994118
12	9/11/2020	3.6047688	8.878476	14.31387	26.7971148
13	10/11/2020	2.9531288	8.950333	14.891356	26.7948148
14	11/11/2020	2.3014888	9.02219	15.468842	26.7925208
15	12/11/2020	1.6498488	9.094047	16.046328	26.7902238
16	13/11/2020	0.9982088	9.165904	16.623814	26.7879268

s denotes the susceptible person at the time t

i denotes the infected person at the time t

r denotes the recovered person at the time t

g denotes transmission coefficient

f denotes recovery.

Now we will solve these equations with the help of HPM.

By the homotopy method (3.1–3.3) we get

$$\left(1-p\right)\frac{dS}{dt}+p\left(\frac{dS}{dt}+si\right)$$

$$\frac{dS}{dt}=p\left(-0.00031800si\right) \tag{3.22}$$

$$\left(1-p\right)\frac{di}{dt}+p\left(\frac{di}{dt}-0.00031800si+0.0714i\right)$$

$$\frac{di}{dt}=p\left(0.00031800si-0.0714i\right) \tag{3.23}$$

$$\frac{dr}{dt}=p0.0714i \tag{3.24}$$

$$s=s_0+p^1s_1+p^2s_2+$$

$$i = i_0 + p^1 i_1 + p^2 i_2 + \ldots$$

$$r = r_0 + p^1 r_1 + p^2 r_2 + \ldots$$

Putting value s, i, r in Equations (3.4–3.6) we get

$$\frac{ds}{dt} = p\left(-0.00031800\{s_0 + p^1 s_1 + p^2 s_2 + \ldots\}\{i_0 + p^1 i_1 + p^2 i_2 + \ldots\}\right) \tag{3.25}$$

$$\frac{di}{dt} = p\left(\begin{array}{l} 0.00031800\left[\{s_0 + p^1 s_1 + p^2 s_2 + \ldots\}\{i_0 + p^1 i_1 + p^2 i_2 + \ldots\}\right] \\ -0.0714\{i_0 + p^1 i_1 + p^2 i_2 + \ldots\} \end{array}\right) \tag{3.26}$$

$$\frac{dr}{dt} = p\,0.0714\{i_0 + p^1 i_1 + p^2 i_2 + \ldots\} \tag{3.27}$$

Both side comparing the coefficient of p in Equations (3.7–3.9) we get

$$s_0 = 14.0379976$$

$$i_0 = 9.432075$$

$$r_0 = 8.983490$$

$$\frac{ds_1}{dt} = -0.00031800(14.0379976)(9.432075)$$

$$s_1 = -0.042105t$$

$$\frac{di_1}{dt} = 0.00031800(14.0379976)(9.432075) - 0.0714(9.432075)$$

$$i_1 = -0.631344t$$

$$\frac{dr_1}{dt} = 0.0714\{9.432075\}$$

$$r_1 = 0.673450t$$

So, by the HPM we get the solution of the above equations

$$s(t) = 14.0379976 - 0.042105t + \ldots$$

$$i(t) = 9.432075 - 0.631344t + \ldots$$

$$r(t) = 8.983490 + 0.673450t + \ldots$$

We have (Figure 3.9) Table 3.3.

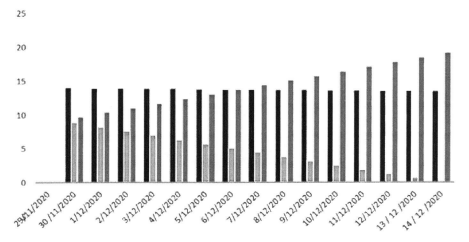

FIGURE 3.9 Number of susceptible (S), infected (I), and recovered (R) COVID-19 cases to December 12, 2020.

TABLE 3.3
SIR Methods Simulation and Results of HPM from November 29 to December 12, 2020

S.NO.	Date	S	I	R
1	29/11/2020	14.0379976	9.432075	8.983490
2	30 /11/2020	13.9958926	8.800731	9.65694
3	1/12/2020	13.9537876	8.169387	10.33039
4	2/12/2020	13.9116826	7.538043	11.00384
5	3/12/2020	13.8695776	6.906699	11.67729
6	4/12/2020	13.8274726	6.275355	12.35074
7	5/12/2020	13.7853676	5.644011	13.02419
8	6/12/2020	13.7432626	5.012667	13.69764
9	7/12/2020	13.7011576	4.381323	14.37109
10	8/12/2020	13.6590526	3.749979	15.04454
11	9/12/2020	13.6169476	3.118635	15.71799
12	10/12/2020	13.5748426	2.487291	16.39144
13	11/12/2020	13.5327376	1.855947	17.06489
14	12/12/2020	13.4906326	1.224603	17.73834

3.5 CONCLUSION

In this chapter we have solved three-dimensional mathematical models of Novel Coronavirus differential equations by using HPM, and we have found that HPM is a very useful method to solve Novel Coronavirus equations. In all the models, to predict the number of corona cases, we can derive a solution using HPM.

Different scientific mathematical models were studied to predict the endemical tendency including severity, reproduction, and herd immunity. It is still a tedious assignment at the current stage of the pandemic. Due to the unique characteristics of the virus, its epidemiological attributes cannot yet be evaluated for certain.

We have tried to predict the infected cases and recovered cases by taking the data for two different months—namely, October and December. The data is presented earlier in the chapter in the forms of tables. We have then plotted the curve based on the data in the table after applying HPM. A few techniques have underlying restrictions, due to lack of information. To fill these research gaps, supplementary research is the need of the hour. An absolute model can be depicted, once this pandemic withers away and sufficient data is collected for study.

REFERENCES

Aristovnik, A., Keržič, D., Ravšelj, D., Tomaževič, N., Umek, L. (2020). Impacts of the COVID-19 Pandemic on Life of Higher Education Students: A Global Perspective *Sustainability* 12 (20), 8438. doi:10.3390/su12208438.

Batista, M. (2020). Estimation of the final size of the COVID-19 epidemic, medRxiv, 2020.2002.2016.20023606.

Brauer, F. (2017). Mathematical epidemiology: Past, present, and future, *Infectious Disease Modelling*, 2 113–127.

COVID-19 March (2020). Educational Disruption and Response. UNESCO. 4 March 2020. Retrieved 28 March 2020.

Dhariwal, M., Fatima, N. (2020). *Homotopy Perturbation Method for Solving Mathematical Model of Novel Coronavirus Differential Equations*. SSRN: https://ssrn.com/abstract=3627481 or http://dx.doi.org/10.2139/ssrn.3627481

Hai, N. D., George, Y., Chao, Z. (2020). Long-term analysis of a stochastic SIRS model with general incidence rates, *SIAM Journal on Applied Mathematics*, 80 (2), 814–838. doi: 10.1137/19M1246973.

He, J. H. (1999). Homotopy perturbation technique, *Computer Methods in Applied Mechanics and Engineering*, 178, 257–262.

He, J. H. (2005a). Application of Homotopy Perturbation Method to nonlinear wave equations, *Chaos, Solitons & Fractals*, 26, 695–700.

He, J. H. (2005b). Homotopy perturbation method for bifurcation of nonlinear problems, *International Journal of Nonlinear Sciences and Numerical Simulation*, 6 (2), 207–208.

He, J. H. (2009). An elementary introduction to the Homotopy Perturbation Method, *Computers and Mathematics with Applications*, 578 (3), 410–412.

Pell, B., Kuang Y., Viboud, C., Chowell, G., (2019). Using phenomenological models for forecasting the 2015 Ebola challenge, *Epidemics*, 22 (2018) 62–70.

Sonnino, G. (2020). Dynamics of the COVID-19–Comparison between the Theoretical Predictions and Real Data. arXiv preprint arXiv: 2003.13540.

Yu, Y. et al. (2020). Dynamical analysis of a diffusive SIRS model with general incidence rate, *Discrete Continuous Dyn Syst-B 2020*, 25 (7), 2433–2451.

4 Predictive Models of Hospital Readmission Rate Using the Improved AdaBoost in COVID-19

Arash Raftarai, Rahemeh Ramazani Mahounaki, Majid Harouni, Mohsen Karimi, and Shakiba Khadem Olghoran

CONTENTS

4.1 INTRODUCTION

The COVID-19 virus is a respiratory infection known as coronavirus. At the beginning of the pandemic, it was thought that the virus could be transmitted between humans and animals and that it could mutate. It has also adapted to the possibility of pathogenesis among humans (Afza et al., 2020; Khan et al., 2021). The outbreak of this deadly disease began in China, and its devastating effects are widely felt in all areas. The severity of the disease caused by this virus has been observed from asymptomatic or mild to high severity and death. Also, a significant number of patients who have obvious evidence of clinical infection need to be hospitalized. The prevalence of this disease has been such that it has posed a great challenge to medical staff and health and hospital equipment. Challenges such as lack of personnel and medical staff, loss of life to medical personnel, lack of hospital equipment and even medication for treatment against the COVID-19 virus have been raised. Maintaining medical staff and maintaining equipment for later use is important. At the same time, existing costs should be reduced as much as possible. One of the challenges facing health organizations is to reduce costs in the face of this deadly disease while increasing the quality of services provided, or at worst, not degrading the quality of services provided, because improving the quality of health services is directly related to human lives. Therefore, reducing costs should not lead to endangering human lives (Abbas et al., 2019a, b, c; Adeel et al., 2020). Quality improvement in the health industry is defined by the motivating forces and their impact. One of the parameters for improving health services' quality in the health industry is to reduce the rate of hospitalization and readmission. In other words, the rate of readmission in a hospital is a determining indicator for measuring the quality of services provided in care (Afza et al., 2019; Al-Ameen et al., 2015; Abbas et al., 2018). As mentioned, inpatient medical care resources such as doctors, nurses, and used equipment, are expensive and limited. Lack of this equipment can cause unwanted and harmful crises. This limitation is clearly palpable in the COVID-19 pandemic. Various studies have concluded that, if a patient needs to be readmitted, the health unit will incur exorbitant costs (Amin et al., 2019a, 2020). Decreasing costs while increasing quality is one of the important challenges of health organizations. Quality of services is important in these organizations because it is directly related to human life. Therefore, cost reduction should not endanger human life. Quality is improved in the health industry by effective factors such as health data. Health data is one of the main properties of each organization associated with the health industry (Ejaz et al., 2018a, 2018b, 2019, 2020, Adabavazeh et al., 2020; Amin et al., 2018, 2019a, 2019b, 2019c; Javed et al., 2019a, 2019b, 2020a, 2020b; Khan et al., 2017, 2019a, 2019b, 2019c, 2019d, 2019e, 2020a, 2020b, 2020d; Marie-Sainte et al., 2019a, 2019b; Mughal et al., 2018a, 2018b; Rahim et al., 2017a, 2017b, 2018a, 2018b, 2018c, 2020a, 2020b; Saba, 2018a, 2018b, 2019a, 2019b, 2019c, 2020a, 2020b; Yousaf et al., 2019a, 2019b). The successful performance of health organizations depends on the procedure of data gathering, storing, and analysis. Useful employment of these data can be turned into a financial source and followed by cost reduction. One practical way of using these data is data mining-based methods and algorithms (Fahad et al., 2018; Husham et al., 2016; Hussain et al., 2020; Iftikhar et al., 2017). Readmission rate reduction is one of the parameters of quality improvement in the health industry. In other words, the hospital readmission rate is a determining index to evaluate the quality of health and care services (Iqbal et al., 2017, 2018, 2019). Medical care resources in medical centers, such as doctors, nurses, and the equipment employed, are limited and expensive; if a patient needs readmission, the respective health unit undergoes extra and heavy costs. In this regard, the United States has introduced various hospital readmission rules through which people requiring readmission pay lower costs and the respective unit has to pay a fine (Jamal et al., 2017; Javed et al., 2018, 2019a, 2020a). The process of identifying patients who may need to be readmitted is very complex, due to such reasons as patients' health status, type of disease, quality of caring for patients during admission, mental and psychological factors after discharge, quality of life after discharge, and social factors (Khan et al., 2018, 2019b, 2020b). Especially in COVID-19, it is very complicated because the virus and its effects have not been discovered yet. Presenting models for determining feasible readmission patterns requires complex methods (Ejaz et al., 2018b; Khan et al., 2020; Ramzan et al., 2018a, 2018b, 2018c). One of these methods is that based on data mining (Liaqat et al., 2020; Lung et al., 2014; Majid et al., 2020; Marie-Sainte et al., 2019a). Using data mining-based methods can reveal hidden patterns between data. Accordingly, the feasible readmission pattern is obtained by high accuracy. The technology of useful data employment to discover knowledge and relation between patterns and data is called data mining. Extracting information and discovering hidden patterns from large databases is complex (Harouni et al., 2014); so, data mining-based methods are used to extract information. There are different steps in a data-mining process, including preprocessing, feature extraction, feature selection, and classification (Agbehadji et al., 2020, Harouni et al., 2010, 2012a). The aim of preprocessing data is to reduce noise and to omit the outlier data. Preprocessing of data is done using normalization, whitening, or similar methods. Data imputation-based methods is another type of data preprocessing, in which missing value is imputed using such methods as regression. In feature extraction and selection, since not all data may be necessary for processing, a part of the data is used by size reduction or feature selection methods. In the classifying stage, machine learning-based methods such as support vector machines (SVM) (Mashood Nasir et al., 2020), K-nearest neighbors (KNN) (Mittal et al., 2020; Mughal et al., 2017, 2018a), artificial neural networks (ANN)) (Rehman et al., 2020; Nazir et al., 2019; Perveen et al., 2020), deep neural networks (DNN) (Perveen et al., 2020; Qureshi et al., 2020; Ramzan et al., 2020a), and ensemble-based classifiers (Rahim et al., 2017b) have been used. Recently, deep learning methods in machine learning have been widely used in data mining (Rehman et al., 2018a, 2019, 2020a). AdaBoost is the most well-known ensemble classifier algorithm. It is considered one of the top ten data-mining algorithms. AdaBoost, which stands for adaptive boosting, is a boosting algorithm. AdaBoost is rooted in probably approximately correct (PAC) learning. Those who introduced PAC proved that a combination of simple learners, which have better results from random selection, can lead to a good final classification (Rehman et al., 2021; Saba, 2019, 2020). This is the main idea of boosting. After several relatively successful algorithms with some limitations, AdaBoost was introduced as the practical algorithm of boosting theory (Saba, 2017; Saba et al., 2018a, 2019a, 2020a). Firstly, boosting algorithms were considered due to their efficacy in working with low-noise data. However, the primary algorithms for noisy data had bad results due to the problem of overfitting. Accordingly, boosting was applied less. AdaBoost can be regarded as a gradient method with constraint of error function limiting as much as margin. AdaBoost classifies with interconnected upright lines forming a curve, so that it focuses on a few problem samples similarly to SVM (Sadad et al., 2018; Ullah et al., 2019). There is a close relationship between boosting and optimization theory. The definition of boosting algorithms for regression, multi-class problems, unsupervised learning, and convergence proof for boosting algorithms using optimization theory confirms this relation (Yousaf et al., 2019a; Atalla et al., 2020). The research studied the readmission rate of heart failure patients for 30 days. It proposed a method including data imputation replacement in preprocessing, feature selection and, finally, naïve Bayes classification and SVM. The accuracy of the proposed method was 79% (Meadem et al., 2013). Also, another study used SVM supervised classifier in three database classifications related to readmission (Ramzan et al., 2020a, 2020b). They employed area under curve (AUC) criterion and the best result was reported as 86% (Yu et al., 2015). Shameer et al. (2017) used machine learning-based methods to predict readmission rates. In their proposed method, principal component analysis has been used for feature selection (Harouni et al., 2012b). The best results obtained in AUC criterion and accuracy criterion were 78% and 84%, respectively (Shameer et al., 2017). To predict readmission rate in heart failure patients at intervals of 30, 60, and 90 days, a study employed SVM classifiers, decision trees, and ANNs. In addition to accuracy criteria of AUC, F1 score and recall have been used. The best result for readmission rate prediction of 30 days in AUC criterion was 78%, while the accuracy of prediction in 90 days was less than 63% (Sohrabi et al., 2019). A study reported a perceptron neural network with an error backpropagation training pattern for readmission rate prediction (Nallamothu et al., 2019). In a South Korean study, a group of 7,590 people infected with the coronavirus was considered the target population. This target population's statistical characteristics were obtained, and the resources used for treatment were monitored to analyze the readmission rate. Logistic regression analysis was used to analyze the factors influencing the readmission rate. From the statistical population, 328 patients have been referred to the hospital again. The readmission rate of men and elderly patients with COVID-19 was higher than the others. Concerns about the factors affecting readmission by logistic regression were divided into two categories: the high risk of readmission and the management of individuals before and after treatment. The results of this study have shown that hospital resource management to control COVID-19 virus disease is the most important factor in controlling the rate of return to hospital (Jeon et al., 2020). In another study, 1,087 patients with coronary artery disease in Wuhan, China, underwent reassessment studies. This study showed that 7.6% of coronary patients were readmitted (Chen et al., 2020). The main purpose of this chapter is to propose a new data mining-based method to design a precise pattern of readmission rate prediction in COVID-19, to help the health system perform more effectively and increase its productivity. In this study, after imputing the database used in (Strack et al., 2014) through iterative robust model-based imputation (IRMI), it is classified using an ensemble classifier based on AdaBoost. In the following, ensemble classifiers are introduced in Section 4.2 and the proposed method is introduced in Section 4.3. Then, the proposed method is evaluated in Section 4.4 and, finally, the conclusion is reported in Section 4.5.

4.2 ENSEMBLE CLASSIFIERS

Here, the standard of AdaBoost algorithm is investigated and shown as the convergence theory of the proposed algorithm. AdaBoost is an ensemble-based training method that creates strong classification from a weak hypothesis coalition by reiteration. One simple learning algorithm is applied (called the base classifier) to gain a learner for that reiteration. The final ensemble classifier is the linear-weight combination of the base classifier in which each of them imposes their own weight. This weight relates to the classification accuracy; that is, classification with lower error causes higher weight. The base classifier should be a little better than a random classifier; hence, they are called weak classifiers. Simple learners like root decision often show a good performance for AdaBoost. Assume that characteristics in datasets have true value; so, we need three root decision parameters:

1. Feature list for test (j).
2. The numerical value of threshold test.
3. Test registration $\{-1, +1\}$. For example in Equation (4.1):

$$h_{j,0,+}(x) = \begin{cases} +1 \text{ if } x^j < \theta \\ -1 \text{ otherwise} \end{cases}$$ (4.1)

where x_j indicates the xth data. Given the aforementioned, the root decision is used as the principal learner, although each weak learner-generated decision making as a true value can be appropriate in the proposed parallel algorithms. According to Equation (4.2), the null hypothesis h can be defined as follows:

$$-h_{j,0,+}(x) = h_{j,0,-}(x) = \begin{cases} +1 \text{ if } x^j < \theta \\ -1 \text{ otherwise} \end{cases}$$ (4.2)

Pseudo codes for AdaBoost are explained in Figure 4.1. If data sets are as Equation (4.3):

$$D_n = \{(x_1, y_1), (x_1, y_1), \dots, (x_n, y_n)\}$$ (4.3)

$x_i = (x_i^1, x_i^2, \dots, x_i^d)$ is the vector with determined d and $y_i \in \{+1, -1\}$. $w^t = \{w_1^t, w_2^t, \dots, w_n^t\}$ is the algorithm which determines weight for all samples in D_n which is $t \in \{1, 2, \dots, T\}$

Input: Training set of n samples (D_n)
Number of boosting iterations (T)
Output: The final classifier (H)
Procedure:
1: $w^1 \leftarrow \left(\frac{1}{n}, \dots, \frac{1}{n}\right)$
2: for $t \leftarrow 1\, To\, T\, do$
3:     $h^{(t)} \leftarrow LEARNWEAKCLASSIFIER(W)$
4:     $\varepsilon_- \leftarrow \sum_{i=1}^{n} w_i^t I\{h^{(t)}(x_i) \neq y_i\}$
5:     $\alpha^t \leftarrow \frac{1}{2} ln\left(\frac{1-\varepsilon_-}{\varepsilon_-}\right)$
6:     for $i \leftarrow 1\, To\, n\, do$
7:         if $h^{(t)} \neq y_i$ then
8:             $w_i^{t+1} \leftarrow \frac{w_i^t}{2\varepsilon_-}$
9:         else
10:            $w_i^{t+1} \leftarrow \frac{w_i^t}{2(1-\varepsilon_-)}$
11:         end if
12:     end for
13:     end for
14:     return $H = \sum_{t=1}^{T} \alpha^t h^{(t)}$

FIGURE 4.1   AdaBoost algorithm.

and T is the total number of boosting reiterations. Before the first reiteration, the weights are evenly set up (line 1) and updated in each successive reiteration (line 7–10). It is important to consider that for all t, we have $\sum_{i=1}^{n} w_i^t = 1$. In each reiteration, the weak learner function is imposed on data weight version and then returned to the weak desired hypothesis (line 3). The weak hypothesis of weight error, i.e. Equation (4.4), is minimized concerning Equation 4.1:

$$\varepsilon_- = \sum_{i=1}^{n} w_i^t I\left\{h^{(t)}(x_i) \neq y_i\right\} \tag{4.4}$$

Here, $1\{A\}$, if A is correct, the characteristic function shows the value of 1; otherwise, it shows 0. The weak learner function always warrants finding an optimal h with $\varepsilon < \frac{1}{2}$. If there is each $h^{(t)}$ with $> \frac{1}{2}$, according to (2), there will be the weight error $(1 - \varepsilon)$, which will be less than ½. Therefore, the desired weak learner always induces -h instead of h. This feature increases the value below 1/s of the sample weight with inappropriate classification. The proper classified sample weight is decreased. Thus, for the next reiteration, classification focuses more on those samples which have been previously improperly classified. In each reiteration, the weight of (a^t) is allocated to the weak classification (line 5). At the end of reiteration T, the algorithm returns to the last H classification, which has weighted the mean of all classifications. The sign of H is used for the final prediction (Freund, 2009).

4.3 THE PROPOSED METHOD

The main purpose of the research is to propose a new method in hospital data classification in readmission rate for COVID-19. The dataset used in the study is (Al-Okaily et al., 2020). This dataset contains 53 features for 100,000 cases referred to healthcare centers. Some of these features are shown in Table 4.1. The lost values in the dataset shown by zero are replaced with new values using the IRMI algorithm (Strack et al., 2014). Ensemble AdaBoost classification has been designed based on random forest, SVM, and KNN single classifiers. This classifier is trained based on the dataset's main features and its objective is to predict and classify readmission rates. A single classifier like random forest, SVM, or KNN in the employed dataset classification has negative points such as low accuracy and high error. The designed classifier has been calculated to overcome the mentioned weaknesses, including low accuracy and sensitivity. Also, the way of appropriate preprocessing to replace lost values is presented.

The three classifiers employed in this research are SVM, KNN, and random forest. Each classifier is trained with a subset of training data. The outputs resulting from each of the classifiers are used for maximum voting. Figure 4.2 shows the block diagram of the proposed method.

TABLE 4.1
The Investigated Features

Row	Feature name	Description
1	Encounter ID	Unique ID of an encounter
2	Patient number	Unique earmarks of a patient
3	Race	Values: Caucasian, Asian, African-American, Spanish, etc.
4	Gender	Values: male, female, unknown/invalid
5	Age	Grouped in 10-year intervals: 0–10, 10–20, …, 90–100
6	Weight	Weight in pounds
7	Admission type	Integer ID pertained to 9 certain values, e.g. emergency, selective, infant, unavailable
8	Discharge disposition	Correct ID pertained to 29 separated values, e.g. referring to doctor, emergency room and transferring from a hospital
9	Admission source	Correct ID pertained to 21 separated values, e.g. discharging, expired, and unavailable
10	Time in hospital	The correct number of days between admission and discharge
11	Payer code	Correct number ID pertained to 33 separated values, e.g. Blue, Blue shield/Cross, Medicare, and ATM
12	Medical specialty	The correct ID of an accepted physician's specialty pertained to 8 separated values, e.g. heart, internal medicine, family/general, and surgery
13	Number of lab procedures	The number of laboratory procedures
14	Number of procedures	The number of procedures (other than laboratory procedures)
15	Number of medications	The number of separate medications during encounter
16	Number of outpatient visits	The number of outpatient visits one year before encounter
17	Number of emergency visits	The number of emergency visits one year before encounter
18	Number of inpatient visits	The number of inpatient visits one year before encounter
19	Diagnosis 1	Primary diagnosis (as the first three digits of ICD9), 848 distinctive values
20	Diagnosis 2	Secondary diagnosis (as the first three digits of coded ICD9), 923 distinctive values
21	Diagnosis 3	Extra secondary diagnosis (as the first three digits of coded ICD9), 954 distinctive values
22	Number of diagnoses	The number of diagnoses entered into the system
23	A1c test result	Shows the results range or if the test was not done. the values "> 200, 300 ">, "normal" and "none" if not measured
24	COVID-19 medications	Shows any change in COVID-19 medications
25	Readmitted	If the patient is readmitted in less than 3 days, >" 3 if the patient is readmitted over 3 days and "no" for no readmission

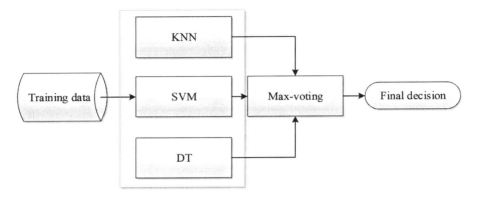

FIGURE 4.2   The block diagram of the proposed method.

The process of AdaBoost ensemble classifier based on the block diagram is performed during the three following phases:

- Feature selection for dividing the main dataset into various fine datasets for each classifier. Selection is performed based on maximum voting law.
- Ten reiterations train the classifier to compute the accuracy of classification. Each classification is selected for validating the subset and the maximum voting law is also trained.
- The output of each classifier for the final decision making is performed based on the maximum voting. Figure 4.3 shows the block diagram of using AdaBoost in readmission data training and classification.

To train each classifier in AdaBoost, a boosting algorithm is used. One of the ways of improving accuracy and sensitivity is the technique of boosting. In combining classifiers through the boosting method, the distribution of certain training sets is shown in series by the patterns. Each classifier in this series set can identify accurately. Each classifier is trained individually and training is applied hierarchically for more training. After each training process, a vector is formed. This vector includes the results obtained by each classifier in each stage of training. Selecting the definite result indicating that the person is healthy or suffering from diabetes is performed based on the maximum voting.

In each stage of AdaBoost-based ensemble classifier training, one of the three classifiers is selected for training in the proposed method. Of course, duplicate selection is not allowed. After training the classifier, cross-validation pertained to the classifier training is computed. The output Y_i includes one of the two classes of readmission and no readmission.

- If $Y_i = 1$, the considered person will be readmitted.
- If $Y_i = -1$, the considered person will not be readmitted.

The results are stored in a vector. Since the number of employed single classifiers is three, the output vector's size will be smaller than 3 in this stage. In the next stage,

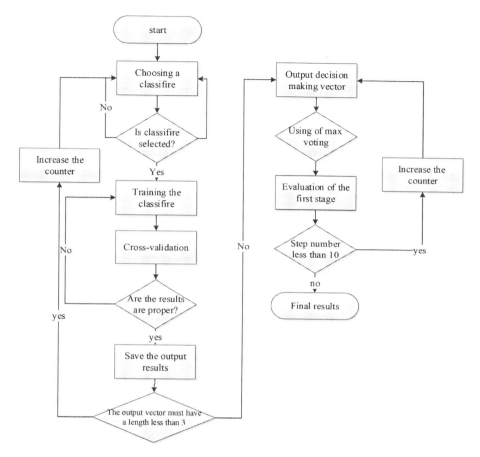

FIGURE 4.3 The block diagram of the proposed method in detail.

the output decision-making vector is formed. The final decision will be made on this vector for each person. The maximum voting law will be used for the final result selection. Thus, the evaluation of the first stage will be repeated.

4.4 EVALUATING THE PROPOSED METHOD

Here, the results obtained by simulating the proposed method based on the AdaBoost ensemble classifier are discussed. To evaluate the proposed method, different tests have been designed. Firstly, the data employed in the chapter are classified through SVM, KNN, and decision tree classifiers. Then, AdaBoost ensemble classification is first tested by SVM. Next, the same process is checked by the decision tree. Simulations are performed using MATLAB software. Finally, the proposed ensemble classification is evaluated based on each of the three classifiers. In the end, the obtained data are compared. Evaluation is the final stage in a classification system (Harouni & Baghmaleki, 2020). In this chapter, accuracy, sensitivity, positive predictive values (PPV), and negative predictive values (NPV) have been employed.

PPV and NPV are truly positive and negative cases reported by a test. In other words, these values indicate, if a person or a test is tested, how much the positive or negative true possibility in the readmission diagnosis is. The relations 5 to 8 show the accuracy, sensitivity, PPV, and NPV (Harouni & Baghmaleki, 2020).

$$\text{Accuracy} = \frac{TP + TN}{TP + TN + FPP + FN} \quad (4.5)$$

$$\text{Sensitivity} = \frac{TP}{TP + FN} \quad (4.6)$$

$$\text{Positive predictive value} = \frac{TP}{TP + FP} \quad (4.7)$$

$$\text{Negative predictive value} = \frac{TP}{TP + FN} \quad (4.8)$$

Where TN indicates true negative, the cases of readmission which are properly diagnosed through test; FN indicates false negative, the cases of readmission which are wrongly diagnosed through test; TP indicates true positive, the cases of readmission which are properly diagnosed through test; FP indicates false positive, the cases of readmission which are wrongly diagnosed through test.

4.4.1 Databases

Datasets play an important role in experiments and results evaluation (Rad et al., 2016, 2013). Data-mining methods in identifying the causes of readmission in COVID-19 virus require standard benchmarked datasets. Table 4.2 shows some of these databases with their access addresses.

4.4.2 The Results Obtained by Single Classifiers Simulation

SVM, KNN, and random forest were used as classifiers. For classification, data was first preprocessed through IRMI. Then, the mentioned classifiers were used to classify data. For further comparison, naïve Bayes was also employed. Table 4.3 presents

TABLE 4.2
Training Databases on COVID-19 Virus

Web Link Address	Name
https://github.com/beoutbreakprepared/nCoV2019	nCoV-2019
https://github.com/CSSEGISandData/ COVID-19	COVID-19
https://www.kaggle.com/allen-institute-for-ai/CORD-19-research-challenge	COVID-19
https://www.who.int/emergencies/diseases/novel-coronavirus-2019/ global-research-on-novel-coronavirus-2019-ncov	WHO COVID-2019
https://github.com/echen102/ COVID-19-TweetIDs	COVID-19 Tweet IDs

TABLE 4.3
The Results Obtained by Simulation Using the Single Classifiers

Evaluation Index	KNN (%)	SVM (%)	Random Forest	Naïve Bayes
Accuracy	86.26	79.123	75.65	73.85
Sensitivity	51.38	68.57	51.72	41.50
PPV	90.33	83.75	76.27	70.96
NPV	77.21	85.89	75.43	74.59

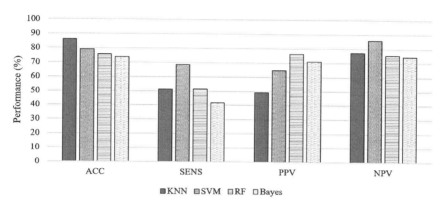

FIGURE 4.4 Comparison of the classifiers in the data classification.

the obtained results. Moreover, Figure 4.4 depicts the comparative diagram of the simulation and classification.

According to Figure 4.4 and Table 4.3, it is observed that the obtained data is at a similar range. However, the KNN classifier outperformed with the best-calculated accuracy value and lower sensitivity and higher PPV and lower NPV. Random forest classifiers, as well as SVM and naïve Bayes, show similar results. However, the obtained results did not reach 90%.

4.4.3 The Results Obtained by Ensemble Classifier with Repeating Classifiers

After preparing data through preprocessing, all three single classifiers are considered as SVM. Also, the same test was repeated for KNN, random forest, and naïve Bayes. Table 4.4 shows the results. The comparison between these ensemble classifiers is presented in Figure 4.5 as well.

As shown in Table 4.4, it is obvious that the KNN classifier outperformed in classification. Notably, PPV and NPV and the obtained sensitivity show acceptable results. For better comparison, the comparative diagrams have been shown in Figures 4.4–4.6. As observed in the figures, combining these classifiers led to better results. Thus, it proves the second hypothesis of the study, indicating the improvement of combining classifiers' results. This superiority can still be understood in all employed criteria.

TABLE 4.4

The Results Obtained by Simulation Using the Reiteration of Single Classifiers in an Ensemble Classifier

Evaluation Index	KNN (%)	SVM (%)	Random Forest	Naïve Bayes
accuracy	85.81	82.10	80.81	77.50
Sensitivity	71.51	75.37	76.25	79.13
PPV	54.54	41.14	36.38	28.30
NPV	81.86	80.81	8.80	74.32

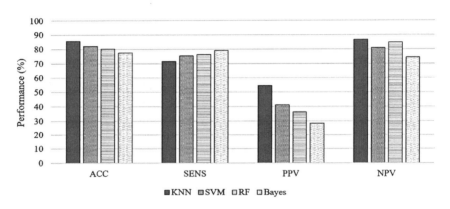

FIGURE 4.5   Comparison of the single classifiers repeated in the ensemble classifier.

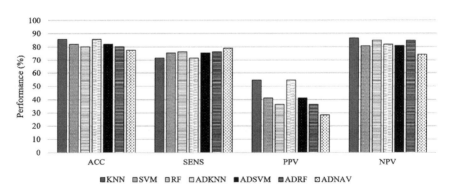

FIGURE 4.6   The comparison of single classifier compared to single classifier repetitions in AdaBoost.

4.4.4   THE PROPOSED COMBINATION IN ADABOOST FOR THE IMPROVEMENT OF THE RESULTS

The main purpose of the present study was to present a new method of improving the evaluation criteria obtained by hospital readmission prediction and classification. In this regard, a triple combination of ensemble classifiers of AdaBoost was proposed.

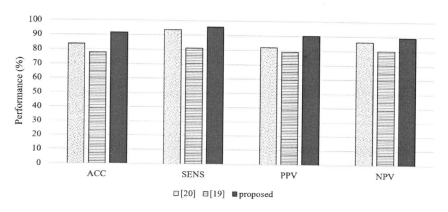

FIGURE 4.7 The comparison of the proposed method (Shameer et al., 2017) and (Sohrabi et al., 2019).

This ensemble combination worked, as Figure 4.7 shows. All three classifiers were alternatively trained ten times. In each stage, a classification result vector and a classification matrix were constructed at the end of the training and testing process. And using the maximum voting law, the output result was selected. The obtained results were classified and compared to the results reported in (Shameer et al., 2017) and (Sohrabi et al., 2019) shown in Table 4.5. Further, Figure 4.8 presents the results more comprehensively.

As observed in Figure 4.8, the proposed method is superior to the base article's method of accuracy, sensitivity, PPV and NPV. Figure 4.8 shows a further comparison between the proposed method and the considered articles.

As shown in Figure 4.8, the results obtained by the proposed method are better than other methods. The main reason for this superiority can be selecting the type of classifiers, the method of boosting training, and its ten times reiteration. Also, using a strong preprocessing method has had a considerable effect on replacing the lost data in improving the results.

TABLE 4.5
The Results Obtained by Simulation Using the Proposed Method in an Ensemble Group

Evaluation Criterion	The Proposed Method (%)	The Method Presented in (Shameer et al., 2017) (%)	The Method Presented in (Sohrabi et al., 2019) (%)
accuracy	91.61	78	86.6
Sensitivity	95.80	81.2	93.07
PPV	90.25	79.12	82.11
NPV	89.31	80.33	86.17

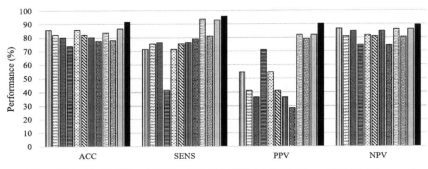

FIGURE 4.8 The comparison of the results obtained by the proposed method and other methods.

4.5 CONCLUSION

Using ensemble classifiers has resolved the challenges and problems of single classifiers. Although classifiers and their reiteration in ensemble classification improve the efficacy and evaluation criteria, it cannot provide a considered response to improving the results. In this study, the proposed method was stimulated and implemented for readmission prediction based on KNN, SVM, and random forest. For further evaluation and comparison, various tests were designed. In the designed tests, the study's data was first classified separately and readmission was predicted. Then, the criteria of sensitivity, accuracy, PPV and NPV were calculated. The same results were tested for AdaBoost classification without the classifiers' reiteration. Finally, the obtained results were compared with the base article. The diagrams and figures reveal the superiority of the proposed method such that the accuracy, sensitivity, NPV, and PPV are 91%, 95%, 90%, and 89%, respectively.

REFERENCES

Abbas, N. Saba, T. Mohamad, D. Rehman, A. Almazyad, A. S., & Al-Ghamdi, J. S. (2018) Machine aided malaria parasitemia detection in Giemsa-stained thin blood smears, *Neural Computing and Applications.*, 29(3), 803–818, doi:10.1007/s00521-016-2474-6.

Abbas, N., Saba, T., Mehmood, Z., Rehman, A., Islam, N., & Ahmed, K. T. (2019a). An automated nuclei segmentation of leukocytes from microscopic digital images. *Pakistan Journal of Pharmaceutical Sciences*, 32(5), 2123–2138.

Abbas, A., Saba, T., Rehman, A., Mehmood, Z., Javaid, N., Tahir, M., Khan, N. U., Ahmed, K. T., & Shah, R. (2019b) Plasmodium species aware based quantification of malaria, parasitemia in light microscopy thin blood smear, *Microscopy Research and Technique*, 82(7), 1198–1214. doi: 10.1002/jemt.23269.

Abbas, N., Saba, T., Rehman, A., Mehmood, Z., Kolivand, H., Uddin, M., & Anjum, A. (2019c). Plasmodium life cycle stage classification-based quantification of malaria parasitaemia in thin blood smears. *Microscopy Research and Technique*, 82(3), 283–295. doi:10.1002/jemt.23170

Adabavazeh, N., Nikbakht, M., & Amirteimoori, A. 2020. Envelopment analysis for global response to novel 2019 coronavirus-SARS-COV-2 (COVID-19). *Journal of Industrial Engineering and Management Studies*, 7, 1–35.

Adeel, A., Khan, M. A., Akram, T., Sharif, A., Yasmin, M., Saba, T., & Javed, K. (2020). Entropy-controlled deep features selection framework for grape leaf diseases recognition. *Expert Systems*, 2020, 1–17.

Afza, F., Khan, M. A., Sharif, M., & Rehman, A. (2019). Microscopic skin laceration segmentation and classification: A framework of statistical normal distribution and optimal feature selection. *Microscopy Research and Technique*, 82(9), 1471–1488.

Afza, F., Khan, M. A., Sharif, M., Saba, T., Rehman, A., & Javed, M. Y. (2020, October). *Skin Lesion Classification: An Optimized Framework of Optimal Color Features Selection.* In *2020 2nd International Conference on Computer and Information Sciences (ICCIS)* (pp. 1–6). IEEE.

Agbehadji, I. E., Awuzie, B. O., Ngowi, A. B., & Millham, R. C. 2020. Review of big data analytics, artificial intelligence and nature-inspired computing models towards accurate detection of COVID-19 pandemic cases and contact tracing. *International Journal of Environmental Research and Public Health*, 17, 5330.

Al-Ameen, Z. Sulong, G. Rehman, A., Al-Dhelaan, A. Saba, T., & Al-Rodhaan, M. (2015) An innovative technique for contrast enhancement of computed tomography images using normalized gamma-corrected contrast-limited adaptive histogram equalization. *EURASIP Journal on Advances in Signal Processing*, 32, 1–12. doi:10.1186/s13634-015-0214-1

Al-Okaily, M., Alqudah, H., Matar, A., Lutfi, A., & Taamneh, A. 2020. Dataset on the acceptance of e-learning system among universities students' under the COVID-19 pandemic conditions. *Data in Brief*, 32, 106176.

Amin, J., Sharif, M., Raza, M., Saba, T., & Anjum, M. A. (2019a). Brain tumor detection using statistical and machine learning method. *Computer Methods and Programs in Biomedicine*, 177, 69–79.

Amin, J., Sharif, M., Raza, M., Saba, T., & Rehman, A. (2019b). *Brain Tumor Classification: Feature Fusion.* In *2019 International Conference on Computer and Information Sciences (ICCIS)* (pp. 1–6). IEEE.

Amin, J., Sharif, M., Raza, M., Saba, T., Sial, R., & Shad, S. A. (2020). Brain tumor detection: A long short-term memory (LSTM)-based learning model. *Neural Computing and Applications*, 32, 15965–15973.

Amin, J., Sharif, M., Rehman, A., Raza, M., & Mufti, M. R. (2018). Diabetic retinopathy detection and classification using hybrid feature set. *Microscopy Research and Technique*, 81(9), 990–996.

Amin, J., Sharif, M., Yasmin, M. Saba, T., & Raza, M. (2019c). Use of machine intelligence to conduct analysis of human brain data for detection of abnormalities in its cognitive functions. *Multimedia Tools and Applictions*, 79(15), 10955–10973. doi:10.1007/s11042-019-7324-y

Chen, J., Xu, X., Hu, J., Chen, Q., Xu, F., Liang, H., Liu, N., Zhu, H., Lan, J., & Zhou, L. 2020. *Clinical Course and Risk Factors for Recurrence of Positive SARS-CoV-2 RNA: A Retrospective Cohort Study from Wuhan*, China, Aging (Albany, NY), 12(17), 16675.

Ejaz, K., Rahim, M. S. M., Bajwa, U. I., Chaudhry, H., Rehman, A., & Ejaz, F (2020) Hybrid segmentation method with confidence region detection for tumor dentification, *IEEE Access*, 10.1109/ACCESS.2020.3016627

Ejaz, K., Rahim, D. M. S. M., Rehman, D. A., & Ejaz, E. F. (2018b). An image-based multimedia database and efficient detection though features. *VFAST Transactions on Software Engineering*, 14(1), 6–15.

Ejaz, K., Rahim, M. S. M., Bajwa, U. I., Rana, N., & Rehman, A. (2019). *An Unsupervised Learning with Feature Approach for Brain Tumor Segmentation Using Magnetic Resonance Imaging.* In *Proceedings of the 2019 9th International Conference on Bioscience, Biochemistry and Bioinformatics* (pp. 1–7).

Ejaz, K., Rahim, M. S. M., Rehman, A., Chaudhry, H., Saba, T., & Ejaz, A. (2018a). Segmentation method for pathological brain tumor and accurate detection using MRI. *International Journal of Advanced Computer Science and Applications*, 9(8), 394–401.

Fahad, H. M., Khan M. U. G., Saba, T., Rehman, A., & Iqbal, S. (2018) Microscopic abnormality classification of cardiac murmurs using ANFIS and HMM. *Microscopy Research and Technique*, 81(5), 449–457. doi: 10.1002/jemt.22998.

Freund, Y. 2009. A more robust boosting algorithm. arXiv preprint arXiv:0905.2138.

Harouni, M., & Baghmaleki, H. Y. 2020. Color image segmentation metrics. *Encyclopedia of Image Processing*, 2018, 95.

Harouni, M., Mohamad, D. & Rasouli, A. 2010. *Deductive method for recognition of on-line handwritten Persian/Arabic characters. The 2nd International Conference on Computer and Automation Engineering (ICCAE)*, IEEE, 791–795.

Harouni, M., Mohamad, D., Rahim, M. S. M., & Halawani, S. M. 2012b. Finding critical points of handwritten persian/arabic character. *International Journal of Machine Learning and Computing*, 2, 573.

Harouni, M., Mohamad, D., Rahim, M. S. M., Halawani, S. M., Afzali, M. J. I. J. O. M. L. & Computing 2012a. Handwritten Arabic character recognition based on minimal geometric features. 2, 578.

Harouni, M., Rahim, M., Al-Rodhaan, M., Saba, T., Rehman, A. & Al-Dhelaan, A. J. T. I. S. J. 2014. Online Persian/Arabic script classification without contextual information. *The Imaging Science Journal*, 62, 437–448.

Husham, A., Alkawaz, M. H., Saba, T., Rehman, A., & Alghamdi, J. S. (2016) Automated nuclei segmentation of malignant using level sets. *Microscopy Research and Technique*, 79(10), 993–997, doi. 10.1002/jemt.22733.

Hussain, N., Khan, M. A., Sharif, M., Khan, S. A., Albesher, A. A., Saba, T., & Armaghan, A. (2020). A deep neural network and classical features based scheme for objects recognition: An application for machine inspection. *Multimed Tools and Applications*. doi: 10.1007/s11042-020-08852-3.

Iftikhar, S. Fatima, K. Rehman, A. Almazyad, A. S., & Saba, T. (2017) An evolution based hybrid approach for heart diseases classification and associated risk factors identification, *Biomedical Research*, 28 (8), 3451–3455

Iqbal, S. Ghani, M. U. Saba, T., & Rehman, A. (2018). Brain tumor segmentation in multispectral MRI using convolutional neural networks (CNN). *Microsc Research and Technique*, 81(4), 419–427. doi: 10.1002/jemt.22994.

Iqbal, S., Khan, M. U. G., Saba, T., & Rehman, A. (2017). Computer assisted brain tumor type discrimination using magnetic resonance imaging features. *Biomedical Engineering Letters.*, 8(1), 5–28. doi. 10.1007/s13534-017-0050-3.

Iqbal, S., Khan, M. U. G., Saba, T. Mehmood, Z. Javaid, N., Rehman, A., & Abbasi, R. (2019) Deep learning model integrating features and novel classifiers fusion for brain tumor segmentation. *Microscopy Research and Technique*, 82(8), 1302–1315, doi:10.1002/jemt.23281

Jamal, A., Hazim Alkawaz, M., Rehman, A., & Saba, T. (2017). Retinal imaging analysis based on vessel detection. *Microscopy Research and Technique*, 80 (17), 799–811. doi:10.1002/jemt.

Javed, R., Rahim, M. S. M., Saba, T., & Rashid, M. (2019a). Region-based active contour JSEG fusion technique for skin lesion segmentation from dermoscopic images. *Biomedical Research*, 30(6), 1–10.

Javed, R., Rahim, M. S. M., & Saba, T. (2019b) An improved framework by mapping salient features for skin lesion detection and classification using the optimized hybrid features. *International Journal of Advanced Trends in Computer Science and Engineering*, 8(1), 95–101.

Javed, R., Rahim, M. S. M., Saba, T., & Rehman, A. (2020a) A comparative study of features selection for skin lesion detection from dermoscopic images. *Network Modeling Analysis in Health Informatics and Bioinformatics*, 9 (1), 4.

Javed, R., Saba, T., Shafry, M., & Rahim, M. (2020b). *An Intelligent Saliency Segmentation Technique and Classification of Low Contrast Skin Lesion Dermoscopic Images Based on Histogram Decision*. In *2019 12th International Conference on Developments in eSystems Engineering (DeSE)* (pp. 164–169).

Jeon, W.-H., Seon, J. Y., Park, S.-Y. & Oh, I.-H. 2020. Analysis of risk factors on readmission cases of covid-19 in the republic of korea: Using nationwide health claims data. *International Journal of Environmental Research and Public Health*, 17, 5844.

Khan, M. A., Akram, T., Sharif, M., Javed, K., Raza, M., & Saba, T. (2020a). An automated system for cucumber leaf diseased spot detection and classification using improved saliency method and deep features selection. *Multimedia Tools and Applications*, 79(1), 1–30.

Khan, M. A., Javed, M. Y., Sharif, M., Saba, T., & Rehman, A. (2019a). *Multi-Model Deep Neural Network Based Features Extraction And Optimal Selection Approach For Skin Lesion Classification*. In *2019 International Conference On Computer And Information Sciences (ICCIS)* (pp. 1–7). IEEE.

Khan, M. A., Sharif, M. I., Raza, M., Anjum, A., Saba, T., & Shad, S. A. (2019b). Skin lesion segmentation and classification: A unified framework of deep neural network features fusion and selection. *Expert Systems*, e12497.

Khan, M. W., Sharif, M., Yasmin, M., & Saba, T. (2017). CDR based glaucoma detection using fundus images: A review. *International Journal of Applied Pattern Recognition*, 4(3), 261–306.

Khan, M. Z., Jabeen, S., Khan, M. U. G., Saba, T., Rehmat, A., Rehman, A., & Tariq, U. (2020). A realistic image generation of face from text description using the fully trained generative adversarial networks. *IEEE Access*, 9, 1250–1260.

Khan, M. A. Kadry, S., Zhang, Y. D., Akram, T., Sharif, M., & Rehman, A., Saba, T. (2021) Prediction of COVID-19- Pneumonia based on selected deep features and one class kernel extreme learning machine. *Computers & Electrical Engineering*, 90, 1–19.

Khan, M. A., Lali, I. U. Rehman, A. Ishaq, M. Sharif, M. Saba, T., Zahoor, S., & Akram, T. (2019c) Brain tumor detection and classification: A framework of marker-based watershed algorithm and multilevel priority features selection, *Microscopy Research and Technique*, 82(6), 909–922. doi:10.1002/jemt.23238

Khan, M. A.,Sharif, M. Akram, T., Raza, M., Saba, T., & Rehman, A. (2020b) Hand-crafted and deep convolutional neural network features fusion and selection strategy: An application to intelligent human action recognition. *Applied Soft Computing*, 87, 105986

Khan, M. A.; Akram, T. Sharif, M., Saba, T., Javed, K., Lali, I. U., Tanik, U. J., & Rehman, A. (2019d). Construction of saliency map and hybrid set of features for efficient segmentation and classification of skin lesion, *Microscopy Research and Technique*, 82(5), 741–763, doi:10.1002/jemt.23220

Khan, M. A., Ashraf, I., Alhaisoni, M., Damaševičius, R., Scherer, R., Rehman, A., & Bukhari, S. A. C. (2020d) Multimodal brain tumor classification using deep learning and robust feature selection: a machine learning application for radiologists. *Diagnostics*,10, 565.

Khan, S. A., Nazir, M., Khan, M. A., Saba, T., Javed, K., Rehman, A., … & Awais, M. (2019e). Lungs nodule detection framework from computed tomography images using support vector machine. *Microscopy Research and Technique*, 82(8), 1256–1266.

Liaqat, A., Khan, M. A., Sharif, M., Mittal, M., Saba, T., Manic, K. S., & Al Attar, F. N. H. (2020). Gastric tract infections detection and classification from wireless capsule endoscopy using computer vision techniques: A review. *Current Medical Imaging*, 16(10), 1229–1242.

Lung, J. W. J., Salam, M. S. H., Rehman, A., Rahim, M. S. M., & Saba, T. (2014) Fuzzy phoneme classification using multi-speaker vocal tract length normalization, *IETE Technical
Review*, 31 (2), 128–136, doi. 10.1080/02564602.2014.892669.

Majid, A., Khan, M. A., Yasmin, M., Rehman, A., Yousafzai, A., & Tariq, U. (2020).
Classification of stomach infections: A paradigm of convolutional neural network along
with classical features fusion and selection. *Microscopy Research and Technique*, 83(5),
562–576.

Marie-Sainte, S. L. Aburahmah, L., Almohaini, R., & Saba, T. (2019a). Current techniques for
diabetes prediction: Review and case study. *Applied Sciences*, 9(21), 4604.

Marie-Sainte, S. L., Saba, T., Alsaleh, D., Alotaibi, A., & Bin, M. (2019b). An improved strategy for predicting diagnosis, survivability, and recurrence of breast cancer. *Journal of
Computational and Theoretical Nanoscience*, 16(9), 3705–3711.

Mashood Nasir, I., Attique Khan, M., Alhaisoni, M., Saba, T., Rehman, A., & Iqbal, T. (2020).
A hybrid deep learning architecture for the classification of superhero fashion products:
An application for medical-tech classification. *Computer Modeling in Engineering &
Sciences*, 124(3), 1017–1033.

Meadem, N., Verbiest, N., Zolfaghar, K., Agarwal, J., Chin, S.-C. & Roy, S. B. 2013. *Exploring
Preprocessing Techniques for Prediction of Risk of Readmission For Congestive Heart
Failure Patients. Data Mining And Healthcare (Dmh), At International Conference On
Knowledge Discovery And Data Mining (Kdd)*, WA, USA.

Mittal, A., Kumar, D., Mittal, M., Saba, T., Abunadi, I., Rehman, A., & Roy, S. (2020).
Detecting pneumonia using convolutions and dynamic capsule routing for chest X-ray
images. *Sensors*, 20(4), 1068.

Mughal, B. Muhammad, N. Sharif, M. Saba, T. Rehman, A. (2017) Extraction of breast border
and removal of pectoral muscle in wavelet domain, *Biomedical Research*, 28(11),
5041–5043.

Mughal, B., Muhammad, N., Sharif, M., Rehman, A., & Saba, T. (2018a). Removal of pectoral
muscle based on topographic map and shape-shifting silhouette. *BMC Cancer*, 18(1),
1–14.

Mughal, B., Sharif, M., Muhammad, N., & Saba, T. (2018b). A novel classification scheme to
decline the mortality rate among women due to breast tumor. *Microscopy Research and
Technique*, 81(2), 171–180.

Nallamothu, B., Liu, W., Singh, K., Ryan, A., Sukul, D., Mahmoudi, E., Waljee, A., Stansbury,
C., & Zhu, J. 2019. Predicting 30-day hospital readmissions using artificial neural networks with medical code embedding. *bioRxiv*, 741504.

Nazir, M., Khan, M. A., Saba, T., & Rehman, A. (2019). *Brain Tumor Detection from MRI
images using Multi-level Wavelets*. In *2019, IEEE International Conference on Computer
and Information Sciences (ICCIS)* (pp. 1–5).

Perveen, S., Shahbaz, M., Saba, T., Keshavjee, K., Rehman, A., & Guergachi, A. (2020).
Handling irregularly sampled longitudinal data and prognostic modeling of diabetes
using machine learning technique. *IEEE Access*, 8, 21875–21885.

Qureshi, I., Khan, M. A., Sharif, M., Saba, T., & Ma, J. (2020) Detection of glaucoma based
on cup-to-disc ratio using fundus images, *International Journal of Intelligent Systems
Technologies and Applications*, 19(1), 1–16, 10.1504/IJISTA.2020.105172

Rad, A. E., Rahim, M. S. M., Rehman, A. Altameem, A. Saba, T. (2013) Evaluation of current
dental radiographs segmentation approaches in computer-aided applications *IETE
Technical Review*, 30(3), 210–222

Rad, A. E., Rahim, M. S. M., Rehman, A. Saba, T. (2016) Digital dental X-ray database for
caries screening. *3D Research*, 7(2), 1–5, doi. 10.1007/s13319-016-0096-5

Rahim, M. S. M., Norouzi, A. Rehman, A., & Saba, T. (2017a) 3D bones segmentation based
on CT images visualization, *Biomedical Research*, 28(8), 3641–3644

Rahim, M. S. M., Rehman, A., & Kurniawan, F. Saba, T. (2017b) Ear biometrics for human classification based on region features mining, *Biomedical Research*, 28 (10), 4660–4664

Ramzan, F., Khan, M. U. G., Iqbal, S., Saba, T., & Rehman, A. (2020b). Volumetric segmentation of brain regions from MRI scans using 3D convolutional neural networks. *IEEE Access*, 8, 103697–103709.

Ramzan, F., Khan, M. U. G., Rehmat, A., Iqbal, S., Saba, T., Rehman, A., & Mehmood, Z. (2020a). A deep learning approach for automated diagnosis and multi-class classification of alzheimer's disease stages using resting-state fMRI and residual neural networks. *Journal of Medical Systems*, 44(2), 37.

Rehman, A., Abbas, N., Saba, T., Mahmood, T., & Kolivand, H. (2018c) Rouleaux red blood cells splitting in microscopic thin blood smear images via local maxima, circles drawing, and mapping with original RBCs, *Microscopic Research and Technique*, 81(7), 737–744. doi: 10.1002/jemt.23030

Rehman, A., Abbas, N., Saba, T., Rahman, S. I. U., Mehmood, Z., & Kolivand, K. (2018a) Classification of acute lymphoblastic leukemia using deep learning, *Microscopy Research & Technique*, 81(11), 1310–1317. doi: 10.1002/jemt.23139

Rehman, A. Abbas. N. Saba, T. Mehmood, Z. Mahmood, T., & Ahmed, K. T. (2018b) Microscopic malaria parasitemia diagnosis and grading on benchmark datasets, *Microscopic Research and Technique*, 81(9), 1042–1058. doi: 10.1002/jemt.23071

Rehman, A., Khan, M. A., Mehmood, Z., Saba, T., Sardaraz, M., & Rashid, M. (2020a) Microscopic melanoma detection and classification: A framework of pixel-based fusion and multilevel features reduction, *Microscopy Research and Technique*, 83(4), 410–423, doi: 10.1002/jemt.23429

Rehman, A., Harouni, M., Saba, T. (2020). Cursive multilingual characters recognition based on hard geometric features. *International Journal of Computer Vision and Robotics*, 10, 213–222.

Rehman, A., Khan, M. A., Saba, T., Mehmood, Z., Tariq, U., & Ayesha, N. (2021). Microscopic brain tumor detection and classification using 3D CNN and feature selection architecture. *Microscopy Research and Technique*, 84(1), 133–149, doi:10.1002/jemt.23597.

Saba, T. (2017) Halal food identification with neural assisted enhanced RFID antenna, *Biomedical Research*, 28(18), 7760–7762.

Saba, T. (2019). Automated lung nodule detection and classification based on multiple classifiers voting. *Microscopy Research and Technique*, 82(9), 1601–1609.

Saba, T. (2020). Recent advancement in cancer detection using machine learning: Systematic survey of decades, comparisons and challenges. *Journal of Infection and Public Health*, 13(9), 1274–1289.

Saba, T., Bokhari, S. T. F., Sharif, M., Yasmin, M., & Raza, M. (2018b). Fundus image classification methods for the detection of glaucoma: A review. *Microscopy Research and Technique*, 81(10), 1105–1121.

Saba, T., Haseeb, K., Ahmed, I., & Rehman, A. (2020b). Secure and energy-efficient framework using Internet of Medical Things for e-healthcare. *Journal of Infection and Public Health*, 13(10), 1567–1575.

Saba, T., Khan, M. A., Rehman, A., & Marie-Sainte, S. L. (2019a). Region extraction and classification of skin cancer: A heterogeneous framework of deep CNN features fusion and reduction. *Journal of Medical Systems*, 43(9), 289.

Saba, T., Khan, S. U., Islam, N., Abbas, N., Rehman, A., Javaid, N. and Anjum, A., (2019b). Cloud-based decision support system for the detection and classification of malignant cells in breast cancer using breast cytology images. *Microscopy Research and Technique*, 82(6), 775–785.

Saba, T., Mohamed, A. S., El-Affendi, M. Amin, J., & Sharif, M. (2020a) Brain tumor detection using fusion of hand crafted and deep learning features. *Cognitive Systems Research*, 59, 221–230

Saba, T., Rehman, A. Mehmood, Z., Kolivand, H., & Sharif, M. (2018a) Image enhancement and segmentation techniques for detection of knee joint diseases: A survey. *Current Medical Imaging Reviews*, 14(5), 704–715, doi. 10.2174/15734056136661 70912164546

Saba, T., Sameh, A., Khan, F., Shad, S. A., & Sharif, M. (2019c). Lung nodule detection based on ensemble of hand crafted and deep features. *Journal of Medical Systems*, 43 (12), 332.

Sadad, T. Munir, A. Saba, T. Hussain, A. (2018) Fuzzy C-means and region growing based classification of tumor from mammograms using hybrid texture feature. *Journal of Computational Science*, 29, 34–45

Shameer, K., Johnson, K. W., Yahi, A., Miotto, R., Li, L., Ricks, D., Jebakaran, J., Kovatch, P., Sengupta, P. P. & Gelijns, S. 2017. *Predictive modeling of hospital readmission rates using electronic medical record-wide machine learning: A case-study using Mount Sinai heart failure cohort. Pacific Symposium on Biocomputing 2017. World Scientific*, Hawaii, United States, 276–287.

Sohrabi, B., Vanani, I. R., Gooyavar, A., & Naderi, N. 2019. Predicting the readmission of heart failure patients through data analytics. *Journal of Information & Knowledge Management*, 18, 1950012.

Strack, B., Deshazo, J. P., Gennings, C., Olmo, J. L., Ventura, S., Cios, K. J. & Clore, J. N. 2014. Impact of HbA1c measurement on hospital readmission rates: analysis of 70,000 clinical database patient records. *BioMed Research International*, 2014.

Ullah, H., Saba, T., Islam, N., Abbas, N., Rehman, A., Mehmood, Z., & Anjum, A. (2019). An ensemble classification of exudates in color fundus images using an evolutionary algorithm based optimal features selection. *Microscopy Research and Technique*, 82(4), 361–372.

Yousaf, K. Mehmood, Z. Saba, T. Rehman, A. Munshi, A. M. Alharbey, R., & Rashid, M. (2019a). Mobile-health applications for the efficient delivery of health care facility to people with dementia (PwD) and support to their carers: A survey. *BioMed Research International*, 2019, 1–26

Yousaf, K., Mehmood, Z., Awan, I. A., Saba, T., Alharbey, R., Qadah, T., & Alrige, M. A. (2019b). A comprehensive study of mobile-health based assistive technology for the healthcare of dementia and Alzheimer's disease (AD). *Health Care Management Science*, (23), 287–309.

Yu, S., Farooq, F., Van Esbroeck, A., Fung, G., Anand, V., & Krishnapuram, B. 2015. Predicting readmission risk with institution-specific prediction models. *Artificial Intelligence in Medicine*, 65, 89–96.

5 Nigerian Medical Laboratory Diagnosis of COVID-19; from Grass to Grace

Uchejeso M. Obeta, Nkereuwem S. Etukudoh, and Chukwudinma C. Okoli

CONTENTS

5.1 INTRODUCTION

The medical laboratory component is a very important public health practice and operation, especially in cases of an outbreak or pandemic such as COVID-19. Medical laboratory science practice involves the analysis of human specimens such as bodily fluids, excretions, and various body swabs for the purpose of diagnosis, treatment and research (Obeta et al., 2019). In Nigeria, the Medical Laboratory Science Council of Nigeria (MLSCN) regulates medical laboratory science practice while the Nigerian Center for Disease Control (NCDC) coordinates all aspects of public health challenges, especially in disease outbreaks such as COVID-19, following the stipulations of the laws establishing them in Nigeria (MLSCN, 2003; NCDC, 2018).

COVID-19, which emanated from Wuhan, China, in late 2019 (Nassiri, 2020) has been said to be zoonotic and spread through traveling (Etukudoh et al., 2020a) and has became a nightmare to the global community. COVID-19 (SARS-CoV-2) genome sequencing shares up to 88% identity with bat SARS-like CoVs (bat-SL-CoVZC45 and bat-SL-CoVZXC21), up to 79% identity with SARS-CoV, and up to 50% identity with MERS-CoV (Lu et al., 2020).

COVID-19 entered Nigeria on February 27, 2020, and as at September 4, 2020 had reached 54, 743 cases (Figure 5.1), according to states. The pandemic revealed the weakness of the health systems of developing countries like Nigeria and humbled

PREVENT THE SPREAD OF COVID-19   4-09-2020   #TakeResponsibility #COVID19Nigeria

S/N	STATE	Number of Cases
1	Lagos	18,255
2	FCT	5,273
3	Oyo	3,168
4	Plateau	2,710
5	Edo	2,594
6	Kaduna	2,184
7	Rivers	2,172
8	Delta	1,756
9	Kano	1,727
10	Ogun	1,676
11	Ondo	1,550
12	Enugu	1,184
13	Ebonyi	1,005
14	Kwara	976
15	Katsina	812
16	Abia	807
17	Osun	795
18	Borno	741
19	Gombe	723
20	Bauchi	669
21	Imo	532
22	Benue	460
23	Nasarawa	437
24	Bayelsa	391
25	Jigawa	322
26	Ekiti	281
27	Akwa Ibom	280
28	Niger	244
29	Adamawa	228
30	Anambra	219
31	Sokoto	159
32	Kebbi	93
33	Taraba	87
34	Cross River	83
35	Zamfara	78
36	Yobe	67
37	Kogi	5
	TOTAL	54,743

@NCDCgov   COVID19.NCDC.GOV.NG   NCDC

FIGURE 5.1 Highest cases of COVID-19 in 36 states of Nigeria and Federal Capital Territory (FCT).

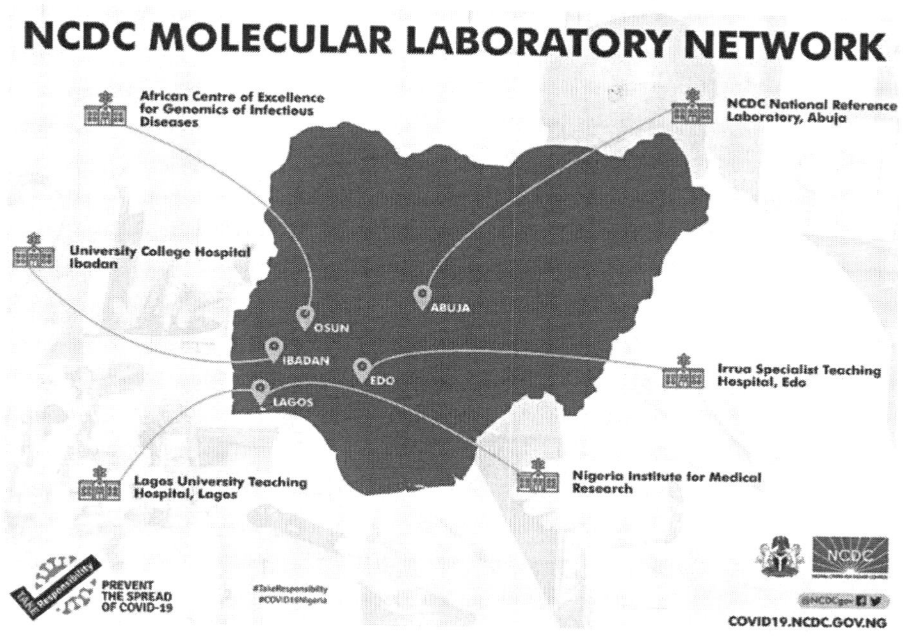

FIGURE 5.2 The number of molecular laboratories (6) available in Nigeria.

countries with strong health systems like the USA. In the battle against COVID-19, medical laboratory diagnosis remained one component of health systems that was considered vital by all, from WHO to local disease control bodies, as it gives the status of the pandemic at any point.

To minimize COVID-19 and associated morbidity and mortality in any country, taking Nigeria as an example, Abdullahi et al. (2020) advised that public health authorities should have a well-coordinated medical laboratory testing/case detection and contacts tracing system in addition to other local infection control measures. Though the number of tests performed in Nigeria as at September 6, 2020, was about 422, 000, it is still not enough, considering the Nigerian population of about 20 million. However, success has been achieved with the number of medical laboratories that has grown from 6 (Figure 5.2) in 5 states to 71 (Figure 5.4) in 35 states and federal capital territory (FCT).

This chapter thoroughly examines the role of the medical laboratory diagnosis of COVID-19 in Nigeria and how the government increased the medical laboratory network from the first day of the pandemic's discovery to the time of preparing this manuscript.

5.2 COVID-19 AND MEDICAL LABORATORY TESTING

Medical laboratory testing is an integral part of public health strategies in fighting any public health challenge or pandemic such as COVID-19.

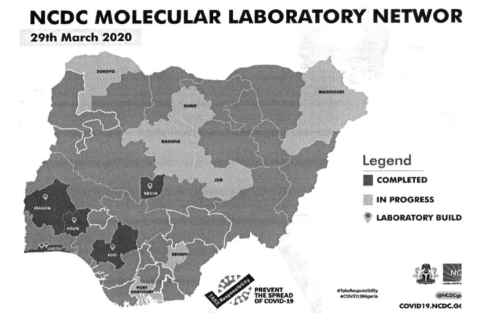

FIGURE 5.3   COVID-19 Testing Laboratories Expansion.

Naidoo and Ihekweazu (2020), while considering the importance of medical laboratories in public health disease outbreaks and control, posit that NCDC after its establishment worked toward strengthening of diagnostic laboratories and networking capabilities for diseases of public health importance, especially lassa fever, monkeypox, yellow fever, cerebrospinal meningitis, influenza, cholera, and other enteric pathogens. On the arrival of COVID-19, NCDC's attention was shifted to testing for coronavirus across the laboratory network (Figure 5.3).

COVID-19 pandemic testing involves viral culture which takes more time, nucleic acid testing (NAT) which can be more rapid using reverse transcription polymerase chain reaction (RT-PCR), and serological testing detecting IgM and IgG antibodies in vivo as a supplement to molecular diagnostic methods. Plasmonic photothermal biosensor and genomic sequencing have been playing an excellent role in medical laboratory testing.

5.3   METHODOLOGIES AND TECHNIQUES FOR COVID-19 TESTING

5.3.1   NEUTRALIZATION/VIROLOGICAL CELL CULTURE TEST

This method is the gold standard for virus discovery, pathogenesis research, and strategy evaluation. This technique is about incubating serum or plasma with live viruses, mainly to perform virus neutralization tests or pseudo virus neutralization tests. The neutralization test is more specific with COVID-19 antibodies (Abdullahi et al., 2020; Li et al., 2020).

The challenges with this method of testing is that it is time consuming, taking not less than three days in turnaround time; there is less availability of permissive cell lines; its capital- and labor-intensive nature; it requires great expertise; and there is less availability of commercial antisera for confirmation of cultures. It is also performed in a model laboratory equipped to biosafety levels (BSL) 2 or 3. When all conditions are achieved, the technique is authoritative, simple, low in cost and of high sensitivity which serve as an advantage (Li et al., 2020).

5.3.2 GENOME SEQUENCING

This analyzes virus evolution, genetic relations with diseases, and pandemic outbreak tracking, and enables the development of new vaccines and therapy. The genomic sequencing for COVID-19 can be nanopore targeted, amplicon sequencing and hybrid capture-based sequencing (Li et al., 2020; Sheikhzadeh et al., 2020).

The demerits of this technique are mainly its high cost, the requirement for sophisticated equipment and a high level of technical skills, and insufficient coverage and development; it also needs a BSL-2 laboratory in which to operate. But it has a very rapid turnaround time, low viral load, high accuracy, and high sensitivity as its advantages.

5.3.3 IMMUNOLOGICAL TESTING

This method relies on the quantification and/or detection of antigen and/or detection of an antigen and-or antibody interactions existing in COVID-19 cases as early as possible. COVID-19-encoded proteins, such as S, N, E, and M proteins, have COVID-19 antibodies on S and N as their antigenic targets.

Mainly, COVID-19 antigens are detected using immunochromatographic assays, possibly found in RapiGEN, Liming Bio, Savant, and Bioeasy lateral flow antigenic detection kits.

The antibodies involved in COVID-19 are IgG, IgM, and IgA which can be determined using various immunological tests as highlighted by Li et al. (2020) such as:

enzyme-linked immunosorbent assay (ELISA) which detects IgG in N and S-proteins;

immunoflorescence assay (IFA) which detects IgG and IgM;

chemiluminescence immunoassay (CLIA) e.g. magnetic chemiluminescence enzyme immunoassay (MCLIA) for COVID-19 IgG and IgM;

lateral flow immunoassay (LFIA) for both IgM and IgG in blood.

The disadvantages of immunological tests are high turnaround time, low sensitivity, cross-reactivity, non-specificity, sophisticated equipment requirements, and low throughput, depending on the particular example. The advantages are convenience, rapidity, small sample size, stable results, and portability of equipment. Importantly, laboratory design can be small and less sophisticated equipment can be incorporated (Li et al., 2020).

5.3.4   BIOSENSORS

These are biological devices that use the selectivity features of a biomolecule and the sensitivity of a physicochemical transducer in combination to provide diagnosis, real-time detection, and routine measurement of COVID-19.

Examples of biosensors in COVID-19 as given by Sheikhzadeh et al. (2020) are:

localized surface plasmon resonance sensor which is applied in detection of RdRp-COVID, ORF1ab-COVID, and E-gene sequence with aid of photo-thermal units;

field effect transistor which is decorated with COVID spike S1 subunit or angiotensin converting enzyme-2 to detect COVID spike protein S1;

cell-based potentiometric biosensor which detects COVID-19 S1 antigen.

5.3.5   NUCLEIC ACID TESTING (NAT)/AMPLIFICATION TESTING

Nucleic acid testing (NAT) is basically a polymerase chain reaction (PCR) which has been the gold standard for COVID-19 since the outbreak. The COVID-19 targets used in the testing are open reading frame ORF1ab region (ORF1ab), RNA-dependent RNA polymerasegene (RdRp), nucleocapsid protein gene (N-gene) and envelope protein gene (E-gene) and spike proteins.

PCR as described by Li et al. (2020) may be:

quantification real-time PCR (qRT-PCR) where COVID-19 is confirmed when all three target genes are positive;

nested RT-PCR where COVID-19 low copy is detected;

nested RT-PCR can be one-step nested RT-PCR (OSN-qRT-PCR);

droplet digital PCR (ddPCR);

loop-mediated isothermal amplification (LAMP);

nanoparticle-based amplification;

portable benchtop-sized analyzers such as Xpert Xpress point of care (GeneXpert systems).

This method of molecular testing using PCR is challenged by complex pretreatment steps, availability of skilled manpower, equipment sophistication, and availability of BSL-2 laboratories.

Simplified machines and rapid test kits having been developed from this method, its general advantages are high sensitivity, specificity, suitability, time saving, and automation systems where a little training can be enough to operate the system.

COVID-19 medical laboratory testing in Nigeria has up to now been based on the gold standard (PCR) using NAT, as recommended by WHO and domesticated by NCDC. Though numerous samples such as nasopharyngeal swabs, oropharyngeal swabs, throat swabs, sputum, saliva, bronchoalveolar lavage fluid, conjunctival swabs, whole blood, serum/plasma, stool, urine, and rectal swabs can be used, Nigeria mostly uses nasopharyngeal and oropharygeal swabs but is expanding to using sputum and serum (blood). The Nigerian protocol depends on the availability of PCR kits; however, the target genes are E, N, PdRp and ORF1ab, while using the available PCR machines including the GeneXpert instrument (Etukudoh et al., 2020b; Enitan et al., 2020; Ikeagwulonu et al., 2020).

Currently, the MLSCN that regulates the practice of medical laboratory science in Nigeria has not approved/validated any rapid testing kit for COVID-19 in Nigeria, but NCDC, following the WHO announcement for emergency use of two antigen-based rapid diagnostic test kits—SD Biosensor and Abbott for COVID-19—have released a guide for their usage in Nigeria, following approval by NAFDAC and validation by MLSCN. This approach would increase the amount of testing and aid surveillance, point-of-care testing and patient management (Obeta et al., 2020; NCDC, 2020; MLSCN, 2020).

5.4 MEDICAL LABORATORY TESTING OF COVID-19 EXPERIENCE IN NIGERIA

After confirmation of COVID-19 from the index case in Nigeria and subsequent detection, NCDC realized that there was an urgent need to set up more molecular laboratories across the country for more testing and coverage of the Nigerian population, leading to the release of the laboratory building structures (Figure 5.3) within the network.

There was multiple online training of medical laboratory scientists and other health professionals for sample collection. There was also recruiting of experienced medical laboratory scientists who are experienced in molecular techniques and retraining for testing for COVID-19 in the upcoming centers (Figure 5.4).

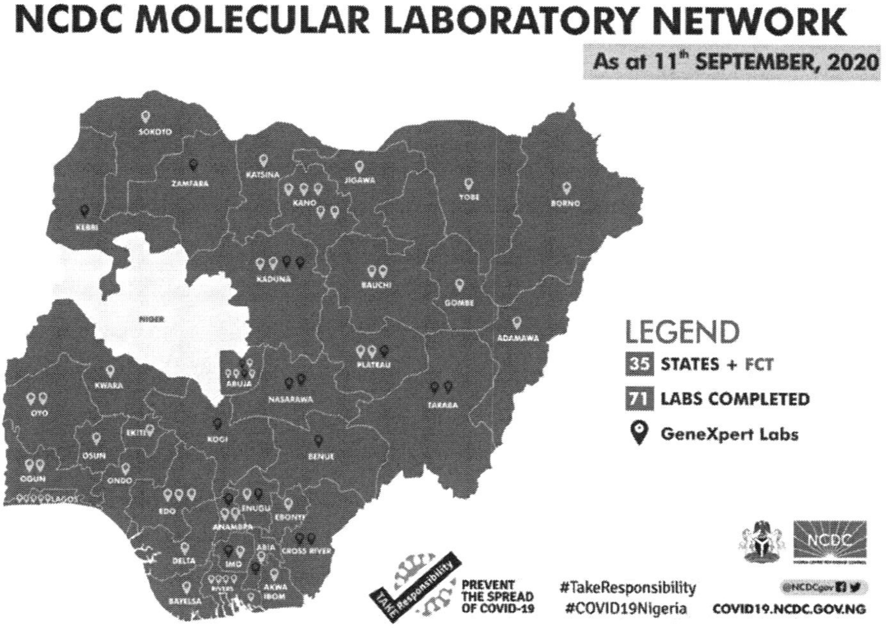

FIGURE 5.4 Molecular Laboratories (PCR) and GeneXpert-activated Laboratories in 35 States and FCT.

5.5   CHALLENGES OF MEDICAL LABORATORY TESTING FOR COVID-19 IN NIGERIA

There are operational differences between the laboratory-developed protocols and commercial consumable (ELISA) protocols which came out following the COVID-19 pandemic. Therefore, the implementation of these new protocols is challenging and requires continuous training for the laboratory staff to meet them. Such training and retraining and new products pose challenges (Bamidele & Rana, 2020).

It is also a challenge that there is limited access to COVID-19 reagents. This restricted the testing of citizens to only those who had symptoms or were at risk, as there is a fear of exhausting the limited supply of COVID-19 reagents and consumables (Akinyemi et al., 2020).

In the testing proper, there is delayed outbreak detection and reporting of COVID-19 cases; this may be due to distances from the testing and collection sites and the technicalities involved.

Also, there is limited access to clinically validated or regulatorily approved molecular and serologic tests either through the WHO network or through commercial manufacturers. Obviously, there are no validated rapid kits approved in Nigeria, as reported by MLSCN (Tang et al., 2020).

There is a lack of availability of local testing kits in Nigeria. The kits can be produced locally but resources and motivation to explore the capacity are not available.

There is a paucity of funding and lack of training of professionals, especially during COVID-19, toward the procurement of diagnostic test kits.

As COVID-19 ravages the world, there are shortages and difficulties in importing large quantities of diagnostic kits into Nigeria.

There is also a lack of political will on the part of the political elites concerning the development of good medical laboratories and hospitals, considering the fact that they can easily embark on medical tourism with their families, making the best health facilities in the country impossible to attain (Etukudoh et al., 2020b).

There is also poor knowledge and research capacity on COVID-19 in Nigeria, as there exist few grants or facilities.

5.6   POST-COVID-19 MEDICAL LABORATORY DIAGNOSIS

Most of the laboratories are temporary or established under emergency conditions, without adequate laboratory organization and plans.

The post-COVID-19 conditions in medical laboratory services, especially at the established COVID-19 testing sites under the NCDC laboratory network, if not followed with the political will of the government, will be in a sorry state. While the GeneXpert laboratories will continue to function optimally for COVID-19 and tuberculosis testing (Bayot & Sanchez, 2020) because they are well built under the hospital system, the molecular laboratories outside the hospital environment will not function optimally, due to the hasty manner of their construction and commencement of operation with support from federal government. Post-COVID-19, they may face a withdrawal of government funding and the facilities may fail to fund the purchase of consumables for such services or other related services because of the capital-intensive nature of molecular machine maintenance and services.

While NCDC had a plan to have one molecular laboratory in its network in every state, some states have more than one. This may increase the budgetary provisions of NCDC and, in a bid to cut down the cost, some laboratories may be closed down, waiting for another outbreak before reconsideration.

5.7 CONCLUSION

Medical laboratory testing is vital in public health emergencies (Kiros et al., 2020) and in COVID-19 in particular. Medical laboratory testing has employed various techniques and methods, such as virological cell culture, genomic sequencing, immunological testing, biosensors, and NAT with PCR, using various samples including nasopharyngeal swabs, oropharyngeal swabs, throat swabs, sputum, saliva, bronchoalveolar lavage fluid, conjunctival swabs, whole blood, serum/plasma, stool, urine, and rectal swabs.

In Nigeria, the protocol of COVID-19 testing is based on NAT using PCR and GeneXpert, and currently rapid test kits are about to be deployed that use biosensor or immunological techniques, subject to MLSCN validation.

In a bid to curtail COVID-19, Nigeria increased its number of molecular (PCR) laboratories, which were not initially meant for COVID-19, from 5 to 71 molecular laboratories including PCR and GeneXpert machines.

It is pertinent to address all the challenges faced in the optimal and functional testing for COVID-19 in such laboratories, to ensure adequate maintenance of such facilities even after the COVID-19 pandemic.

ACKNOWLEDGMENTS

The authors acknowledge the medical laboratory scientists in Nigeria who have played and are still playing major roles in the installation of testing services for COVID-19.

REFERENCES

Abdullahi IN, Emeribe AN, Akande AO, Ghamba PE, Adekola HA, Ibrahim Y, Dangana A. (2020). Challenges of COVID-19 testing services in Africa. *J Infect Dev Ctries*, 14(7), 691–695.

Akinyemi KO, Fakorede CO, Anjorin AAA, Abegunrin RO, Adunmo O, Ajoseh SO and Akinkunmi FM. (2020). Intrigues and challenges associated with COVID-19 pandemics in Nigeria. *Health*, 12, 954–971. doi:10.4236/health.2020.128072

Bamidele JA and Rana KM. (2020). Challenges of healthcare delivery in the context of COVID-19 Pandemic in Sub-Saharan Africa. *Iberoamerican J Med*, 02 100–109. doi:10.5281/zenodo.3755414

Bayot ML and Sanchez RS. (2020). Coronavirus disease (COVID-19) testing using the GeneXpert System: A technical guide on laboratory systems strengthening for COVID-19 pandemic response. 1st Edition. *Coach MLB Consulting*. Metro Manila, Philippines.

Enitan SS, Akele RY, Agunsoye CJ, Olawuyi KA, Nwankiti AJ, Oluremi AS, Ofodile CA, Olayanju AO, Alaba OEG, and Enitan CB. (2020). Molecular diagnosis of COVID-19 in Nigeria: Current practices, challenges and opportunities. *J Infect Dis & Case Rep*, SRC/JIDSCR 116-129. doi:10.47363/JIDSCR/2020(1)116.

Etukudoh NS, Ejinaka O, Obeta U, Utibe E, Lote-Nwaru I, Agbalaka P and Shaahia D. (2020a). Zoonotic and parasitic agents in bioterrorism. *J Inf Dis Trav Med*, 4(2), 000139. doi:10.23880/jidtm-16000139

Etukudoh NS, Ejinaka RO, Olowu FA, Obeta MU, Adebowale OM and Udoudoh M P. (2020b). Coronavirus (COVID-19); Review from A Nigerian perspective. *Am J Biomed Sci & Res*, 9(1). doi:10.34297/AJBSR.2020.09.001347

Ikeagwolunu RC, Obeta MU and Ugwu IN. (2020). Systematic review of laboratory parameters predicting severity and fatality of COVID-19 hospitalised patients New Zealand *J Med Lab Sci*, 74, 165–180

Kiros T, Kiros M, Andalem H, Hailemichael W, Damite S, Eyayu T, Getu S and Tiruneh T. (2020). Laboratory diagnosis of COVID-19: Role of laboratory medicine. *J Clin Chem Lab Med*, 3, 142. doi:10.35248/clinical-chemistry-laboratory-medicine.20.3.142

Li C, Zhao C, Bao J, Tang B, Wang Y and Gu B. (2020). Laboratory diagnosis of coronavirus disease-2019 (COVID-19). *Clinica Chimica Acta*, 510: 25–46

Lu R, Zhao X, Li J, Niu P, Yang B, Wu H, Wang W, Song H, Huang B, Zhu N, Bi Y, Ma X, Zhan F, Wang L, Hu T, Zhou H, Hu Z, Zhou W, Zhao L, Chen J, Meng Y, Wang J, Lin Y, Yuan J, Xie Z, Ma J, Liu WJ, Wang D, Xu W, Holmes EC, Gao GF, Wu G, Chen W, Shi W and Tan W. (2020). Genomic characterization and epidemiology of 2019 novel coronavirus: implications for virus origins and receptor binding. *Lancet*, 395(10224), 565–574

MLSCN, (2003). *Medical laboratory science council of Nigeria, Act, 2003. No: 11 Federal Official Gazette*, 2004 www.mlscn.gov.ng

MLSCN (2020). *Press Release on the pre-market validation of COVID-19 test kits*. Accessed on 15th October, 2020 from www.mlscn.gov.ng

Naidoo D and Ihekweazu C. (2020). Nigeria's efforts to strengthen laboratory diagnostics – Why access to reliable and affordable diagnostics is key to building resilient laboratory systems. *Afr J Lab Med 2020*, 9(2), a1019 doi: 10.4102/ajlm. v9i2.1019

Nassiri N (2020) Perspective on Wuhan Viral Pneumonia. *Adv in Pub. Health, Com and Trop Med: APCTM-106*. Kosmos Publishers.

NCDC (2018) *Nigeria Centre for Disease Control and Prevention (Establishment) Act*, 2018. www.ncdc.gov.ng

NCDC (2020). *Guidance on the Use of Antigen Rapid Diagnostic Kits for Diagnosis of SARS-COV-2 Infection in Nigeria*. Accessed on 28th October, 2020 from www.ncdc.gov.ng

Obeta MU, Ejinaka RO, Ofor IB, Ikeagwulonu RC, Agbo EC and Abara US. (2020). Nigerian COVID-19 (Coronavirus) patients update, the realities with medical laboratory diagnostic sites. *Am J Epidemiol Infect Dis*, 8 (1), 13–15. doi:10.12691/ajeid-8-1-3.

Obeta MU, Maduka KM, Ofor IB and Ofojekwu NM. (2019). Improving quality and cost diminution in modern healthcare delivery: The role of the medical laboratory scientists in Nigeria. *Inte J Business Management Invention (IJBMI)*, 08(03) 08–19.

Sheikhzadeh E, Eissa S, Ismail A, and Mohammed Zourob M. (2020). Diagnostic techniques for COVID-19 and new developments. *Talanta*, 220. doi:10.1016/j.talanta.2020.121392

Tang YW, Schmitz JE, Persing DH and Stratton CW. (2020). Laboratory diagnosis of COVID-19: current issues and challenges. *Journal of Clinical Microbiology*, 58, e00512–e00520. doi:10.1128/JCM.00512-20.

6 COVID-19 CT Image Segmentation and Detection
Review

Zahra Nourbakhsh

CONTENTS

6.1 INTRODUCTION

Deep learning (DL) is a kind of machine learning (ML) that can learn images through supervised learning, unsupervised learning, and semi-supervised learning. DL also identifies as a deep neural network (DNN). The convolutional neural network (CNN) is a method of DNN that uses preprocessing less than other networks (Rehman et al., 2020b; Rehman et al., 2021a; Iftikhar et al., 2017; Hussain et al., 2020). The methods are used in text, sound, images, etc. The DL method includes segmentation or detection to diagnose COVID-19. In this study, we work with a CNN in images. Image segmentation is also used for image analysis and processing, such as automated driving, medical images, etc. This method works with pixel values (Iqbal et al., 2018a, 2018b, 2019). Images can be classified using automatic segmentation and manual segmentation (Wang, 2016). COVID-19 has symptoms such as fever, coughing, and sneezing (Landry et al., 2020; Khan et al., 2020d), and sometimes breathlessness and loss of taste (Sharifi-Razavi et al., 2020). At first, the disease was taken to be pneumonia, but in time it was found to be a new disease. DL methods can detect many diseases, so has been used to detect COVID-19. Some studies have already used the normal patient, COVID-19, and types of pneumonia such as bacterial and viral to search the model. Then the new images to the diagnostic data are given to the model. Some researchers have used X-rays and computed tomography (CT) images for training and diagnosis (Rehman, 2020, 2021). For example, Guan et al. (2020a, 2020b), Gozes et al. (2020), Chung et al. (2020), and Rehman et al. (2021a, 2021b) have used CT images to detect COVID-19. Image segmentation is a challenge in image analysis, so a meta-heuristic algorithm can be used, such as a genetic algorithm (Elsayed et al., 2014), cuckoo search (Gaál et al., 2020; Agrawal et al., 2013), whale optimization algorithm (WOA) (Abd El Aziz et al., 2017), honey-bee mating optimization (Horng, 2010), particle swarm optimization (PSO) (Di Martino & Sessa, 2020; Harouni et al., 2014; Habibi & Harouni, 2018), or Harris Hawks optimization (HHO) (Das & Padhy, 2018), WOA having been used for dividing liver images (Mostafa et al., 2017). Shi et al. (2020) used an artificial intelligence (AI) method for image data diagnosis, segmentation, etc. Some authors have reported that X-ray images are better than CT images for diagnosis of COVID-19 (Latif et al., 2020; Khan et al., 2020a, 2020b). There are limited images in some model training (Ho et al., 2019; Harouni et al., 2010; Rehman et al., 2020a; Khan et al., 2019b). Sethy and Behera (2020) used ResNet-50 and support vector machines (SVM) in X-ray images for detecting COVID-19. Abbas et al. (2020) used CNN in X-ray images for diagnosis of COVID-19. Alqudah et al. (2019) used CNN, SVM, and random forest to classify COVID-19. Salman et al. (2020) created a CNN on chest X-ray (CXR) for COVID-19 infection. Karim et al. (2020) produced a DeepConvExplainer on CXR images for detecting COVID-19. Jamil and Hussain (2020) implemented a deep CNN on CXR to identify COVID-19. Asif and Wenhui (2020) created Inception v3 on CXR images. Loey et al. (2020) applied a GANjaved model (including GoogLeNet, AlexNet, and ResNet-18) on CXR images. Apostolopoulos and Mpesiana (2020) used a CNN model on X-ray images to detect COVID-19. Jaiswal et al. (2020) presented a deep transfer learning (DTL) method using DenseNet-201 for COVID-19 classification. Zhang et al. (2020) developed a confidence-aware anomaly detection model on CXR images to diagnose COVID-19. Singh et al. (2020) used a CNN model on CXR images to classify COVID-19. Adhikari (2020) proposed an automatic diagnostic network-based dense net on both X-ray and CT images for detecting infections to diagnose COVID-19. Ghoshal and Tucker (2020) created a Bayesian CNN-based DL model on CXR images to detect COVID-19. Ucar and Korkmaz (2020) presented a SqueezeNet model with Bayesian optimization on X-ray images to diagnose COVID-19. Barstugan et al. (2020) implemented a local directional pattern, gray level size zone matrix, gray level co-occurrence matrix, discrete wavelet transform algorithms, gray level run length matrix, and SVM (Mohammadi Dashti & Harouni, 2018; Javed et al., 2021) on X-ray images to diagnose COVID-19 (Khan et al., 2021a, 2021b). Punn and Agarwal (2020) used ResNet and Inception v3 on X-ray images for detecting COVID-19. Sarker et al. (2020) utilized DenseNet-121 on X-ray images to classify COVID-19. Wang et al. (2020c) used pre-training DL on lung images to detect COVID-19 infection. Accuracy in this model was 87%. Barstugan et al. (2020) created an SVM model on chest CT images to classify COVID-19 patients. Accuracy in this model was 99.68%. Yu et al. (2019) used 3D CNN on CT images to detect COVID-19 infection and used ResNet for feature extraction (Rehman et al., 2021c). Accuracy in this model was 86.70%. Gozes et al. (2020) utilized 2D and 3D DL models on CT features for the classification of COVID-19. Accuracy in this model was 99.60%. Zhao et al. (2020) implemented a COVID-CT dataset comprising 397 normal patients and 349 COVID-19 images. Corman et al. (2020) used real-time reverse transcription polymerase chain reaction (RRT-PCR) to detect COVID-19. Wang et al. (2020d) created a deep CNN on chest radiology images to detect COVID-19. Narin et al. (2020) developed a DNN to diagnose COVID-19. Various challenges in these studies include small datasets, the overfitting problem in a model, and the noise of the data, and most of the studies are looking for new methods or a combination of several methods to increase the accuracy and performance of the detection, segmentation or classification of COVID-19, and finally compare their models with other methods and show the proposed model of their study performs better than other models.

Now, we talk about the methods used in disease images and bring some studies in this section. Images that we can work with DL methods for the diagnosis of disease include X-ray images, CT images, MRI, and ultrasound (US) (Haque & Neubert, 2020). To detect digestive disease, ResNet-101 and the SVM method have 99.13% accuracy (Khan et al., 2020e). The detection method for the diagnosis of disease can be used for coronary artery disease (Wang et al., 2017a). Işın et al. (2016) used segmentation brain tumor with CNN for detecting cancer. This study reported that CNN could automatically receive complex features in multi-mode MRI images. According to 2013 Brats test dataset, the author finds that this study shows that this architecture is better than others. Bauer et al. (2011) used SVM for segmentation of brain tissue in ten datasets of patients. Islam et al. (2013) used the AdaBoost algorithm in MRI images to segment tumors. Ardila et al. (2019) used 3D DL to detect the lung automatically. Suzuki (2017) explains the application of CNNs and massive-training artificial neural networks (ANNs) for detecting medical images. Coudray et al. (2018) used an Inception v3 model to detect lung tumors automatically. Yıldırım et al. (2018) used 1D CNN for detecting cardiac arrhythmia. Hannun et al. (2019) created a DNN model to detect arrhythmia and classification ambulatory in 12 rhythm classes. This article used 59,543 patients for this model. Acharya et al. (2017) implemented a CNN with nine layers to classify heartbeats. Esteva et al. (2017) used a DNN model to classify skin cancer. This study first describes related work in DL method for disease detection and segmentation and then the studies that have investigated DL methods such as CNN, segmentation, and detection on coronavirus diagnoses.

In part 2, we talk about the datasets used in the studies. In part 3, we introduce the evaluation parameters used in the studies, and the part 3 discussion and part 4 are the conclusion sections.

6.2 DEEP LEARNING (DL)

DL is a subset of AI and ML that uses a human brain to find the pattern in input data and data processing to allow a high-precision model for prediction, detection, and classification, etc. in data (Khan et al., 2019a, 2020b, 2020c; Mashood Nasir et al., 2020; Saba et al., 2021b). DL is often used in complex tasks such as image processing, computer vision, speech processing, and image recognition (Amin et al., 2019; Rehman et al., 2018;

Iqbal et al., 2018a; Mughal et al., 2018b). DL has various types such as CNN, long short-term memory, and recurrent neural network (Mittal et al., 2020; Javed et al., 2019a, 2019b, 2020, Khan et al., 2019c, 2020f; Husham et al., 2016).

6.2.1 CONVOLUTIONAL NEURAL NETWORKS (CNNs)

CNN is a type of DL (Iqbal et al., 2018a; Adeel et al., 2020; Saba, 2020). This network has a three-layer convolutional layer, pooling layer, and fully connected layer, and is used in image processing, pattern recognition, and computer vision (Al-Ameen et al., 2015; Afza et al., 2020; Abbas et al., 2019a). This network can use activation functions, such as ReLU, sigmoid, softmax, and tanh. If these filters used to repeat in the model, it created an activation map that was also named a feature map. This CNN can automatically detect important features and reduce the characteristic. (Khan et al., 2020a)) created a DL model based on Xception architecture that trained end-to-end on ImageNet and was named the coronet method on 290 samples of CXR images to identify COVID-19. This model is done on four classes (normal, COVID-19, bacterial pneumonia, and viral pneumonia) and three classifications (normal, COVID-19, and pneumonia). This method can achieve 89.60% accuracy and 94.59% for three classes, 89.5% for four classes, and 99% for binary classes. The disadvantage is that this model only used a small dataset. This study tested the model in the dataset of Ozturk et al. (2020) and normal and pneumonia images CXR dataset. The extreme version of the inspection (Xception) method was developed in a study by (Das et al. (2020) on CXR images to detect COVID-19. The accuracy of the method is 97.40%. The advantage is that the Xception method can work with large datasets and train the weights for large and small datasets. The disadvantage is that the data in this method is limited. In this study, the activation function is shown in Equation (6.1):

$$R^l = F\left(\sum_{k \in L_r} a^k \otimes x^l + S^l\right) \tag{6.1}$$

In this formula, F means activation function, a^k is feature of l the former layer, x^l is coefficients and S^l is biases value.

Equation (6.2) shows the ReLU activation function:

$$F(\alpha) = \max(0, \alpha) \tag{6.2}$$

In this formula, $\alpha = 0$ if $F(\alpha)$ is smaller than zero.
Loss function is defined as in Equation (6.3):

$$T(ZP) = -\sum_j Z_j \log(P_j) \tag{6.3}$$

In here, Z_j shows label i of the image, P_j define the results parameter in the softmax activation function. Hemdan et al. (2020) implemented several DL models named COVIDX-Net (Abbas et al., 2019b, 2019c) (for example, VGG19, Inceptionv3, InceptionResNetv2, MobileNet, Xception, ResNetv2, and DenseNet-201) on 50 images comprised of 25 normal and 25 COVID-19 for 2D X-ray images for detecting COVID-19. The accuracy of this model is 90% in VGG19 and DenseNet-201 methods. VGG19 and DenseNet-201 performed better than the other three methods. Figure 6.1 shows the proposed COVIDX-Net. DenseNet-121 architecture is based on DenseNet architecture with 121 layers.

A CNN model named COVIDX-Net is used in a study by Sandler (Sandler et al., 2018). This model used a DenseNet method on CXR images. This method is a mobile application that used a CNN model. The accuracy of this model was 85%. This model was trained on the ImageNet dataset.

6.2.2 SEGMENTATION

Image segmentation divides images into parts (Harouni et al., 2012b; Mughal et al., 2017, 2018a). Image segmentation is used in fingerprint images, medical image processing, and self-driving cars, etc. These include behavioral, psychographic, demographic, and geographic applications (Harouni et al., 2012a; Lung et al., 2014; Nazir et al., 2019; Saba, 2021). In this section, we review image segmentation in the diagnosis of COVID-19. Amyar et al. (2020) performed multi-task learning methods on CXR images to detect coronavirus. The proposed method used a dataset of 1,369 patients consisting of 449 COVID-19, 425 normal, 98 lung cancer, and 397 different kinds of pathology data collected from a study by Liang and Zheng (2020) and medical segmentation. This method achieved 94% accuracy. The authors compare the performance of this method with the U-Net method. In this study, the author claimed that this method could improve segmentation results. This model addressed two

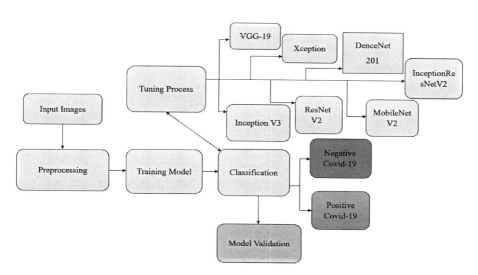

FIGURE 6.1 General COVIDX-Net detection process.

challenges: (1) limited data; (2) the use of a multi-task model instead of ResNet-50 for improving performance and preventing overfitting. The disadvantages were that this method only used CT images, and its data is limited (Saba et al., 2021a). DL techniques (DenseNet) created in a study by Zhang et al. (2020) on CT images from segmentation (Corman et al., 2020) of COVID-19, pneumonia, and normal and DeepLabv3 data were collected: 219 images from 110 patients with COVID-19, 224 from patients with influenza-A viral pneumonia, and 175 healthy cases. The accuracy of the proposed method was 86.7%. This research can show different Novel Coronavirus Pneumonia (NCP) from pneumonia and normal. This model used from lung segmentation and ResNet-18 from accuracy. This method's architecture was 92.49% accuracy for different NCP of the other two classes (pneumonia and normal), 94.93% sensitivity, and 91.13% specificity. The advantage of this method is that it can assist physicians with an inaccurate diagnosis. The 2,246 images used in this study include 759 NCP, 797 pneumonia, and 697 normal. U-Net, DRUMSET-FCN, SEGMENT, and DeepLabv3 were used to study the performance of the model. Cohen et al. (2020) developed CXR images with a path-based CNN for small training images. ImageNet parameters for weight values in the training network and the Adam algorithm for optimization network were used in this method. In this proposed method, the classifications were local and global. The accuracy of the method is 91.9%. The advantage of this method is that it has little complexity, and is therefore suitable for increased data. The disadvantage is that it is difficult to perform DL in large datasets. This method first trained data with FC-DenseNet-103 to prevent overfitting and complexity and to learn with ImageNet to increase data, and then the Adam optimizer was used. The data collection for segmentation was the JSRT and SCR datasets and for XCR it was UNSLM. Wang et al. (Guan et al., 2020a) used a DNN and U-Net model on chest CT images to recognize COVID-19. This research used 419 images from training and 131 images from testing. The author could achieve 90% accuracy. The method in this study is 3D deep CNN (deco net). In this study the first segmentation was of the lung with U-Net, then it was sent to 3D DNN.

6.2.3 DETECTION

Image detection is used to detect objects in the images and processing that image (Harouni et al., 2012c; Javed et al., 2021). For example, in medical images, from the number of symptoms in an image, we can identify whether that person has a certain disease or not (Perveen et al., 2020; Ullah et al., 2019; Rad et al., 2013). In image detection, there are algorithms such as single shot detectors, region-based convolutional neural networks (R-CNN), and faster R-CNN, for image detection applications in various fields of computer vision such as security, image retrieval, and edge identification. In this section we review image detection in the diagnosis of COVID-19. Abdel-basset et al. (2020) reported a hybrid method of COVID-19 diagnosis model on an improved marine predators algorithm (IMPA) and a ranking-based diversity reduction (RDR) on CXR images to detect COVID-19. The results were compared with five algorithms (WOA, sine-cosine algorithm, SAIP swarm algorithm, HHA, and equilibrium optimizer (EO)). The result showed that the performance of the IMPA algorithm is better than others. The disadvantage is that this

TABLE 6.1

Number of Images for Each Dataset (Narin et al., 2020)

Class or Datasets	Bacterial Pneumonia	COVID-19	Normal	Viral Pneumonia
Dataset-1	—	341	2800	—
Dataset-2	—	341	—	1493
Dataset-3	2772	341	—	—

algorithm only works in grayscale images. Narin et al. (2020) created a CNN on 100 CXR images (50 COVID and 50 normal) to detect COVID-19. The authors used five CNN models, including Inceptionv3, ResNet-50, and Inception-ResNetv2 and ResNet-101, and achieved 97% accuracy with Inceptionv3 and 87% accuracy with the Inception-ResNetv2 method. This study used three different COVID-19 datasets comprising bacterial pneumonia, normal, and viral pneumonia. In Table 6.1, dataset-1, dataset-2, and dataset-3 are binary datasets. This model trained on ImageNet for two classes by the Softmax classifier and achieved best performance. This study used five parameters for this model's performance—namely, accuracy, specificity, recall, precision, and F1-score. The highest accuracy of this model is 96.1% for ResNet-50 and ResNet-101. The results of accuracy of this study for COVID-19/normal is 96.1%, for COVID-19/viral pneumonia 99.5%, and for COVID-19/bacterial pneumonia 99.7%.

Wang et al. (2020b) trained 13,975 CXR images with the COVID-Net method to detect COVID-19. This method can achieve 93.3% accuracy. In Figure 6.2, the COVID-Net architecture is shown. According to this study, we understand COVID-Net to be the best method.

Ozturk et al. (2020) proposed a model for automatic exploration of COVID-19 with different CXR images (including 500 pneumonia, 125 COVID-19, and 500 normal). In this method, the author first trained the DarkCovidNet and classification into three categories: pneumonia, COVID, and No-Findings. The proposed model achieved an accuracy of 98.8% in detecting COVID-19. This model achieved 98.08% for binary images and 87.02% for multi-class images. The disadvantage is that this method worked in few X-ray images. The activation function of this model is leaky ReLu with 17 convolution layers. Alom et al. (2020) applied the Inception Recurrent Residual Convolutional Neural Network (IRRCNN) and NABLA-3 network models on CXR and CT images. The test accuracy of the model is 84.67% for X-ray and 98.88 for CT images. The disadvantages of this model are that the experiment is on small data, and there is a false positive in the model's detection. Jaiswal et al. (2020) proposed a DenseNet201 model on chest CT images. The accuracy of the proposed training model is 99.82% and that of the proposed testing model 96.25%. Chen et al. (2020) present a DL-based model on 51 CT images for COVID-19 and 55 CT images for other diseases. The accuracy of this model for each patient was 95.24% and for each image 98.85%. The advantage of this model is that it can decrease reading time by 95% and thereby enhance radiologists' performance. This model used a BrightSpeed CT scanner, Optima CT680, and Revolution CT implementation and the

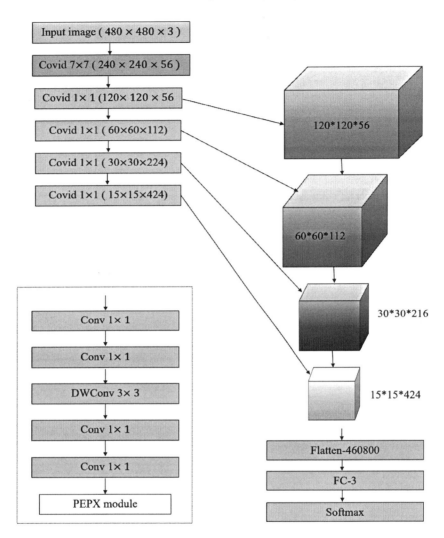

FIGURE 6.2   Samples of X-ray for two classes: normal & COVID-19-infected.

unit++ training algorithm. The authors used 289 CT images for learning and 600 CT images for testing. This model used accuracy, sensitivity, specificity, positive predictive value (PPV) (Kenari et al., 2010), and negative predictive value (NPV) for evaluation performance. Disadvantage: this study does not take normal patient images. Brunese et al. (2020) developed a DL model on X-ray images. To predict COVID-19, the Softmax activation function was used. The authors used VGG-16 by 16 layers that trained on ImageNet dataset. The accuracy of the model was 97%, sensitivity was 96%, specificity was 98%, and F-measure was 94%. Apostolopoulos and Mpesiana (2020) conducted TL with a CNN model on CXR images. Data classification was COVID-19, pneumonia, and normal for automatically identifying COVID-19.

Accuracy in this model was 96.78%, sensitivity was 98.66%, and specificity was 96.46%. This study used 1,427 images to train and test. Several datasets were used in this model for an example for analysis, X-ray images for Cohen and Radiopaedia, and the Italian Society of Medical Interventional Radiology (SIRM) and the Radiological Society of North America websites. This study used the ReLu activation function and the Adam optimizer and two different datasets. Elasnaoui and Chawki (2020) produced a hybrid DL model using VGG-16, VGG-19, DenseNet201, Inception-ResNet-v2, Inception-v3, ResNet50, and MobileNet-v2 on 6,087 images (1,583 normal, 1,493 COVID, and 2,780 bacterial pneumonia), CXR, and CT images. The result of 88.09% accuracy for DenseNet201 and 82.18% for Inception-ResNet-v2 shows Inception-ResNet-v2 and DenseNet201 are better evaluated from another model. This model used F1-score, accuracy, specificity, sensitivity, and precision parameters for evaluation (Harouni and Baghmaleki, 2020). Finally, we understand that the Inception-ResNet-v2 model with 92.07% for F1-score, 92.11% for sensitivity, 92.38% for precision, 92.18% for accuracy, and 96.06 for specificity is better than other models for the detection of COVID-19. Wang et al. (2020a) exhibited a COVID-19 pneumonia lesion segmentation network (COPLE-NET) model for reducing noise on CT images of COVID-19. For the evaluation model, an MAE parameter was used. Sethy and Behera (2020) developed a ResNet50 plus SVM model on CXR images. This study used accuracy, FPR, Kappa, and MCC for evaluation. This method achieved 95.38% of accuracy, 95.52% of FPR, 90.76% of Kappa, and 91.41% of MCC. This study used 2 datasets and an open dataset that give 51 images of the first dataset, 133 images of the second dataset, and 133 images of the open dataset. Farooq and Hafeez (2020) created a COVIDResNet method that used ResNet-50 method on CXR images. This method can achieve 93.96% of accuracy that showed COVIDResNet was a better performance than ResNet50 because the accuracy of ResNet50 was 83.50% in this study. This study's classification data was divided into four classes, normal, viral, bacterial, and Covid-19, and compared these four classes for sensitivity, F-1 score, and precision and understand that Covid-19 in another class for better sensitivity, F-1 score, and precision. Gozes et al. (2020) conducted an INF-NET method on CT images for detecting COVID-19. This study used the INF-NET method to solve diversity problems in infection and high data collection at a low time and used semi-supervised learning for reducing the label data. This method used the MAE parameter for evaluation (Harouni and Baghmaleki, 2020). Guan et al. (2020a) produced a weakly supervised DL on CT images for detecting COVID-19. This study used accuracy, precision, sensitivity, and specificity for evaluation. The accuracy of this method was 95%. Jamshidi et al. (2020) proposed an ANN model including GAN, LSTM, RNN, and ELM on CT images and X-ray images. This method has five layers with each one responsible for doing the work. We see each layer's performance: layer 1 receives inputs database, layer 2 was a selection, layer 3 was image-based techniques, layer 4 was optimization, and layer 5 was output diagnoses.

6.3   DATASETS

In this section, we introduce the duties used in different studies and explain each other in the sample and show the samples of the data used (Rad et al., 2016;

TABLE 6.2
Summary Dataset

Disease	Number of Images
Normal	310
Bacterial pneumonia	330
Viral pneumonia	327
COVID-19	284

Rahim et al., 2017a, 2017b). Here are some of the studies on CT images. Some studies used X-ray images, and others used both and have tried to implement their models on those datasets and, with the correct setting of the parameters, can achieve a good result. The dataset used in the above studies are: Khan et al. (2020a), Narin et al. (2020), Jaiswal et al. (2020) using the Kaggle[1] dataset. Khan et al. utilized ImageNet dataset of GitHub containing 290 X-ray images and Kaggle. This dataset contains 1,203 normal images, 931 viral pneumonia, and 660 bacterial pneumonia that we used 1300 images of 2 datasets. Dataset summary in this model includes normal, bacterial pneumonia, viral pneumonia, and COVID-19, as shown in Table 6.2. Narin et al. used Cohen et al. (2020) dataset, 2,800 normal chest images for Wang et al. (2017b) dataset, and 2,772 bacterial and 1,493 viral pneumonia for this dataset. This image was binary and 224 × 224-pixel size and Jaiswal used a dataset with 1,262 COVID-19 and 1,230 normal images. This dataset divided total images into two classes— COVID-19 and normal—and trained and tested the model with these classes.

The GitHub[2] dataset was utilized in Hemdan et al. (2020), Narin et al. (2020), Ozturk et al. (2020), Elasnaoui and Chawki (2020), and Farooq and Hafeez (2020) studies. In the article of Hemdan et al., in the step of preprocessing, the images taken this data in the size of 224*224 and then divided them into two groups of COVID-19 or not. In the training and validation step, the model also divides 20% of the data to the test and 80% of the rest in training and validation. The classification step classifies test data, the Narin et al. dataset, with eight images—X1, X2, X3, X4, X5, X6, X7, and X8. This data has four classes of images— normal, bacterial, viral pneumonia, and COVID-19 images. This dataset utilizes 341 COVID-19 patients. Ozturk et al. used 2 datasets with 127 X-ray images of COVID-19 including 43 women and 82 men for the GitHub dataset and 500 pneumonia, and 500 non-finding of 2 classes— normal and pneumonia—in DOI[3] dataset. Elasnaoui and Chawki utilized Cohen et al. (2020) for CXR images. This article used 6,087 images: 1,493 of coronavirus, 2,780 bacterial pneumonia, 1,583 normal, and 231 of COVID-19. Farooq used the GitHub[4] dataset that exists through COVID-Net and is available to people. This dataset includes 4 classes—viral, normal, COVID-19, and bacterial pneumonia—of 2,839 patients from 5,941 images of chest radiography, including 931 bacterial pneumonia, 45 COVID-19, 660 viral pneumonia, and 1,203 normal images.

Amyar et al. (2020) used three datasets with 347 COVID-19 images and 397 non-COVID images in Zhao et al. (2020) dataset, 100 COVID-19 CT scan images in Covid19[5] dataset, and 425 CT scan images of normal patients and 98 lung cancer

TABLE 6.3
Classification Dataset

Dataset	Class	#	Bits	Lung Mask	Heart Mask
JSRT/SCR	Normal	20	12	O	O
NLM (MC)	Normal	73	8	O	—
CoronaHack	Normal	98	24	—	—
NLM (MC)	Tuberculosis	57	8	O	—
CoronaHack	Pneumonia (Bacterial)	21	24	—	—
Cohen et al.	Pneumonia (Bacterial)	33	24	—	—
	Pneumonia (Virus)	20	20	—	—
	Pneumonia (COVID-19)	180	24	—	—

TABLE 6.4
Segmentation Dataset (Guan et al., 2020a)

Dataset	Class	#	Bits	Lung Mask	Heart Mask
Training JSRT/ SCR	Normal/ Nodule	197	2	O	O
Validation JSRT/ SCR	Normal/ Nodule	57	12	O	O
NLM (MC)	Normal	73	8	O	—

TABLE 6.5
Classification Method

Dataset	Normal	Bacterial	Tuberculosis	Viral	COVID-19	Total
Training	134	39	41	14	126	354
Validation	19	5	5	2	18	49
Test	38	10	11	4	36	99
Total	191	54	57	20	180	502

images in Henri Becquerel Cancer Center in Rouen, France, dataset. A CXR dataset was used in Guan et al. (2020a). This dataset is shown in Tables 6.3–6.5. This study compared results with the COVID-Net algorithm (Shan et al., 2020).

In Table 6.5, JCR is the Japanese Society of Radiological Technology (Shiraishi et al., 2000) and used 247 CXR images, mask classification has collected SCR dataset (Van Ginneken et al., 2006), USNLM was the US National Library of Medicine, and MC was collected Montgomery County (Jaeger et al., 2014).

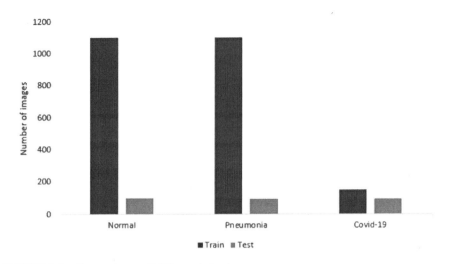

FIGURE 6.3   Comparisons of different infections.

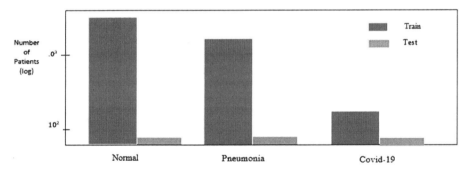

FIGURE 6.4   Comparisons of different infections on number of patients.

Figure 6.3 depicts the COVIDx dataset that was used in Shan et al. (2020) article. Distribution for COVIDx is shown in Figure 6.2. According to Figure 6.3, the number of normal images in the training data is more than 1,000 images and for testing data is less than 100, for pneumonia images for testing less than 100, and training images more than 1,000 and less than normal images, and for COVID-19, it is also more than 100 images for testing and more than 100 for training images.

In another study, Brunese et al. (2020) used 3 datasets—Cohen et al. (2020), Ozturk et al. (2020), and Wang et al. (2017b) with CXR or CT images. Figure 6.4 illustrates a type of label. This dataset contains 6,523 CXR images comprising 250 COVID-19, 2753 pulmonary diseases, and 3,529 normal patient images.

Guan et al. (2020a) and Elasnaoui and Chawki (2020) used DOI[6] dataset. Guan et al. used this dataset for training model that was public access and Elasnaoui and Chawki used DOI (Kermany et al., 2018) for CT images. Sethy and Behera (2020) used 3 datasets—GitHub[7], Kaggle[8], and Open[9]—that used 50 images of the first

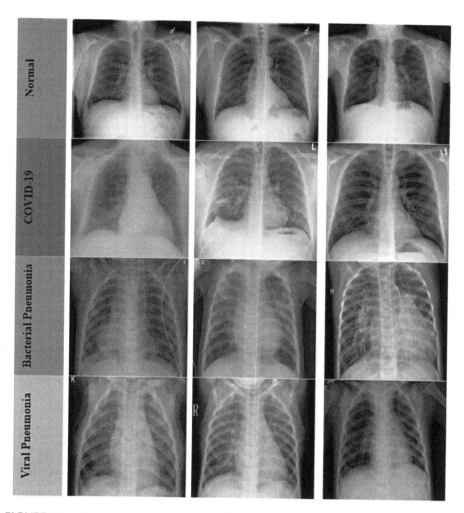

FIGURE 6.5 X-ray image comparisons of different viruses.

TABLE 6.6
Detail Dataset (Sethy and Behera, 2020)

Samples	Number	Repository
Covid-19+ Without MERS, SARS, ARDS	25	http://www.github.com/
Covid-19-	25	http://www.kaggle.com/

dataset, 133 images of the second dataset, and 133 images of the third dataset. Figure 6.5 shows a sample of X-ray used in the model. Table 6.6 shows a detailed dataset.

Jamshidi et al. (2020) used Zhao et al. (2020) dataset for this study. This article used big data in the model. Das et al. (2020) used Ozturk et al. (2020) dataset that includes COVID-19, normal, and pneumonia classes. In this dataset, three classes—COVID-19, pneumonia, and various types of infections—have 70% of them for training.

COVID[10] dataset was used in Gozes et al. (2020) article for labeled CT images and Cohen et al. (2020) dataset was used for unlabeled CT images. This article utilizes 157 international patients of China and USA.

6.3.1 EVALUATION

In this study, the above articles' parameter evaluation is accuracy, sensitivity, specificity, PPV, NPV, F1-score, recall, precision, MSE (Amyar et al., 2020), and MAE (Wang et al., 2020a).

The formulas of this parameter are (Saba et al., 2018, 2019a, 2020; Fahad et al., 2018) (Table 6.7).

TABLE 6.7
Metric Parameter

$$\text{Accuracy} = \frac{TN + TP}{TN + TP + FN + FP} \tag{6.4}$$

$$\text{Sensitivity} = \frac{TP}{Positive} \tag{6.5}$$

$$\text{Specificity} = \frac{TN}{Negative} \tag{6.6}$$

$$\text{PPV} = \frac{TN}{\left(TP + FP\right)} \text{ (Harouni and Baghmaleki)} \tag{6.7}$$

$$\text{Recall} = \frac{TP}{\left(TP + FP\right)} \tag{6.8}$$

$$\text{Precision} = \frac{TP}{\left(TP + FP\right)} \text{ (Harouni and Baghmaleki)} \tag{6.9}$$

$$\text{NPV} = \frac{TN}{\left(TN + FN\right)} \tag{6.10}$$

$$\text{F1} - \text{Score} = \frac{2 * \left(\text{Precision} * \text{Recall}\right)}{\left(\text{Precision} + \text{Recall}\right)} \tag{6.11}$$

$$\text{MSE} = \frac{1}{n} \sum_{t=1}^{n} \left(y_{true} - y_{predict}\right)^2 \tag{6.12}$$

$$\text{MAE} = \frac{\sum_{i=1}^{n} |y_i - x_i|}{n} \tag{6.13}$$

TABLE 6.8
Confusion Matrix

	Predicted Positive	Predicted Negative
Actual Positive	True Positive (TP)	False Negative (Mostafa et al.)
Actual Negative	False Positive (FP)	True Negative (TN)

The above parameters are used for evaluation between several methods (Harouni & Baghmaleki, 2020; Khan et al., 2020d; Saba et al., 2019b; Jamal et al., 2017; Ejaz et al., 2018). Accuracy is shown with %, and each method has higher accuracy, which is better. This parameter is used in every above article to evaluate the performance. In this formula TP represents true positive, FP represents false positive, TN represents true negative, FN represents false negative, PPV means positive predictive value, and NPV means negative predictive value. Positive means COVID-19, and negative means normal or another disease, true positive is the number of COVID patients who are correctly detected, false positive the number of people who are mistakenly recognized, true negative the number of people properly detected as healthy, and false negative the people who have been mistakenly recognized as healthy. In MSE, y-true is the main value and y-predict is the predicted value that the author predicts. TP, FP, TN, and TP are shown in Table 6.8.

6.4 DISCUSSION

COVID-19 came into being in the last year and rapidly grew into a worldwide epidemic. The diagnosis of this disease is difficult because of its being new. Also, due to the diversity of the type of disease, it cannot easily be detected with manual methods, hence the use of artificial intelligence methods, such as DL, segmentation, detection, classification, and so on. To detect this high-precision disease, many studies have been written in this regard using various DL methods. We use better parameters such as accuracy, specificity, sensitivity, F-1 score, PPV, NPV, etc. to show better performance. The summary of these studies is presented in Table 6.9.

According to Table 6.9, we understand these studies used different CNN methods such as CoroNet, Xception, COVIDX-Net (for example, VGG19, ResNet50, InceptionV3, InceptionResNetV2, and DensNet201), COVID-MobileXpert, segmentation methods such as multi-task learning, DenseNet, Path-Based Method and DeCovNet, and detection methods such as IMPA+RDR, DTL (InceptionV3, ResNet50, Inception-ResNetV2, ResNet101), DTL (InceptionV3, ResNet50, Inception-ResNetV2, ResNet101), DTL (InceptionV3, ResNet50, Inception-ResNetV2, ResNet101), (5 CNN Models Include InceptionV3, ResNet50, Inception-ResNetV2, ResNet101), COVID-Net, DarkCovidNet, IRRCNN, DenseNet201 Model, VGG16, TL with CNN, VGG16, (VGG19, DenseNet201, Inception-ResNet-V2, Inception-V3, Resnet50, and MobileNet-V2), ResNet50 plus SVM model, COVIDResNet, Inf-Net Method, Weakly Supervised DL (an artificial neural network model including GAN, LSTM, RNN, and ELM) for diagnosis of COVID-19 (Khan et al., 2017, Ramzan et al., 2020b, Ramzan et al., 2020a). According to accuracies, we understand that

TABLE 6.9
Summary of Methods

Study	Type of Image	Method Used	Accuracy
Khan et al. (2020a)	CXR	CoroNet	89.6%
Narayan Das et al. (2020)	CXR	Xception	97.40%
Hemdan et al. (2020)	CXR	COVIDX-Net (for Example VGG19, ResNet50, InceptionV3, InceptionResNetV2, and DenseNet201)	90%
Li and Zhu product (Jamshidi et al., 2020)	CXR	COVID-MobileXpert	
Amyar et al. (2020)	Chest CT	Multi-Task Learning	94%
Zhang et al. (Gozes et al., 2020)	Chest CT	DenseNet	92.49%
Oh et al. (Cohen et al., 2020)	CXR	Path-Based Method	91.9%
Wang et al. (Guan et al., 2020a)	Chest CT	DeCovNet	90%
Abdel-Basset et al. (2020)	CXR	IMPA+RDR	
Narin et al. (normal) (Narin et al., 2020)	CXR	DTL (Inception V3, ResNet50, Inception-ResNetV2, ResNet101)	96.1%
Narin et al. (viral pneumonia) (Narin et al., 2020)		DTL (InceptionV3, ResNet50, Inception-ResNetV2, ResNet101)	99.5%
Narin et al. (bacterial pneumonia) (Narin et al., 2020)	CXR	DTL (InceptionV3, ResNet50, Inception-ResNetV2, ResNet101)	99.7%
Narin et al. (2020)	CXR	5 CNN Models Include InceptionV3, ResNet50, Inception-ResNetV2, ResNet101	97%
Wang et al. (Shan et al., 2020)	CXR	COVID-Net	93.3%
Ozturk et al. (2020)	CXR	DarkCovidNet	98.8%
Alom et al. (2020)	CXR and CT	IRRCNN	84.67% for X-ray and 98.78 for CT
Jaiswal et al. (2020)	Chest CT	DenseNet-201 model	99.82% in Train model and 96.25% in Testing model
Chen et al. (2020)	Chest CT	VGG16	97%
Brunese et al. (2020)	CXR	Transfer learning with CNN	96.78%
Elasnaoui and Chawki (2020)	CT and X-ray	VGG16, VGG19, DenseNet201, Inception_ResNet_V2, Inception_V3, Resnet50, and MobileNet_V2	92.18%
Sethy and Behera (2020)	CXR	ResNet50 plus SVM model	95.38%
Farooq and Hafeez (2020)	CXR	COVIDResNet	93.26%
Fan et al. (Gozes et al., 2020)	CT	Inf-Net Method	

(Continued)

TABLE 6.9
(CONTINUED)

Study	Type of Image	Method Used	Accuracy
Guan et al. (Guan et al., 2020a)	CT	Weakly Supervised DL	95%
Jamshidi et al. (2020)	CT and X-ray	Proposed an Artificial Neural Network model include GAN, LSTM, RNN, and ELM	

DenseNet201 Model with 99.82% of accuracy in training model, DarkCovidNet with 98.8% of accuracy, DTL (InceptionV3, ResNet50,Inception-ResNetV2, ResNet101) with 99.7% of accuracy, DTL (InceptionV3, ResNet50,Inception-ResNetV2, ResNet101 (Khan et al., 2019b, Khan et al., 2019d) method with 99.5% of accuracy, Xception method with 97.40% of accuracy, and 5 CNN models including InceptionV3, ResNet50, Inception-ResNetV2, ResNet101, and VGG16 with 97% of accuracy is better performance than other methods. The purpose of these studies is to solve the challenge of increasing the accuracy of detection, segmentation, classification, and prediction. Thus, different parameters are used to model for the evaluation performance. In all studies using the precision parameter, it has shown which method is better than the other methods of action.

6.5 CONCLUSION

In this chapter, we set out to review DL methods to recognize COVID-19. Various studies in this field have been investigated. They have tried to recognize with higher accuracy and in less time by artificial intelligence methods for detection and classification of images to help radiologists and doctors. Each of the studies has tried to compare its methods with previous studies and show that their method has higher performance than other methods. They have evaluated their method according to accuracy, sensitivity, specificity, PPV, NPV, recall, precision, F-1 score, MSE, and MAE, that most studies used accuracy, sensitivity, and specificity for evaluation and compared them with other methods and have shown that their selective method has a superior performance than other methods.

NOTES

1 https://www.kaggle.com/paultimothymooney/chest-xray-pneumonia
2 https://github.com/ieee8023/covid-chestxray-dataset
3 doi:10.1101/2020.03.12.20027185
4 https://github.com/lindawangg/COVID-Net
5 http://medicalsegmentation.com/covid19/
6 doi:10.7937/K9/TCIA.2017.3r3fvz08
7 http://www.github.com/

8 http://www.kaggle.com/
9 https://openi.nlm.nih.gov/
10 HTTPS:// Medical segmentation.com/ COVID19/

REFERENCES

Abbas, A., Abdelsamea, M. M. & Gaber, M. M. 2020. Classification of COVID-19 in chest X-ray images using DeTraC deep convolutional neural network. arXiv preprint arXiv: 2003.13815.

Abbas, N., Saba, T., Mehmood, Z., Rehman, A., Islam, N. & Ahmed, K. T. 2019a. An automated nuclei segmentation of leukocytes from microscopic digital images. *Pakistan Journal of Pharmaceutical Sciences*, 32(5).

Abbas, N., Saba, T., Rehman, A., Mehmood, Z., Javaid, N., Tahir, M., Khan, N. U., Ahmed, K. T. & Shah, R. 2019b. Plasmodium species aware based quantification of malaria parasitemia in light microscopy thin blood smear. *Microscopy Research and Technique*, 82, 1198–1214.

Abbas, N., Saba, T., Rehman, A., Mehmood, Z., Kolivand, H., Uddin, M. & Anjum, A. 2019c. Plasmodium life cycle stage classification based quantification of malaria parasitaemia in thin blood smears. *Microscopy Research and Technique*, 82, 283–295.

Abd El Aziz, M., Ewees, A. A. & Hassanien, A. E. 2017. Whale optimization algorithm and moth-flame optimization for multilevel thresholding image segmentation. *Expert Systems with Applications*, 83, 242–256.

Abdel-Basset, M., Mohamed, R., Elhoseny, M., Chakrabortty, R. K. & Ryan, M. 2020. A hybrid COVID-19 detection model using an improved marine predators algorithm and a ranking-based diversity reduction strategy. *IEEE Access*, 8, 79521–79540.

Acharya, U. R., Oh, S. L., Hagiwara, Y., Tan, J. H., Adam, M., Gertych, A. & San Tan, R. 2017. A deep convolutional neural network model to classify heartbeats. *Computers in Biology and Medicine*, 89, 389–396.

Adeel, A., Khan, M. A., Akram, T., Sharif, A., Yasmin, M., Saba, T. & Javed, K. 2020. Entropy-controlled deep features selection framework for grape leaf diseases recognition. *Expert Systems*, 1(2), 1–21.

Adhikari, N. C. D. 2020. Infection severity detection of CoVID19 from X-Rays and CT scans using artificial intelligence. *International Journal of Computer (IJC)*, 38, 73–92.

Afza, F., Khan, M. A., Sharif, M., Saba, T., Rehman, A. & Javed, M. Y. *Skin Lesion Classification: An Optimized Framework of Optimal Color Features Selection. 2020 2nd International Conference on Computer and Information Sciences (ICCIS)*, 2020. IEEE, 1–6.

Agrawal, S., Panda, R., Bhuyan, S. & Panigrahi, B. K. 2013. Tsallis entropy based optimal multilevel thresholding using cuckoo search algorithm. *Swarm and Evolutionary Computation*, 11, 16–30.

Al-Ameen, Z., Sulong, G., Rehman, A., Al-Dhelaan, A., Saba, T. & Al-Rodhaan, M. 2015. An innovative technique for contrast enhancement of computed tomography images using normalized gamma-corrected contrast-limited adaptive histogram equalization. *EURASIP Journal on Advances in Signal Processing*, 2015, 1–12.

Alom, M. Z., Rahman, M., Nasrin, M. S., Taha, T. M. & Asari, V. K. 2020. COVID_MTNet: COVID-19 detection with multi-task deep learning approaches. arXiv preprint arXiv: 2004.03747.

Alqudah, A. M., Qazan, S., Alquran, H., Qasmieh, I. A. & Alqudah, A. 2019 COVID-2019 Detection using X-ray images and artificial intelligence hybrid systems. *Biomedical Signal and Image Analysis and Project; Biomedical Signal and Image Analysis and Machine Learning Lab:* Boca Raton, FL.

Amin, J., Sharif, M., Raza, M., Saba, T. & Rehman, A. *Brain Tumor Classification: Feature Fusion. 2019 International Conference on Computer and Information Sciences (ICCIS)*, 2019. IEEE, 1–6.

Amyar, A., Modzelewski, R., Li, H., & Ruan, S. 2020. Multi-task deep learning based CT imaging analysis for COVID-19 pneumonia: Classification and segmentation. *Computers in Biology and Medicine*, 126, 104037.

Apostolopoulos, I. D. & Mpesiana, T. A. 2020. Covid-19: Automatic detection from x-ray images utilizing transfer learning with convolutional neural networks. *Physical and Engineering Sciences in Medicine*, 43(2), 635–640.

Ardila, D., Kiraly, A. P., Bharadwaj, S., Choi, B., Reicher, J. J., Peng, L., Tse, D., Etemadi, M., Ye, W., & Corrado, G. 2019. End-to-end lung cancer screening with three-dimensional deep learning on low-dose chest computed tomography. *Nature Medicine*, 25, 954–961.

Asif, S. & Wenhui, Y. 2020. Automatic detection of COVID-19 using X-ray images with deep convolutional neural networks and machine learning. medRxiv.

Barstugan, M., Ozkaya, U. & Ozturk, S. 2020. Coronavirus (covid-19) classification using ct images by machine learning methods. arXiv preprint arXiv: 2003.09424.

Bauer, S., Nolte, L.-P. & Reyes, M. *Fully automatic segmentation of brain tumor images using support vector machine classification in combination with hierarchical conditional random field regularization. International Conference on Medical Image Computing and Computer-Assisted Intervention*, 2011. Springer, 354–361.

Brunese, L., Mercaldo, F., Reginelli, A. & Santone, A. 2020. Explainable deep learning for pulmonary disease and coronavirus COVID-19 detection from X-rays. *Computer Methods and Programs in Biomedicine*, 196, 105608.

Chen, J., Wu, L., Zhang, J. Zhang, L., Gong, D., Zhao, Y., Chen, Q., Huang, S., Yang, M. & Yang, X. 2020. Deep learning-based model for detecting 2019 novel coronavirus pneumonia on high-resolution computed tomography. *Scientific Reports*, 10, 1–11.

Chung, M., Bernheim, A., Mei, X., Zhang, N., Huang, M., Zeng, X., Cui, J., Xu, W., Yang, Y. & Fayad, Z. A. 2020. CT imaging features of 2019 novel coronavirus (2019-nCoV). *Radiology*, 295, 202–207.

Cohen, J. P., Morrison, P., Dao, L., Roth, K., Duong, T. Q. & Ghassemi, M. 2020. Covid-19 image data collection: Prospective predictions are the future. arXiv preprint arXiv: 2006.11988.

Corman, V. M., Landt, O., Kaiser, M., Molenkamp, R., Meijer, A., Chu, D. K., Bleicker, T., Brunink, S., Schneider, J. & Schmidt, M. L. 2020. Detection of 2019 novel coronavirus (2019-nCoV) by real-time RT-PCR. *Eurosurveillance*, 25, 2000045.

Coudray, N., Ocampo, P. S., Sakellaropoulos, T., Narula, N., Snuderl, M., FENYö, D., Moreira, A.L., Razavian, N. & Tsirigos, A. 2018. Classification and mutation prediction from non–small cell lung cancer histopathology images using deep learning. *Nature Medicine*, 24, 1559–1567.

Das, N. N., Kumar, N., Kaur, M., Kumar, V. & Singh, D. 2020. Automated deep transfer learning-based approach for detection of COVID-19 infection in chest X-rays. *Irbm*.

Das, S. P. & Padhy, S. 2018. A novel hybrid model using teaching–learning-based optimization and a support vector machine for commodity futures index forecasting. *International Journal of Machine Learning and Cybernetics*, 9, 97–111.

Di Martino, F. & Sessa, S. 2020. PSO image thresholding on images compressed via fuzzy transforms. *Information Sciences*, 506, 308–324.

Ejaz, K., Rahim, M. S. M., Rehman, A., Chaudhry, H., Saba, T. & Ejaz, A. 2018. segmentation method for pathological brain tumor and accurate detection using MRI. *International Journal of Advanced Computer Science and Applications*, 9, 394–401.

Elasnaoui, K. & Chawki, Y. 2020. Using X-ray images and deep learning for automated detection of coronavirus disease. *Journal of Biomolecular Structure and Dynamics*, 1–22.

Elsayed, S. M., Sarker, R. A. & Essam, D. L. 2014. A new genetic algorithm for solving optimization problems. *Engineering Applications of Artificial Intelligence*, 27, 57–69.

Esteva, A., Kuprel, B., Novoa, R. A., Ko, J., Swetter, S. M., Blau, H. M. & Thrun, S. 2017. Dermatologist-level classification of skin cancer with deep neural networks. *Nature*, 542, 115–118.

Fahad, H., Ghani Khan, M. U., Saba, T., Rehman, A. & Iqbal, S. 2018. Microscopic abnormality classification of cardiac murmurs using ANFIS and HMM. *Microscopy Research and Technique*, 81, 449–457.

Farooq, M. & Hafeez, A. 2020. Covid-resnet: A deep learning framework for screening of covid19 from radiographs. arXiv preprint arXiv: 2003. 14395.

Gaál, G., Maga, B. & Lukács, A. 2020. Attention u-net based adversarial architectures for chest x-ray lung segmentation. arXiv preprint arXiv: 2003.10304.

Ghoshal, B. & Tucker, A. 2020. Estimating uncertainty and interpretability in deep learning for coronavirus (COVID-19) detection. arXiv preprint arXiv: 2003.10769.

Gozes, O., Frid-Adar, M., Greenspan, H., Browning, P. D., Zhang, H., Ji, W., Bernheim, A. & Siegel, E. 2020. Rapid ai development cycle for the coronavirus (covid-19) pandemic: Initial results for automated detection & patient monitoring using deep learning ct image analysis. arXiv preprint arXiv: 2003.05037.

Guan, C. S., Lv, Z. B., Yan, S., Du, Y. N., Chen, H., Wei, L. G., Xie, R. M. & Chen, B. D. 2020a. Imaging features of coronavirus disease 2019 (COVID-19): Evaluation on thin-section CT. *Academic Radiology*, 27(5), 609–613.

Guan, W.-J., Ni, Z.-Y., Hu, Y., Liang, W.-H., Ou, C.-Q., He, J.-X., Liu, L., Shan, H., Lei, C.-L. & Hui, D. S. 2020b. Clinical characteristics of 2019 novel coronavirus infection in China. MedRxiv.

Habibi, N. & Harouni, M. 2018. Estimation of Re-hospitalization risk of diabetic patients based on radial base function (rbf) neural network method combined with colonial competition optimization algorithm. *Majlesi Journal of Electrical Engineering*, 12, 109–116.

Hannun, A. Y., Rajpurkar, P., Haghpanahi, M., Tison, G. H., Bourn, C., Turakhia, M. P. & Ng, A. Y. 2019. Cardiologist-level arrhythmia detection and classification in ambulatory electrocardiograms using a deep neural network. *Nature Medicine*, 25, 65.

Haque, I. R. I. & Neubert, J. 2020. Deep learning approaches to biomedical image segmentation. *Informatics in Medicine Unlocked*, 18, 100297.

Harouni, M. & Baghmaleki, H. Y. 2020. Color image segmentation metrics. arXiv preprint arXiv:09907.

Harouni, M., Mohamad, D., Rahim, M. S. M. & Halawani, S. M. 2012a. Finding critical points of handwritten Persian/Arabic character. *International Journal of Machine Learning Computing*, 2, 573.

Harouni, M., Mohamad, D., Rahim, M. S. M., Halawani, S. M. & Afzali, M. 2012b. Handwritten Arabic character recognition based on minimal geometric features. *International Journal of Machine Learning Computing*, 2, 578.

Harouni, M., Mohamad, D. & Rasouli, A. *Deductive method for recognition of on-line handwritten Persian/Arabic characters. 2010 The 2nd International Conference on Computer and Automation Engineering (ICCAE)*, 2010. IEEE, 791–795.

Harouni, M., Rahim, M., Al-Rodhaan, M., Saba, T., Rehman, A. & Al-Dhelaan, A. 2014. Online Persian/Arabic script classification without contextual information. *The Imaging Science Journal*, 62, 437–448.

Harouni, M., Rahim, M., Mohamad, D., Rehman, A. & Saba, T. 2012c. Online cursive Persian/Arabic character recognition by detecting critical points. *International Journal of Academic Research*, 4(2).

Hemdan, E. E.-D., Shouman, M. A. & Karar, M. E. 2020. Covidx-net: A framework of deep learning classifiers to diagnose covid-19 in x-ray images. arXiv preprint arXiv: *2003.11055.*

Ho, T. K. K., Gwak, J., Prakash, O., Song, J.-I., & Park, C. M. (2019). *Utilizing Pretrained Deep Learning Models for Automated Pulmonary Tuberculosis Detection Using Chest Radiography. Asian Conference on Intelligent Information and Database Systems,* Springer, 395–403.

Horng, M.-H. 2010. Multilevel minimum cross entropy threshold selection based on the honey bee mating optimization. *Expert Systems with Applications,* 37, 4580–4592.

Husham, A., Hazim Alkawaz, M., Saba, T., Rehman, A. & Saleh Alghamdi, J. 2016. Automated nuclei segmentation of malignant using level sets. *Microscopy Research and Technique,* 79, 993–997.

Hussain, N., Khan, M. A., Sharif, M., Khan, S. A., Albesher, A. A., Saba, T. & Armaghan, A. 2020. A deep neural network and classical features based scheme for objects recognition: An application for machine inspection. *Multimed Tools and Application.* doi:10.1007/s11042-020-08852-3.

Iftikhar, S., Fatima, K., Rehman, A., Almazyad, A. S. & Saba, T. 2017. An evolution based hybrid approach for heart diseases classification and associated risk factors identification, *Biomedical Research,* 28(8), 3451–3455.

Iqbal, S., Ghani Khan, M. U., Saba, T., Mehmood, Z., Javaid, N., Rehman, A., & Abbasi, R. 2019. Deep learning model integrating features and novel classifiers fusion for brain tumor segmentation. *Microscopy Research and Technique,* 82, 1302–1315.

Iqbal, S., Ghani, M. U., Saba, T. & Rehman, A. 2018a. Brain tumor segmentation in multispectral MRI using convolutional neural networks (CNN). *Microscopy Research and Technique,* 81, 419–427.

Iqbal, S., Khan, M. U. G., Saba, T. & Rehman, A. 2018b. Computer-assisted brain tumor type discrimination using magnetic resonance imaging features. *Biomedical Engineering Letters,* 8, 5–28.

Işin, A., Direkoğlu, C. & Şah, M. 2016. Review of MRI-based brain tumor image segmentation using deep learning methods. *Procedia Computer Science,* 102, 317–324.

Islam, A., Reza, S. M. & Iftekharuddin, K. M. 2013. Multifractal texture estimation for detection and segmentation of brain tumors. *IEEE Transactions on Biomedical Engineering,* 60, 3204–3215.

Jaeger, S., Candemir, S., Antani, S., Wáng, Y.-X. J., Lu, P.-X. & Thoma, G. 2014. Two public chest X-ray datasets for computer-aided screening of pulmonary diseases. *Quantitative Imaging in Medicine and Surgery,* 4, 475.

Jaiswal, A., Gianchandani, N., Singh, D., Kumar, V. & Kaur, M. 2020. Classification of the COVID-19 infected patients using DenseNet201 based deep transfer learning. *Journal of Biomolecular Structure and Dynamics,* 1–8.

Jamal, A., Hazim Alkawaz, M., Rehman, A. & Saba, T. 2017. Retinal imaging analysis based on vessel detection. *Microscopy Research and Technique,* 80, 799–811.

Jamil, M. & Hussain, I. 2020. Automatic detection of COVID-19 infection from chest X-ray using deep learning. *medRxiv.*

Jamshidi, M., Lalbakhsh, A., Talla, J., Peroutka, Z., Hadjilooei, F., Lalbakhsh, P., Jamshidi, M., La Spada, L., Mirmozafari, M. & Dehghani, M. 2020. Artificial intelligence and COVID-19: Deep learning approaches for diagnosis and treatment. *IEEE Access,* 8, 109581–109595.

Javed, R., Rahim, M. S. M., Saba, T. & Rashid, M. 2019a. Region-based active contour JSEG fusion technique for skin lesion segmentation from dermoscopic images. *Biomedical Research,* 30, 1–10.

Javed, R., Rahim, M. S. M., Saba, T. & Rehman, A. 2020. A comparative study of features selection for skin lesion detection from dermoscopic images. *Network Modeling Analysis in Health Informatics and Bioinformatics*, 9, 4.

Javed, R., Saba, T., Shafry, M. & Rahim, M. 2019b. *An Intelligent Saliency Segmentation Technique and Classification of Low Contrast Skin Lesion Dermoscopic Images Based on Histogram Decision. 2019 12th International Conference on Developments in eSystems Engineering (DeSE),*. IEEE, 164–169.

Javed, R., Shafry, M., Saba, T., Fati, S.M., Rehman, A. et al. 2021. Statistical histogram decision based contrast categorization of skin lesion datasets dermoscopic images. *Computers, Materials & Continua*, 67, 2, 2337–2352.

Karim, M., Döhmen, T., Rebholz-Schuhmann, D., Decker, S., Cochez, M. & Beyan, O. 2020. Deepcovidexplainer: Explainable covid-19 predictions based on chest x-ray images. *arXiv preprint arXiv: 2004.04582*.

Kenari, A. R., Hosseinkhani, J., Shamsi, M. & Harouni, M. *A robust and high speed E-voting algorithm using elgammel cryptosystem. 2010 The 2nd International Conference on Computer and Automation Engineering (ICCAE)*, 2010. IEEE, 812–816.

Kermany, D., Zhang, K. & Goldbaum, M. 2018. Labeled optical coherence tomography (OCT) and Chest X-Ray images for classification. *Mendeley Data*, 2. doi:10.17632/rscbjbr9sj, 3.

Khan, A. I., Shah, J. L. & Bhat, M. M. 2020a. Coronet: A deep neural network for detection and diagnosis of COVID-19 from chest x-ray images. *Computer Methods and Programs in Biomedicine*, 105581, 196.

Khan, A. R., Khan, S., Harouni, M., Abbasi, R., Iqbal, S., & Mehmood, Z. (2021a). Brain tumor segmentation using K-means clustering and deep learning with synthetic data augmentation for classification. *Microscopy Research and Technique*. doi:10.1002/jemt.23694

Khan, M. A., Akram, T., Sharif, M., Javed, K., Raza, M. & Saba, T. 2020b. An automated system for cucumber leaf diseased spot detection and classification using improved saliency method and deep features selection. *Multimedia Tools and Applications*, 1–30.

Khan, M. A., Ashraf, I., Alhaisoni, M., Damaševičius, R., Scherer, R., Rehman, A. & Bukhari, S. A. C. 2020c. Multimodal brain tumor classification using deep learning and robust feature selection: A machine learning application for radiologists. *Diagnostics*, 10, 565.

Khan, M. A., Javed, M. Y., Sharif, M., Saba, T. & Rehman, A. *Multi-model deep neural network based features extraction and optimal selection approach for skin lesion classification. 2019 International Conference On Computer And Information Sciences (Iccis)*, 2019a. IEEE, 1–7.

Khan, M.A. Kadry, S., Zhang, Y.D., Akram, T., Sharif, M., Rehman, A. & Saba, T. (2021b) Prediction of COVID-19 - pneumonia based on selected deep features and one class kernel extreme learning. *Machine, Computers & Electrical Engineering*, 90, 106960.

Khan, M. A., Kadry, S., Zhang, Y.-D., Akram, T., Sharif, M., Rehman, A. & Saba, T. 2020d. Prediction of COVID-19-pneumonia based on selected deep features and one class kernel extreme learning machine. *Computers & Electrical Engineering*, 90, 106960.

Khan, M. A., Lali, I. U., Rehman, A., Ishaq, M., Sharif, M., Saba, T., Zahoor, S. & Akram, T. 2019b. Brain tumor detection and classification: A framework of marker-based watershed algorithm and multilevel priority features selection. *Microscopy Research and Technique*, 82, 909–922.

Khan, M. A., Sharif, M., Akram, T., Raza, M., Saba, T. & Rehman, A. 2020e. Hand-crafted and deep convolutional neural network features fusion and selection strategy: An application to intelligent human action recognition. *Applied Soft Computing*, 87, 105986.

Khan, M. A., Sharif, M. I., Raza, M., Anjum, A., Saba, T. & Shad, S. A. 2019c. Skin lesion segmentation and classification: A unified framework of deep neural network features fusion and selection. *Expert Systems*, e12497.

Khan, M. W., Sharif, M., Yasmin, M. & Saba, T. 2017. CDR based glaucoma detection using fundus images: a review. *International Journal of Applied Pattern Recognition*, 4, 261–306.

Khan, M. Z., Jabeen, S., Khan, M. U. G., Saba, T., Rehmat, A., Rehman, A. & Tariq, U. 2020f. A realistic image generation of face from text description using the fully trained generative adversarial networks. *IEEE Access*. doi:10.1109/ACCESS.2020.3015656.

Khan, S. A., Nazir, M., Khan, M. A., Saba, T., Javed, K., Rehman, A., Akram, T. & Awais, M. 2019d. Lungs nodule detection framework from computed tomography images using support vector machine. *Microscopy Research and Technique*, 82, 1256–1266.

Landry, M. D., Geddes, L., Moseman, A. P., Lefler, J. P., Raman, S. R. & Van Wijchen, J. 2020. Early reflection on the global impact of COVID19, and implications for physiotherapy. *Physiotherapy*, 107, A1–A3.

Latif, S., Usman, M., Manzoor, S., Iqbal, W., Qadir, J., Tyson, G., Castro, I., Razi, A., Boulos, M. N. K. & Weller, A. 2020. Leveraging data science to combat COVID-19: A comprehensive review. https://qmro.qmul.ac.uk/xmlui/handle/123456789/67218

Liang, G. & Zheng, L. 2020. A transfer learning method with deep residual network for pediatric pneumonia diagnosis. *Computer Methods and Programs in Biomedicine*, 187, 104964.

Loey, M., Smarandache, F. & Mkhalifa, N. E. 2020. Within the lack of Chest COVID-19 X-ray dataset: A novel detection model based on GAN and deep transfer learning. *Symmetry*, 12(4), 651.

Lung, J. W. J., Salam, M. S. H., Rehman, A., Rahim, M. S. M. & Saba, T. 2014. Fuzzy phoneme classification using multi-speaker vocal tract length normalization. *IETE Technical Review*, 31, 128–136.

Mashood Nasir, I., Attique Khan, M., Alhaisoni, M., Saba, T., Rehman, A., & Iqbal, T. 2020. A hybrid deep learning architecture for the classification of superhero fashion products: An application for medical-tech classification. *Computer Modeling in Engineering & Sciences*, 124, 1017–1033.

Mittal, A., Kumar, D., Mittal, M., Saba, T.0, Abunadi, I., Rehman, A. & Roy, S. 2020. Detecting pneumonia using convolutions and dynamic capsule routing for chest X-ray images. *Sensors*, 20, 1068.

Mohammadi Dashti, M., & Harouni, M. 2018. Smile and laugh expressions detection based on local minimum key points. *Signal Data Processing*, 15, 69–88.

Mostafa, A., Hassanien, A. E., Houseni, M. & Hefny, H. 2017. Liver segmentation in MRI images based on whale optimization algorithm. *Multimedia Tools and Applications*, 76, 24931–24954.

Mughal, B., Muhammad, N., Sharif, M., Rehman, A. & Saba, T. 2018a. Removal of pectoral muscle based on topographic map and shape-shifting silhouette. *BMC Cancer*, 18, 778.

Mughal, B., Muhammad, N., Sharif, M., Saba, T. & Rehman, A. 2017. Extraction of breast border and removal of pectoral muscle in wavelet domain. *Biomedical Research*

Mughal, B., Sharif, M., Muhammad, N. & Saba, T. 2018b. A novel classification scheme to decline the mortality rate among women due to breast tumor. *Microscopy Research and Technique*, 81, 171–180.

Narin, A., Kaya, C. & Pamuk, Z. 2020. Automatic detection of coronavirus disease (covid-19) using x-ray images and deep convolutional neural networks. *arXiv preprint arXiv: 2003.10849.*

Nazir, M., Khan, M. A., Saba, T. & Rehman, A. *Brain tumor detection from MRI images using multi-level wavelets. 2019 International Conference on Computer and Information Sciences (ICCIS)*, 2019. IEEE, 1–5.

Ozturk, T., Talo, M., Yildirim, E. A., Baloglu, U. B., Yildirim, O. & Acharya, U. R. 2020. Automated detection of COVID-19 cases using deep neural networks with X-ray images. *Computers in Biology and Medicine*, 121, 103792.

Perveen, S., Shahbaz, M., Saba, T., Keshavjee, K., Rehman, A. & Guergachi, A. 2020. Handling irregularly sampled longitudinal data and prognostic modeling of diabetes using machine learning technique. *IEEE Access*, 8, 21875–21885.

Punn, N. S. & Agarwal, S. 2020. Automated diagnosis of COVID-19 with limited posteroanterior chest X-ray images using fine-tuned deep neural networks. *arXiv preprint arXiv: 2004.11676.*

Rad, A. E., Mohd Rahim, M. S., Rehman, A., Altameem, A., & Saba, T. 2013. Evaluation of current dental radiographs segmentation approaches in computer-aided applications. *IETE Technical Review*, 30, 210–222.

Rad, A. E., Rahim, M. S. M., Rehman, A. & Saba, T. 2016. Digital dental X-ray database for caries screening. *3D Research*, 7, 18.

Rahim, M. S. M., Norouzi, A., Rehman, A. & Saba, T. 2017a. 3D bones segmentation based on CT images visualization. *Biomedical Research*, 28(8), 3641–3644.

Rahim, M. S. M., Rehman, A., Kurniawan, F. & Saba, T. 2017b. Ear biometrics for human classification based on region features mining. *Network Modeling Analysis in Health Informatics and Bioinformatics*, 9(1). doi:10.1007/s13721-019-0209-1.

Ramzan, F., Khan, M. U. G., Iqbal, S., Saba, T. & Rehman, A. 2020a. Volumetric segmentation of brain regions from mri scans using 3D convolutional neural networks. *IEEE Access*, 8, 103697–103709.

Ramzan, F., Khan, M. U. G., Rehmat, A., Iqbal, S., Saba, T., Rehman, A. & Mehmood, Z. 2020b. A deep learning approach for automated diagnosis and multi-class classification of Alzheimer's disease stages using resting-state fMRI and residual neural networks. *Journal of Medical Systems*, 44, 37.

Rehman, A. (2020). *Ulcer Recognition based on 6-Layers Deep Convolutional Neural Network*. In *Proceedings of the 2020 9th International Conference on Software and Information Engineering (ICSIE)* (pp. 97–101). Cairo Egypt.

Rehman, A. (2021) Light microscopic iris classification using ensemble multi-class support vector machine. *Microscopic Research & Technique*, doi:10.1002/jemt.23659

Rehman, A., Abbas, N., Saba, T., Mahmood, T. & Kolivand, H. 2018. Rouleaux red blood cells splitting in microscopic thin blood smear images via local maxima, circles drawing, and mapping with original RBCs. *Microscopy Research and Technique*, 81, 737–744.

Rehman, A., Harouni, M. & Saba, T. 2020a. Cursive multilingual characters recognition based on hard geometric features. *International Journal of Computational Vision Robotics*, 10, 213–222.

Rehman, A., Khan, M. A., Mehmood, Z., Saba, T., Sardaraz, M. & Rashid, M. 2020b. Microscopic melanoma detection and classification: A framework of pixel-based fusion and multilevel features reduction. *Microscopy Research and Technique*, 83, 410–423.

Rehman, A., Khan, M. A., Saba, T., Mehmood, Z., Tariq, U. & Ayesha, N. 2021a. Microscopic brain tumor detection and classification using 3D CNN and feature selection architecture. *Microscopy Research and Technique*, 84, 133–149.

Rehman, A., Saba, T., Ayesha, N., Tariq, U. (2021c) Deep learning-based COVID-19 detection using CT and X-ray images: Current analytics and comparisons, *IEEE IT Professional*, doi:10.1109/MITP.2020.3036820

Rehman, A., Sadad, T., Saba, T., Hussain, A., Tariq, U. (2021b) Real-time diagnosis system of COVID-19 using X-ray images and deep learning, *IEEE IT Professional*, doi:10.1109/ MITP.2020.3042379

Saba, T. 2020. Recent advancement in cancer detection using machine learning: Systematic survey of decades, comparisons and challenges. *Journal of Infection and Public Health*, 13, 1274–1289.

Saba, T. 2021. Computer vision for microscopic skin cancer diagnosis using handcrafted and non-handcrafted features. *Microscopy Research and Technique*. doi:10.1002/jemt.23686.

Saba, T., Abunadi, I., Shahzad, M. N., & Khan, A. R. (2021a). Machine learning techniques to detect and forecast the daily total COVID-19 infected and deaths cases under different lockdown types. *Microscopy Research and Technique*, doi:10.1002/jemt.23702

Saba, T., Abunadi, I., Shahzad, M. N., & Khan, A. R. (2021b). Machine learning techniques to detect and forecast the daily total COVID-19 infected and deaths cases under different lockdown types. *Microscopy Research and Technique*, doi:10.1002/jemt.23702

Saba, T., Khan, M. A., Rehman, A. & Marie-Sainte, S. L. 2019a. Region extraction and classification of skin cancer: A heterogeneous framework of deep CNN features fusion and reduction. *Journal of Medical Systems*, 43, 289.

Saba, T., Khan, S. U., Islam, N., Abbas, N., Rehman, A., Javaid, N. & Anjum, A. 2019b. Cloud-based decision support system for the detection and classification of malignant cells in breast cancer using breast cytology images. *Microscopy Research and Technique*, 82, 775–785.

Saba, T., Rehman, A., Mehmood, Z., Kolivand, H. & Sharif, M. 2018. Image enhancement and segmentation techniques for detection of knee joint diseases: A survey. *Current Medical Imaging*, 14, 704–715.

Salman, F. M., Abu-Naser, S. S., Alajrami, E., Abu-Nasser, B. S. & Alashqar, B. A. 2020. Covid-19 detection using artificial intelligence. *IJAER*, 4, 18–25.

Sandler, M., Howard, A., Zhu, M., Zhmoginov, A. & Chen, L.-C. 2018. Mobilenetv2: Inverted residuals and linear bottlenecks. *Proceedings of the IEEE conference on computer vision and pattern recognition*, 4510–4520.

Sarker, L., Islam, M. M., Hannan, T. & Ahmed, Z. 2020. COVID-DenseNet: A deep learning architecture to detect COVID-19 from chest radiology images. doi:10.20944/preprints202005.0151.v1

Sethy, P. K. & Behera, S. K. 2020. Detection of coronavirus disease (covid-19) based on deep features. *Preprints*, 2020030300, 2020.

Shan, F., Gao, Y., Wang, J., Shi, W., Shi, N., Han, M., Xue, Z. & Shi, Y. 2020. Lung infection quantification of covid-19 in ct images with deep learning. *arXiv preprint arXiv: 2003.04655*.

Sharifi-Razavi, A., Karimi, N. & Rouhani, N. 2020. COVID-19 and intracerebral haemorrhage: Causative or coincidental? *New Microbes and New Infections*, 35, 100669.

Shi, F., Wang, J., Shi, J., Wu, Z., Wang, Q., Tang, Z., He, K., Shi, Y. & Shen, D. 2020. Review of artificial intelligence techniques in imaging data acquisition, segmentation and diagnosis for covid-19. *IEEE Reviews in Biomedical Engineering*, 14, 3–15.

Shiraishi, J., Katsuragawa, S., Ikezoe, J., Matsumoto, T., Kobayashi, T., Komatsu, K.-I., Matsui, M., Fujita, H., Kodera, Y. & Doi, K. 2000. Development of a digital image database for chest radiographs with and without a lung nodule: receiver operating characteristic analysis of radiologists' detection of pulmonary nodules. *American Journal of Roentgenology*, 174, 71–74.

Singh, D., Kumar, V. & Kaur, M. 2020. Classification of COVID-19 patients from chest CT images using multi-objective differential evolution–based convolutional neural networks. *European Journal of Clinical Microbiology & Infectious Diseases*, 39(7), 1379–1389.

Suzuki, K. 2017. Overview of deep learning in medical imaging. *Radiological Physics and Technology*, 10, 257–273.

Ucar, F. & Korkmaz, D. 2020. COVIDiagnosis-net: Deep bayes-squeezenet based diagnostic of the coronavirus disease 2019 (COVID-19) from X-ray images. *Medical Hypotheses*, 140, 109761.

Ullah, H., Saba, T., Islam, N., Abbas, N., Rehman, A., Mehmood, Z. & Anjum, A. 2019. An ensemble classification of exudates in color fundus images using an evolutionary algorithm based optimal features selection. *Microscopy Research and Technique*, 82, 361–372.

Van Ginneken, B., Stegmann, M. B. & Loog, M. 2006. Segmentation of anatomical structures in chest radiographs using supervised methods: A comparative study on a public database. *Medical Image Analysis*, 10, 19–40.

Wang, G. 2016. A perspective on deep imaging. *IEEE access*, 4, 8914–8924.

Wang, G., Liu, X., Li, C., Xu, Z., Ruan, J., Zhu, H., Meng, T., Li, K., Huang, N. & Zhang, S. 2020a. A noise-robust framework for automatic segmentation of covid-19 pneumonia lesions from ct images. *IEEE Transactions on Medical Imaging*, 39, 2653–2663.

Wang, J., Ding, H., Bidgoli, F. A., Zhou, B., Iribarren, C., Molloi, S. & Baldi, P. 2017a. Detecting cardiovascular disease from mammograms with deep learning. *IEEE Transactions on Medical Imaging*, 36, 1172–1181.

Wang, L., Lin, Q. L. & Wong, A. 2020d. A tailored deep convolutional neural network design for detection of covid-19 cases from chest radiography images. *Journal of Network and Computer Applications*, 20, 1–12.

Wang, L., Lin, Z. Q. & Wong, A. 2020b. Covid-net: A tailored deep convolutional neural network design for detection of covid-19 cases from chest x-ray images. *Scientific Reports*, 10, 1–12.

Wang, S., Zha, Y., Li, W., Wu, Q., Li, X., Niu, M., Wang, M., Qiu, X., Li, H., & Yu, H. 2020c. A fully automatic deep learning system for COVID-19 diagnostic and prognostic analysis. *European Respiratory Journal*, 56(2), 2000775.

Wang, X., Peng, Y., Lu, L., Lu, Z., Bagheri, M. & Summers, R. M. *Chestx-ray8: Hospital-scale chest x-ray database and benchmarks on weakly-supervised classification and localization of common thorax diseases. Proceedings of the IEEE Conference on Computer Vision And Pattern Recognition*, 2017b. 2097–2106.

Yildirim, Ö., Pławiak, P., Tan, R.-S. & Acharya, U. R. 2018. Arrhythmia detection using deep convolutional neural network with long duration ECG signals. *Computers in Biology and Medicine*, 102, 411–420.

Yu, X., Zeng, N., Liu, S., & Zhang, Y.-D. 2019. Utilization of DenseNet201 for diagnosis of breast abnormality. *Machine Vision and Applications*, 30, 1135–1144.

Zhang, J., Xie, Y., Li, Y., Shen, C., & Xia, Y. 2020. Covid-19 screening on chest x-ray images using deep learning based anomaly detection. *arXiv preprint arXiv: 2003.12338*.

Zhao, J., Zhang, Y., He, X., & Xie, P. 2020. COVID-CT-dataset: A CT scan dataset about COVID-19. *arXiv preprint arXiv: 2003.13865*.

7 Interactive Medical Chatbot for Assisting with COVID-related Queries

Aayush Gadia, Palash Nandi, and Dipankar Das

CONTENTS

7.1 INTRODUCTION

In general, conversational agents are classified into chatbots and dialogue systems [(Dingli & Scerri, 2013), (Larsson & Traum, 2000)]. However, the main difference between them is that chatbots aim at keeping a conversation going (e.g., for entertainment purposes), without an explicit attempt at understanding the meaning of the utterances. On the other hand, dialogue systems try to establish an effective communication channel between the system and the user, for sharing information, responding to commands, and building common knowledge (Taylor, 2020).

Hirschberg and Manning (Kazi et al., 2012) identify four main factors that have enabled the significant progress in computational linguistics and dialogue systems witnessed in the last years, namely:

   i. the availability of increased computational power;
  ii. the availability of very large amounts of linguistic data in digital form;
 iii. the development of effective machine learning techniques; and
 iv. the increased understanding of the structure and role of natural language, in particular in social contexts.

One of the first chatbots was ELIZA (Weizenbaum, 1966), developed in 1966, which simulated a psychotherapist. After the success of ELIZA, hundreds of chatbots have been developed. For instance, PARRY (Colby et al., 1971) simulates a paranoid person. Chatterbot (Ministry of Communications, Department of Posts, 2017) was one of the first chatbots to pass a restricted version of the Turing test.

Several languages, frameworks, and toolkits for the development of dialogue systems have recently been proposed. The project TRINDI (Mauldin, 1994; Weizenbaum, 1966), for instance, developed the framework TrindiKit for simplifying the implementation of dialogue-processing theories, making the development of dialogue systems easier. Pamini (Purohit et al., 2015) is another framework for defining human-robot interaction strategies based on generic interaction patterns. Recent work on spoken dialogue systems was done by Berg (Berg, 2015) who designed and developed NADIA, a framework for the creation of natural dialogue systems.

Creation and development occur when there is an emergency or urgency. COVID-19 is a global pandemic and it is a completely new urgency in the disease category with no anti-viral treatment. The best way to prevent infection is to avoid being exposed to coronavirus. To prevent people from getting exposed, people have to be made aware of the virus, its symptoms, preventive steps, updates, myths, government helpline numbers, nearby hospitals, medicine shops, etc.; once people are aware they can be alert. Therefore, correct and prompt information delivery is a necessity during the pandemic. In contrast to the above approaches, our present research is primarily intended for the general public who has no medical training. Therefore, the following are some of the crucial challenges which people generally face in finding the correct information in a pandemic situation like COVID-19:

- *Reliability of Online Information (Hirschberg & Manning, 2015):* Being in the WWW with virtual affinity, we are unaware of the fact that the online data which is propagated by different sources have different qualities, with some of them being reliable while some are not. Therefore, it is very difficult to identify the genuine and reliable sources of information, because it might be that the best results returned by the search may not always contain true information. Especially, in the domain of healthcare, the reliability of information becomes essential and critical, and should be considered seriously (Hirschberg & Manning, 2015).
- *Combine information from various sources (Hirschberg & Manning, 2015):* Generally, multiple pieces of information are taken to answer a question which concerns complex health issue. This information needs to be obtained, cleaned, organized, and analyzed, and has to be verified to frame the right answer for user queries. For these very reasons, the process is made more difficult by the

usage of different vocabularies and the possible inconsistencies among these information sources (Hirschberg & Manning, 2015).

- *False Information Circulation:* In this context, the circulation and/or dissemination of false information on the COVID-19 outbreak is growing fast (Chakravorti, 2020; Van der Kooij et al., 2006). False information on different cures for COVID-19, like eating oregano or rinsing the mouth with salty water, are examples of the false information being circulated. Also, myths like "bat soups" being the cause of COVID-19 infection, or myths like the virus was made by the US are spread. This news favors the rise of racist attitudes (Aguilera, 2020) and puts the health of a vast population at risk, thereby impacting the government's capability to effectively tackle the COVID-19 situation (Russo et al., 2019).

To overcome the above challenges, our conversational agent is specifically designed for helping users during the COVID-19 pandemic. It consists of three main modules: *Know-Corona, Self-Assess*, and *Help-Desk*. In this chapter, we especially focus on developing techniques related to two main modules—*Know-Corona* and *Help-Desk*. The *Self-Assess* module integrates medical entity recognition algorithms that can detect whether the user is at risk of COVID-19 infection and recommends similar diseases and symptoms by employing medical concept identification techniques (Mondal et al., 2017; Nuruzzaman & Hussain, 2018; PTI, 2011). The *Know-Corona* module answers any of the user's queries on COVID-19 by doing a local search on the scraped data of trusted websites like Centers for Disease Control and Prevention (https://www.cdc.gov/coronavirus/2019-nCoV/index.html), World Health Organization (WHO) ((https://www.who.int/emergencies/diseases/novel-coronavirus-2019) and National Health Portal of India (https://www.nhp.gov.in/disease/communicable-disease/novel-coronavirus-2019-ncov). We have developed a reverse question clustering algorithm for developing the Question Answering (QA) framework destined for *Know-Corona*. The module also extracts accurate answers by employing local search results. On the other hand, using the *Help-Desk* module, the user can search for hospitals and medicine shops by entering any postal PIN code of a state or country (e.g., the state of West Bengal in India in our case). Therefore, any user living in West Bengal can locate local as well as state-level COVID-19-designated hospitals and local medicine shops respectively. The feature provides information like Name, Address, Contact Numbers, Emergency Numbers, Ambulance Numbers as per the information available. The support of our conversational agent can also be extended to other states of India and regional languages could also be added as a conversation medium.

To build the above QA module along with an interactive dialogue framework for implementing COVID-19 conversational assistant, we have used the Rasa framework (https://rasa.com/docs/rasa/). With the help of the Rasa framework (https://rasa.com/docs/rasa/) documentation, we designed the chatbot system for our required features. Mainly, the Rasa framework is used for intent identification and generation of proper response against the user query. The default user interface of Rasa has been used for the proposed conversational agent, thereby saving our time in building those components and helping us focus on building the core features of our research.

The other sections of the chapter are as follows. Section 7.2 highlights the promising applications and tools developed so far during the short period of the pandemic of say 6 months. Section 7.3 gives an overview of the planning and task ontology required to handle the COVID-19 pandemic. The detailed methodologies or algorithms for developing each of the modules are described in Section 7.4. Section 7.5 mentions the experiments for different individual research components as well as different evaluation metrics. Finally, Section 7.6 concludes with some future roadmaps for the betterment of the agent-based systems.

7.2   RELATED WORK

Previously, several other researchers have also developed conversational agents for the healthcare domain.

- Aarogya Setu (https://play.google.com/store/apps/details?id=nic.goi. aarogyasetu&hl=en_IN) developed by the National Informatics Centre, Government of India, is the Indian COVID-19 tracking mobile application (see Figure 7.1). Currently, Aarogya Setu has four features: Your Status (this feature finds out the risk of a user getting infected by COVID-19); Self-Assess (the feature finds out whether the user is infected with any of the COVID-19 symptoms and what are the risks associated); COVID-19 Update (this tells the user about the updates on local and national COVID-19 cases); and E-pass (this is available to a user if they have applied for it). In our research study, we have also incorporated the "Self-Assess" feature like the Aarogya Setu (https://play. google.com/store/apps/details?id=nic.goi.aarogyasetu&hl=en_IN) but also features like "FirstAid" (which can locate medicine shops and COVID-19-designated local hospitals near the user), and "Know-Corona" (this answers any COVID-19-related query of the user).
- MedChatBot (Klüwer, 2011): in this study a conversational agent was built for medical students. It uses Chatterbean which is open source and based on AIML. Also, the Unified Medical Language System (UMLS) (Bodenreider, 2004) is used as a knowledge source. UMLS(Bodenreider, 2004) data is used for giving responses to the user queries. Here, only UMLS is used as a knowledge source for answering questions and the chatbot has been designed to answer only a certain category of questions. In our research, we have made the conversational agent only for the COVID-19 domain and it can answer any of the user's queries on COVID-19; unlike MedChatbot, it is not restricted to any particular category of questions. The answers to the queries are extracted from trusted websites like WHO (https://www.who.int/ emergencies/diseases/novel-coronavirus-2019), CDC (https://www.cdc. gov/coronavirus/2019-nCoV/index.html), NHP India (https://www.nhp.gov.in/disease/ communicable-disease/novel-coronavirus-2019-ncov) and the dataset remains updated.
- Open-source medical data like UMLS and SNOMED CT are available but we have not used them in our research. UMLS (Bodenreider, 2004) contains more

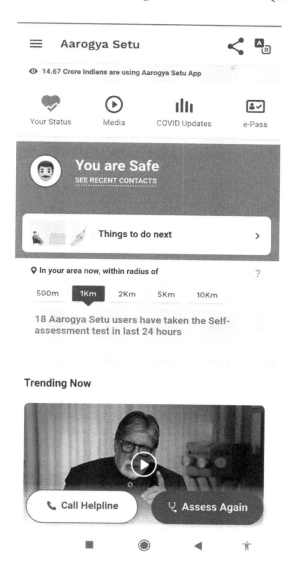

FIGURE 7.1   Aarogya Setu application.

than a million concepts on medical science which cover various medical aspects and it also has 135 semantic types. Every medical concept is mapped to one or more than one semantic type but apart from the medical concepts and semantic types, COVID-19 is a new disease, and consequently people have many general queries about it, and the availability of correct information to those queries is very much required. Therefore, UMLS cannot be used in this case, because it only covers the medical concepts, not the general queries and local data.

- SNOMED CT (https://www.snomed.org/) is a dynamic, scientifically validated clinical health care terminology and infrastructure. By applying SNOMED CT coding, data can be captured, shared, and aggregated in a consistent way across specialties and domains of care. SNOMED CT is best used for electronic medical records, ICU monitoring, and clinical decision support (Van der Kooij et al., 2006). In our research, we have made a conversational agent and here our purpose is to answer user queries correctly rather than capturing terminology. So, we have not used it in our research; we would have used it if we had made a clinical decision support system or software to be used in patient treatment.
- Intent Classification of Short-Text on Social Media (Soares & Parreiras, 2020): here, the research focuses on addressing the problem of intent classification into multiple classes, with the use-case being of the data of the social website which is generated during any crisis. The method used in creating a hybrid feature representation combines the top-down processing which uses the knowledge-guided patterns with the bottom-up processing which uses the bag-of-tokens model. In our research the entire problem of intent classification was handled by the Rasa framework; therefore we just focused on the functionality rather than on the intent classification problem.
- A literature review on question-answering techniques, paradigms, and systems (Traum & Larsson, 2003). Here, a comprehensive study of the QA by its literature is done. This paper focuses on the paradigms, metrics, domains, concepts, and technologies that are used for QA research. Therefore, it is observed how the different approaches fit the different domains and what results are obtained by the implementation of these different approaches. However, in our research, the algorithm that we have used for the Question Answering (QA) is completely new and unique.
- A survey on chatbot implementation in the customer service industry through deep neural networks (Peltason & Wrede, 2010). Here, a survey is done on the existing chatbots and the techniques that are used in them. This paper discusses what the differences and similarities are among the existing chatbots, and their limitations. A total of 11 popular chatbots were considered for this paper and comparisons were made among them on their functionalities and specifications. By the research, it was seen that 75% of customers had experienced bad customer service. It was also seen that the generation of long, informative, and meaningful responses is a challenge. Therefore, people have been using handwritten rules and templates for developing chatbots, but recently these methods have been replaced by neural networks by the development of deep learning algorithms. In our research, we have used neither handwritten rules and templates nor neural networks. Our custom-designed algorithms along with the custom-made dataset just focus on solving the issue of providing information on COVID-19 with maximum customer satisfaction.

7.3   SYSTEM FRAMEWORK

Our conversational agent has been built on the Rasa framework (Rasa is an open-source machine learning framework for building AI assistants and chatbots). Based

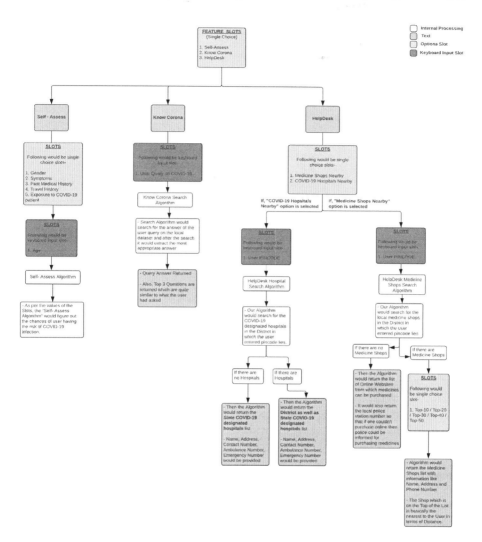

FIGURE 7.2 Working Design of Conversational Agent.

on existing conversational agents for COVID-19 like Aarogya Setu (https://play. google.com/store/apps/details?id=nic.goi.aarogyasetu&hl=en_IN) and based on the overall challenges faced by people during this pandemic, three different modules were incorporated into the conversational agent. The modules are: *Self-Assess, Know-Corona* and *Help-Desk*. In this research paper, two features are discussed in detail: *Know-Corona* and *Help-Desk*. Given in Figure 7.2 is the working design of the conversational agent. The system of the conversational agent has been designed such that, for every query entered, the system can identify which feature should be used to solve that query. Below are some of the examples-

- Example 1—If a user enters "Find hospitals near me," then the system identifies that the *Help-Desk* feature should be used to answer the query.
- Example 2—If a user enters "Find medicine shops near me," then the system identifies that the *Help-Desk* feature should be used to answer the query.
- Example 3—If the user enters "What are the symptoms of COVID-19?" then the system identifies that the *Know-Corona* feature should be used to answer the query.
- Example 4—If a user enters "Why should I wear a mask?" then the system identifies that the *Know-Corona* feature should be used to answer the query.
- Example 5—If a user enters "Help me," then the system shows options of the features incorporated like *Self-Assess*, *Know-Corona*, and *Help-Desk*, and from those users can choose one.

First, the framework shows options of the modules listed *Self-Assess, Know-Corona,* and *Help-Desk* if a user enters too ambiguous a query like "Help me" or "I need help." Then, users must choose one to proceed with. The proposed conversational agent is not restricted to answering queries related to the modules; it can also answer general chitchat like: "Hi," "Hello," "How are you?" "Bye." In the case of irrelevant queries like "Get me a car," "Buy me a pizza," etc., it redirects the user politely to ask a valid question. It is not necessary to enter the same text as enlisted in the database for getting the answers to the queries. A user could enter any query they want to, provided the semantics of the context are the same. For example, a user can ask "How would I know if I have been infected with Cov virus?" instead of "What are the common symptoms of COVID-19?" Therefore, any user can interact with the conversational agent in two ways. First, when the user enters the query, the conversational agent identifies which module the query is intended for and starts acting accordingly. Second, the user chooses the module from the options given when the conversational agent starts up.

The working of the features is the same, irrespective of the method through which the query is requested. The Rasa framework has two main components: "Rasa NLU" which is the Natural Language Understanding component, and "Rasa Core" which is the Dialogue Management component. Rasa NLU is a natural language processing tool for intent classification, response retrieval, and entity extraction (https://rasa.com/docs/rasa/). Rasa Core is the dialogue engine that uses a machine learning model trained on example conversations to decide what to do next. Therefore, configuration files of Rasa NLU are "nlu.md" and specifying a pipeline in "config.yml." Similarly, configuration files of Rasa Core are "stories.md," "actions.py," and "domain.yml," and specifying a policy in "config.yml." Figure 7.3 shows how all the different configuration files co-work together. From the figure, it is absolutely clear that the pipeline and policies in "config.yml" are responsible for building the Rasa NLU and Rasa Core modules respectively. By these configurations along with the data in "nlu.md," the NLU model is trained. Similarly, by the configurations along with the data in "stories.md," the Core model is trained. Thus, the trained NLU model is responsible for identifying the intent of the user input utterance and the trained Core model is responsible for figuring out the response to the particular user input utterance intent. Now, the different intents, entities created in "nlu.md," are also

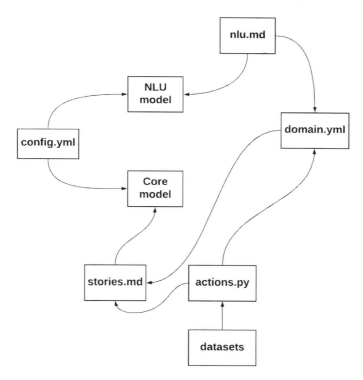

FIGURE 7.3   Co-working of different configuration files in Rasa.

stored in "domain.yml" which is the universe of the system. Similarly, all the custom actions created in "actions.py" are also stored in "domain.yml." All the utterance actions are stored in "domain.yml." Now the stories which are dialogue sessions, in the "stories.md," fetch information from "domain.yml" as it is the universe and contains the response of all the utterances, and if a user enters a query and that has to be passed into the custom actions in "actions.py," then those data are also passed in the "actions.py" file, using which the data is fetched from the datasets, as shown in the figure above. Therefore, the figure gives a brief understanding of how all the different files of the system co-work together.

7.3.1 RASA FRAMEWORK FOR INTENT CLASSIFICATION

As explained above, Rasa NLU is responsible for the intent classification and how all the different configuration files co-work together. Therefore, for enabling Rasa to classify the intents, we set its pipeline configuration in "config.yml" by specifying a pipeline and similarly we specify the different intents for which classification must be made in the "nlu.md" file. To train our model for the intent classification, we prepared different intents in the "nlu.md" file. Figure 7.4 shows the training data that was prepared for training our model so that it can classify the intents correctly. For our research, we created three main intents; as seen from the figure above, they are:

```
## intent:generl_question
- How Corona breakout was started ?
- What are the signs of covid infections ?
- How did this Corona pandemic started? , How this Corona disaster initiated ?
- Can you tell me about any Home diagnosis for Corona? ?
- Can you tell me the treatment of Corona patient in Home ?
- What are the most simple way for prevention of Covid ?
- How does a novel corona virus spreads ?
- What is the story behind corona virus outbreak ?
- Why might someone blame or avoid individuals and groups (create stigma) because of COVID-19 ?
- How to stop false idea about Corona ?
- Why number of affected cases for past days are changing ?
- are patients with high Blood presure prone to Covid ?
- What is the origin of the virus? ?
- How the Covid virus spread from me ?

114
115    ## intent:medicine-shops_search
116    - please look for medicine shops near me
117    - please look for medicine shops near my area
118    - i want to buy medicines
119    - i am looking for medicine shops
120    - where can i buy medicines?
121    - medicine shops near me
122    - suggest some medicine shops to me
123
105    ## intent:hospitals_search
106    - find hospitals near me
107    - share the details of hospitals in my area
108    - hospitals near me
109    - give me the details of hospitals near my area
110    - tell me the hospitals where I can get COVID-19 treatment
111    - i am looking for some hospitals near my area
112    - suggest some hospitals urgently
```

FIGURE 7.4   Training dataset of different intents.

"general query" "hospital search" and *"medicine shops search."* The *"general query"* intent was created for the *Know-Corona* module which is the Question Answering module. The *"hospital search"* and the *"medicine shops search"* intents were created for the *Help-Desk* module, which tells the user about the local COVID-19-designated hospitals and shares information about local medicine shops. Apart from the three intents below, several other intents were also created to enable the chitchat conversations between the conversational agent and the user. Figure 7.5 below shows the different intents for the same.

7.3.2 *KNOW-CORONA* QA MODULE

The *Know-Corona* section module is for letting people clear their doubts relating to the facts around coronavirus. People can easily ask corona-related questions on the chat interface directly or can reach this module via the *"general query"* section. This module works by finding synonyms of the keywords of the user-given query. The intuition behind the use of synonyms is that between two similar questions of the same context there should be a common set of synonyms. For example, "How can I protect myself from covid infection?" and "How to take a shield against coronavirus?" will have more common synonym words like "shield," "save," "defend," "protect," "virus," "infection," "disease," "germs," and "illness" than there are between

```
1    ## intent:bot_challenge          33   | yo
2     - are you a bot?                34   - hola
3     - are you a human?              35   - hi?
4     - am I talking to a bot?        36   - hey bot!
5     - am I talking to a human?      37   - hello friend
6     - Who are you?                  38   - Hi !! Can you help me ?
7     - What do you do?               39   - I need your help
8     - Who made you?                 40   - Could you help me?
9                                     41   - Hi
10   ## intent:greet                  42   - help me urgently
11    - good evening                  43   - help me
12    - Hi bot                        44   - help me bot
13    - Hey bot                       45
14    - Hello                         46   ## intent:goodbye
15    - Good morning                  47   - bye
16    - hi again                      48   - goodbye
17    - hi folks                      49   - see you around
18    - hi Mister                     50   - see you later
19    - hi pal!                       51   - ttyl
20    - hi there                      52
21    - greetings                     53   ## intent:insult
22    - hello everybody               54   - fuck you
23    - hello is anybody there        55   - you are dumb
24    - hello robot                   56   - stupid bot
25    - hallo                         57   - what the hell do you know ?
26    - heeey                         58   - just go away
27    - hi hi                         59   - get lost
28    - hey                           60   - you are stupid
29    - hey hey                       61   - you are foolish
30    - hello there                   62
31    - hi                            63   ## intent:out_of_scope
32    - hello                         64   - Will you marry me?
                                      65   - Can you get me a pizza?
                                      66   - Do my homework ?
                                      67   - Can you do my homework?
```

FIGURE 7.5   Training dataset of different chitchat intents.

the questions "How can I protect myself from covid infection?" and "Does high temperature kill covid infection?" The difference is much wider and more prominent if candidate statements/questions themselves are clusters of similar statements/questions. So the *Know-Corona* (KC) module uses a reverse clustering policy to find a candidate of similar context. We will use an example to demonstrate this.

7.3.2.1   Architecture of *Know-Corona* QA Module

Let's assume a user-given query (UQ) is *"Will being slothful have a good impact on my health?"* and we have the first question cluster (CQ1) as { *"Is lethargy good for my body?"*, *"Are lazy people damaging their body?"*, *"How does inactivity affect us?"*} and second question cluster (CQ2) as { *"Has AI already been applied in the health domain?"*, *"How will pseudo intelligence affect us in future?"*, *"Has AI been applied in the health domain?"*}. It is evident that UQ is very similar to CQ1 in terms of synonyms of their words. So the answer of CQ1 can be returned to the user against UQ. Figure 7.6 shows the proposed architecture of the *Know-Corona* module.

7.3.3   Help-Desk Module

The *Help-Desk* feature was made to provide the details of local medicine shops and local COVID-19-designated government and state-level hospitals to the user. So to understand how the system was built for the feature, let's break the feature into two sub-features and discuss each individually:

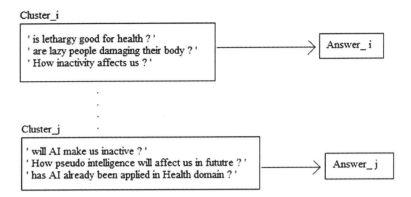

FIGURE 7.6   Intuition behind *Know-Corona* module.

- <u>COVID-19 Hospitals Nearby</u>:

    This sub-feature is just responsible for finding all the local COVID-19-designated government hospitals and all the state-level designated COVID-19 hospitals. Here, custom form action was prepared. Therefore, whenever the conversational agent identifies that the user has selected the "find hospitals" option or the user input is of "hospital search" intent then automatically the custom form action gets triggered. The name of our custom form action is "doctor_hospital_form."
    This particular form action uses the datasets of West Bengal PIN codes—i.e. all the PIN codes in the state mapped with their respective district name; the form action also uses the COVID-19 Government Helpline Numbers dataset and, lastly, the form action uses the dataset containing the COVID-19-designated hospitals in West Bengal. How the dataset was prepared and the overall algorithm of the form action is explained in the Methodology section below.

- <u>Medicine Shops Nearby</u>:

    This sub-feature is just responsible for finding all the medicine shops in or near the PIN code which the user has entered. Here, custom form action was prepared. Therefore, whenever the conversational agent identifies that the user has selected the "medicine shops" option or the user input is of "medicine shops_ search" intent then automatically the custom form action gets triggered. The name of our custom form action is "medicine_shops_form." This particular form action uses the dataset of medicine shops in West Bengal which has medicine shops mapped to the PIN code. How the dataset was prepared and the overall algorithm of the form action is explained in the Methodology section below.

7.4   METHODOLOGY

This section focuses on how the above-designed system works. Here, we discuss the algorithms that are used for each of the features, the datasets that were used for each of the features, how the datasets were prepared, and statistics of the datasets. Also, we demonstrate how our conversational agent will interact with the user, and we also discuss the various ways in which the conversational agent interacts with the user. In this paper, we have discussed two modules in detail: *Know-Corona* and *Help-Desk*.

7.4.1 *KNOW-CORONA* QA MODULE

The dataset used in the *Know-Corona* module consists of FAQ questions from WHO (https://www.who.int/emergencies/diseases/novel-coronavirus-2019), CDC (https://www.cdc.gov/coronavirus/2019-nCoV/index.html) and NHP (https://www.nhp.gov.in/disease/communicable- disease/novel-coronavirus-2019-ncov). Later, with the help of five people, each question is turned into a cluster of similar queries though the answer remains unmodified. Statistics related to the *Know-Corona* dataset are given in Table 7.1.

7.4.1.1 Algorithm of Know-Corona (KC) Module

To find similarity scores against a question cluster "QC" for a user-given query "UQ," the agent must find out the size of three sets for each question of that QC. First, what synonyms are common between UQ and a question of cluster QC, i.e. common synonym set (CSS). Second, what synonyms were expected from a question of QC but are absent, i.e. absent synonym set (ASS), and what synonyms are not in CSS but still present in the question's synonym set of QC, i.e. redundant synonyms set (RSS). Finally, the multiplicative sum of the size of CSS, size of ASS, and size of RSS with their weighted coefficients "coef CSS," "coef ASS," and "coef RSS" respectively for each question of that cluster, the algorithm yields the similarity score of that cluster. This KC module generates optimal results when coef CSS has a positive value and both coef ASS and coef RSS are negative. Based on that approach, we receive a list of similarity scores against the question clusters for a user-given query, i.e. UQ. The QC with the highest similarity score and with a similar context to UQ and its answer can be delivered to the user as a response. The algorithm is given below.

TABLE 7.1
Statistics of the Dataset of *Know-Corona* Module

Total number of different sources	3
Total number of clusters	301
Maximum number of queries in a cluster	5
Minimum number of queries in a cluster	2
Average number of queries in a cluster	3.29
Maximum number of words in a query	45
Minimum number of words in a query	3
Average number of words in a query	10.56
Maximum number of characters in a query	237
Minimum number of characters in a query	16
Average number of characters in a query	60.28

QCS = Set of all available question clusters. for QC_j of QCS:
{
 confidence_QC_j = 0
 ESS = Set of expected synonyms of UQ
 LEN_ESS = Length of ESS
 for question_i of QC_j :
 {
  confidence_i = 0
  QiSS = Set of extracted synonyms.from question_i
  CSS = Set of common synonyms between QiSS and ESS
  RSS = Set of synonyms that are in QiSS but not useful.
  ASS = Synonyms that are in ESS but absent in QiSS
  LEN_QiSS = Length of QiSS
  LEN_CSS = Length of CSS
  LEN_RSS = LEN_QiSS - LEN_CSS
  LEN_ASS = LEN_ESS - LEN_CSS
  confidence_i += coef_CSS * LEN_CSS
  confidence_i += coef_RSS * LEN_RSS
  confidence_i += coef_ASS * LEN_ASS
  confidence_QC_j += confidence_i
 }
 Assign confidence_QC_j as a similarity score of QC_j.
}
    selected_cluster = find the cluster with maximum similarity score return response
of selected cluster

Now let's take a scenario with respect to coef CSS = 6, coef ASS = 2, and coef RSS = 0.5. The user-given query "UQ" is "Ways to protect a kid from covid virus?". Table 7.2 shows the computation for Cluster C1 and Table 7.3 shows the computation for Cluster C2.

TABLE 7.2
KC Computation on Cluster C1

UQ: *Ways to protect a kid from covid virus?*
ESS: *defend, save, protect, safeguard, preserve, virus, germs, illness, microbe, bacteria, kid, child, young, juvenile*
LEN_ESS: 14

	C1_Q1	C1_Q2	C1_Q3
LEN_QiSS	6 { defend, save, shield, safeguard, preserve, protect }	7 { safeguard, germs, protection, save, shield, bacteria, illness }	9 { precaution, virus, safeguard, defense, germs, illness, infection, microbe, bacteria }
LEN_CSS	4 {defend, save, safeguard, preserve }	5 { safeguard, germs, save, bacteria, illness }	6 { virus, safeguard, germs, illness, microbe, bacteria}
LEN_ASS	14–4 = 10	14–5 = 9	14–6 = 8
LEN_RSS	6–4 = 2	7–5 = 2	9–6 = 3
Similarity Score	4*6–10*2–2*0.5 = 3	5*6–9*2–2*0.5 = 11	4*6–8*2–3*0.5 = 6.5

Total similarity score: 3 + 11 + 6.5 = 20.5

TABLE 7.3
KC Computation on Cluster C2

UQ: *Ways to protect a kid from covid virus?*
ESS: *defend, save, protect, safeguard, preserve, virus, germs, illness, microbe, bacteria, kid, child, young, juvenile*
LEN_ESS: 14

	C2_Q1	C2_Q2	C2_Q3
Available Synonyms	13 { hot, warm, mellow, climate, weather, prevent, kill, terminate, virus, germs, illness, microbe, bacteria }	13 { high, tall, big, climate, heat, temperature, kill, eliminate, germs, terminate, infection, bacteria, microbe }	8 { heat, warmth, eliminate, kill, remove, germs, bacteria, microbe }
Size of CSS	5{ virus, germs, illness, microbe, bacteria }	3 { germs, bacteria, microbe }	3 { germs, bacteria, microbe }
Size of ASS	14−5 = 9	14−3=11	14−3=11
Size of RSS	13−5 = 8	13−3 = 10	8−3 = 5
Similarity Score	6*6−9*2−8*0.5 = 14	3*6−11*2−10*0.5 =−9	3*6−1*2−5*0.5 = −6.5

Total similarity score: 14−9−6.5 = -1.5

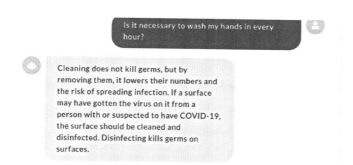

FIGURE 7.7 Sample Response by KC Module against User Query.

As per the algorithm, an answer related to C1 will be selected as a response. Figure 7.7 below shows a response given by the agent to the user against a given query.

7.4.2 HELP-DESK

Referring to Figure 7.2, the *Help-Desk* module comprises two sub-modules, the first being "<u>COVID-19 Hospitals Nearby</u>" and the second being "<u>Medicine Shops Nearby</u>."

If the user chooses the *"COVID-19 Hospitals Nearby"* option then corresponding custom form action would be initiated (as explained in detail in the System section above), and the same is the scenario for the *"Medicine Shops Nearby"* option.

Understanding both of them in detail below,

- COVID-19 Hospitals Nearby
  Here are the steps involved in the algorithm:
  a) PIN code of the user is asked.
  b) After the user enters the PIN code, the algorithm uses the "West Bengal PIN codes" dataset; using the dataset it extracts the district and state name in which the PIN code exists.
  c) Further, the state name extracted from step b is used to find the COVID-19 helpline numbers of the respective state, using the "COVID-19 Government Helpline Numbers" dataset.
  d) Further, the district name extracted from step b is used to find the COVID-19-designated hospitals that exist in that district, using the "COVID-19-designated hospitals in the West Bengal state" dataset.
  e) Further, the state-level designated hospitals are also searched in the "COVID-19-designated hospitals in the West Bengal state" dataset.

Therefore, our algorithm will share the state-level as well as district-level hospitals with the user; if no district-level hospitals exist, then simply the state-level hospitals information will be shared. Information like Hospital Name, Address, General Helplines, Emergency Helpline Number, and Ambulance Contact Number are shared.

Now, we will discuss the datasets used for this feature, how they were prepared, and the statistics of the dataset. The following datasets were used for this feature:

a) "West Bengal PIN codes" dataset which is a CSV file, with all the PIN codes of the state of West Bengal mapped to their respective district names. The dataset was prepared by formatting the All India PIN code Directory available at Open Government Data Platform India (Ministry of Communications, Department of Posts, 2017)

b) "COVID-19 Government Helpline Numbers" dataset which is a CSV File, with all the states and Union Territories in India mapped to their respective COVID-19 helpline numbers. The dataset was prepared from the Helpline Numbers of States & Union Territories file released by the Ministry of Health and Family Welfare, Government of India (Ministry of Health and Family Welfare, Government of India, 2020).

c) "COVID-19-designated hospitals in the West Bengal state" dataset which is an Excel file, containing the COVID-19-designated hospitals in the state of West Bengal. The dataset contains the Hospital Name, Type (Private or Government), Location, District, State, General Helpline Number, Emergency Contact Number, Ambulance Contact Number, Email ID, and Address. The dataset was prepared from the Notification regarding 66 COVID-19-designated hospitals in West Bengal file released by the Department of Health & Family Welfare, Government of West Bengal (Director of Health Service, Government of West Bengal, 2020).

TABLE 7.4
Statistics of the Datasets used in "COVID-19 Hospitals Nearby" Feature

Name of Dataset	Size (in entries)
West Bengal PIN codes	1128
COVID-19 Government Helpline Numbers	37
COVID-19-designated hospitals in the West Bengal state	66

Table 7.4 shows the statistics of the datasets which have been used for the *"COVID-19 Hospitals Nearby"* feature. The "West Bengal PIN codes" dataset consists of 1,128 unique PIN codes of the state of West Bengal. The "COVID-19 Government Helpline Numbers" dataset contains the COVID-19 Helpline Numbers of the Central Government, 28 different states in India and 8 different Union Territories in India. Finally, the "COVID-19-designated hospitals in the West Bengal state" dataset contains information of 66 hospitals (both government and private) which have been designated as hospitals for COVID-19 treatment.

West Bengal has 23 districts. As per the data, only the Kalimpong district does not have a COVID-19-designated hospital. There are four hospitals in the Kolkata district which are also state-level hospitals as per the data released (Director of Health Service, Government of West Bengal, 2020), of which three are government hospitals and one is a private hospital.

- Medicine Shops Nearby
  Here are the steps involved in the algorithm:
  a) PIN code of the user is requested.
  b) After the user enters the PIN code, the algorithm uses the "Medicine Shops in West Bengal" dataset, and the algorithm searches whether medicine shops exist in that PIN code or not.
  c) If no medicine shops exist in that PIN code, then the online websites where the medicines can be bought are shared with the user. Also, the details of local police helpline numbers are shared with the user, whom they can contact if they cannot buy medicines online due to non-availability or delivery issues.
  d) If the medicine shops exist at a given PIN code then the user is asked to choose the option of *Top (10/20/30/40/50)*, where these options are the number of medicine shops' details the user wants.

Here the medicine shops are sorted in ascending order of the distances of the medicine shops with the average GPS coordinate of a PIN code. Information like Name of the Shop, Address, Landline Number (if), Mobile Number (if) are shared with the user.

Figure 7.8 shows how the details of the medicine shops are stored for each PIN code in the dataset. For each PIN code there exists an average latitude and longitude (GPS coordinates); therefore, for the medicine shops which are in that PIN code or

```
{
  "700001": {
    "latitude": "22.5738720000001",
    "longitude": "88.34585",
    "medicine-shops": [
      {
        "distance": 297,
        "eLoc": "7S214E",
        "email": "",
        "entryLatitude": 22.571268,
        "entryLongitude": 88.3469270000001,
        "keywords": [
          "HLTMDS"
        ],
        "landlineNo": "",
        "latitude": 22.571399,
        "longitude": 88.3469610000001,
        "mobileNo": "",
        "orderIndex": 1,
        "placeAddress": "Hare Street, Fairley Place, BBD Bagh, Kolkata, West Bengal, 700001",
        "placeName": "Shreema Medical Stores",
        "type": "POI"
      },
```

FIGURE 7.8   Medicine Shops dataset example.

near that PIN code, each one of them has their GPS coordinates. Therefore, the difference of the GPS coordinates of the store with the GPS coordinates of the PIN code is stored as distance in the *"distance"* variable (in meters). Therefore, in this way for each PIN code, there are various medicine shops' details which are saved and these details are in ascending order of the distances of the shops with the GPS coordinates of the PIN code. So the shop which is closest to the GPS coordinates of the PIN code (here it is "700001") would be ranked 1st when a user searches for medicine shops in the PIN code "700001." In this way, the data is arranged. So, if a user wants the details of the Top 10 medicine shops, then the Top 10 nearest medicine shops' details are shared which are near to a user-entered PIN code's GPS coordinates. Similarly, if a user wants to know the details of the Top 50 medicine shops, then if 50 medicine shops' details are not available and let's say only 35 of them are available, then 35 medicine shops' data are shared with the user, and if all 50 are available then all of them are shared. This feature is for users living in the state of West Bengal, as the data prepared is of West Bengal only.

Now, we will discuss the datasets used for this feature, how they were prepared, and the statistics of the datasets. The following datasets were used for this feature:

a. "Medicine Shops in West Bengal" dataset is a JSON File, containing all the PIN codes of West Bengal, with each of the PIN codes mapped to their average latitude and longitude, and the medicine shops' list which contains each Shop's Name, Address, Mobile Number (if), Landline Number (if) and distance from the coordinates. The dataset was prepared from the MapmyIndia Nearby API (https://www.mapmyindia.com/api/advanced-maps/doc/nearby-api).

Table 7.5 shows the statistics of the datasets which have been used for the *"Medicine Shops Nearby"* feature. The "Medicine Shops in West Bengal" dataset consists of 32,681 medicine shops' details (Name, Address, Mobile, Landline).

TABLE 7.5
Statistics of the Datasets used in "Medicine Shops Nearby" Feature

Name of Dataset	Size (in entries)
Medicine Shops in West Bengal	32,681

7.5 EXPERIMENTAL RESULTS

Here we will discuss the experimental results of the *Know-Corona* and *Help-Desk* features and the methods that were used to evaluate them.

7.5.1 KNOW-CORONA

For evaluating the *Know-Corona* module we used questions of the same dataset for now and have compared the performance of the current synonym-based mode with models of the following architectures:

- Synonym-based matching with answer (SA)
- Synonym-based matching with each question (SQ)
- Synonym-based matching with a single question and its answers (SQSA)
- Synonym-based matching with a cluster of questions and their answers (CQSA)
- SpaCy sentence similarity module is accessed through the homepage or by a query.

We have also chosen the following parameters for performance evaluation:

- Exact Match: If the sequence number of the returned answer and the asked question is the same then it is an exact match.
- Acceptability: If it is not an exact match but the context of the returned answer is the same as the given question then it is in an acceptable state.
- MRR score: Represents how correct the algorithms are.
- Time: Average time taken to answer a single question.

Then their performance scores are compared. The result is given below in Table 7.6.

TABLE 7.6
Comparison of Performance Metrics for *Know-Corona* Module

Model Names	Exact Match	Acceptability	MRR
CQSA	0.0352	0.132	0.0016
SA	0.0332	0.145	0.023
SQSA	0.0332	0.132	0.035
SQ	0.45180	0.498	0.58
KC	0.61794	0.332	0.622
SpaCy	0.99004	0.996	0.765

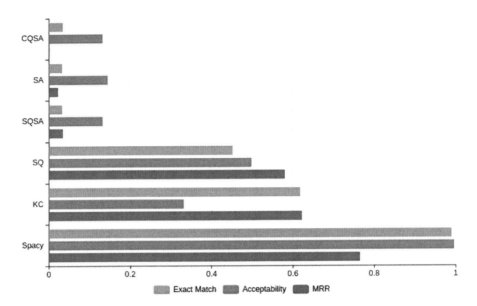

FIGURE 7.9 Performance comparison of different models for **Know-Corona** module.

Figure 7.9 is the visual representation of the performance comparison of different models for the *Know-Corona* module.

7.5.2 HELP-DESK

- **Score:**

  For evaluating the *Help-Desk* feature, we have used the Rasa test's functionality (https://rasa.com/docs/rasa/), and for using that functionality, a "test dataset" was prepared as shown in Figure 7.10. Basically, for this feature, the most important step is correct intent identification, so if the user intent is identified correctly and the correct dialogue utterance is replied by the conversational agent then the feature will work fine, as it is a custom form action; therefore, until all the slots in the form are filled fully, the form will keep iterating among the slots (https://rasa.com/docs/rasa/). Figure 7.11 shows the Intent Prediction Confidence Distribution for the *Help-Desk* feature. From the figure, it is clear that for more than 100 samples our system is predicting the intent with more than 95% confidence whereas for fewer than 5 samples our system is predicting the intent with confidence between 35% to 40%. The samples whose intent has been predicted with more than 95% confidence are the correct predictions as per Figure 7.11, whereas the samples whose intent has been predicted with confidence between 35% to 40% are incorrect predictions as per Figure 7.11.

```
## find hospitals happy path-1
* greet: hi
  - utter_greet
  - utter_ask_Help
* inform: {"Help":"find-hospitals"}
  - doctor_hospitals_form

## find hospitals happy path-2
* hospitals_search: find hospitals near me
  - doctor_hospitals_form

## find medicine shops happy path-1
* greet: hey
  - utter_greet
  - utter_ask_Help
* inform: {"Help":"find-medicine-shops"}
  - medicine_shops_form

## find medicine shops happy path-2
* medicine-shops_search: bot, find medicine shops near me
  - medicine_shops_form
```

FIGURE 7.10   Test dataset for the *Help-Desk* feature.

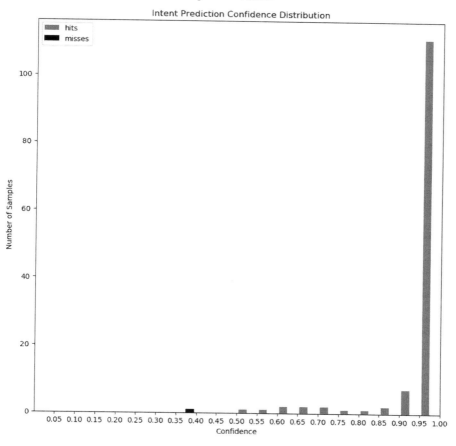

FIGURE 7.11   *Help-Desk* feature's Intent Prediction Confidence Distribution.

```
## find hospitals happy path-1
* greet: hi
    - utter_greet
    - utter_ask_Help
* inform: {"Help":"find-hospitals"}    <!-- predicted: inform: {"Help":"[find-hospitals](Help)"} -->
    - slot{"Help": "find-hospitals"}
    - doctor_hospitals_form    <!-- predicted: action_QA_fallback -->


## find medicine shops happy path-1
* greet: hey
    - utter_greet
    - utter_ask_Help
* inform: {"Help":"find-medicine-shops"}    <!-- predicted: inform: {"Help":"[find-medicine-shops](Help)"} -->
    - slot{"Help": "find-medicine-shops"}
    - medicine_shops_form    <!-- predicted: action_QA_fallback -->
```

FIGURE 7.12 . *Help-Desk* feature failed stories.

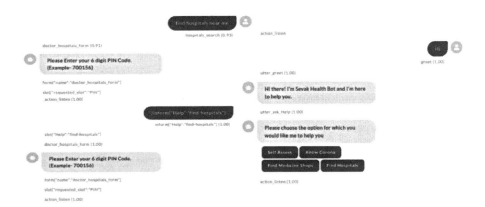

FIGURE 7.13   Choosing "Find Hospitals" option to interact / Typing the query for searching for hospitals to interact.

Figure 7.12 shows the sample instances when the Intents of the *Help-Desk* feature were incorrectly predicted.

- **Output:**

Figures 7.13 and 7.14 show the two ways in which the feature can interact with the user. The user can interact in two methods—first, by choosing the options, and second, by just typing the query. Figures 7.15 and 7.16 show the output of the conversational agent for the "COVID-19 Hospitals Nearby" feature and the "Medicine Shops Nearby" feature.

Here the PIN code given was "700001" and the "Top-10" option was chosen as the number of medicine shops' information needed.

In Figure 7.17, the PIN code entered is "713152." The PIN code is of West Bengal but no medicine shops exist in the dataset at that PIN code, so the online websites and the local police helpline numbers were shared.

FIGURE 7.14   (a) Choosing the "Find medicine shops" option to interact. (b) – Typing the query for searching medicine shops to interact.

7.6   CONCLUSION

In this paper, we have presented a conversational agent made for COVID-19. Our research approach is different from the existing approaches because we have made efforts to bring all the necessary features for COVID-19 into one single application.

Conversational agents are more user-friendly and give a user experience that is much better than normal applications. They also make using technology much easier and interesting.

By our research, a user could search for any query on COVID-19 and could get instant answers; a user can have the information of the COVID-19-designated hospitals in their locality, and/or could easily find the details of medicine shops in their locality.

The biggest advantage of all these features is that they are internet independent. There is no need to have an internet connection to use the features.

The majority of the Indian population lives in rural areas (PTI, 2011) and rural areas have problems with poor internet connectivity. Therefore, our research can be well implemented in rural areas and it would be beneficial for rural people as it would also create awareness among them.

Future work in the research involves adding the data of COVID-19-designated hospitals and local medicine shops of other states and Union Territories in India.

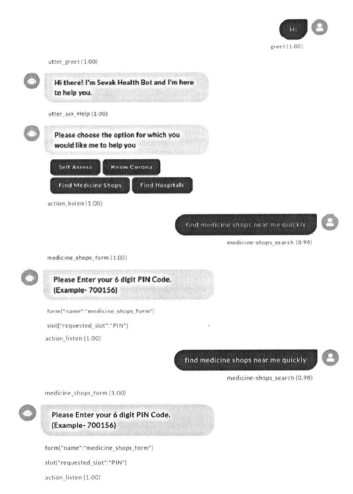

FIGURE 7.15 Output of the conversational agent for the "COVID-19 Hospitals Nearby" feature.

Also, tie-ups could be done with the respective state government on bringing an online consultation service of government hospital doctors on the conversational agent.

These doctors should not only be specialized in helping users with COVID-19-related queries but doctors of different specializations should be available for online consultation so that people can follow the social distancing protocol and stay at home rather than rushing to hospitals, unless it is an emergency.

FIGURE 7.16   Output of the conversational agent for the "Medicine Shops Nearby" feature.

In India, we have different regional languages, so adding the support of those regional languages to the conversational agent and its different features in the future would be very helpful to the local people.

FIGURE 7.17   Output 2 of the conversational agent for the "Medicine Shops Nearby" feature.

REFERENCES

Aguilera, J. (2020, February 03). Xenophobia 'Is A Pre-Existing Condition'. *How Harmful Stereotypes and Racism are Spreading Around the Coronavirus*. Retrieved December 26, 2020, *from* https://time.com/5775716/xenophobia-racism-stereotypes-coronavirus/

Berg M.M. (2015) NADIA: A Simplified Approach Towards the Development of Natural Dialogue Systems. In: Biemann C., Handschuh S., Freitas A., Meziane F., Métais E. (eds) *Natural Language Processing and Information Systems. NLDB 2015. Lecture Notes in Computer Science*, vol 9103. Springer, Cham. https://doi.org/10.1007/978-3-319-19581-0_12

Bodenreider O. (2004). The Unified Medical Language System (UMLS): Integrating biomedical terminology. *Nucleic Acids Research*, 32 (Database issue), D267–D270. doi: 10.1093/nar/gkh061

Chakravorti, B. (2020, February 6). *As Coronavirus Spreads, So Does Fake News*. Retrieved December 26, 2020, from https://www.bloomberg.com/opinion/articles/2020-02-05/as-coronavirus-spreads-so-does-fake-news

Colby, K., Weber, S. & Hilf, F. (1971). Artificial Paranoia. *Artificial Intelligence*, 2. 1–25. 10.1016/0004-3702(71)90002-6.

Dingli, A. & Scerri, D. (2013) Building a Hybrid: Chatterbot – Dialog System. In: Habernal, I., Matoušek, V. (eds) *Text, Speech, and Dialogue. TSD 2013. Lecture Notes in Computer Science*, vol 8082. Springer, Berlin, Heidelberg. doi: 10.1007/978-3-642-40585-3_19

Director of Health Service, Government of West Bengal. (2020). *Notification regarding 66 COVID designated Hospitals in West Bengal [PDF File]*. Retrieved from https://www.wbhealth.gov.in/uploaded_files/go/SPSRC-47.pdf

Hirschberg, J., & Manning, C. D. (2015). Advances in Natural Language Processing. *Science (New York, N.Y.)*, 349(6245), 261–266. https://doi.org/10.1126/science.aaa8685

Kazi, H., Chowdhry, B. & Memon, Z. (2012). MedChatBot: An UMLS based Chatbot for Medical Students. *International Journal of Computer Applications*. 55. 1–5. doi:10.5120/8844-2886.

Klüwer, T. (2011). From Chatbots to Dialog Systems. In Perez-Marin, D., & Pascual-Nieto, I. (eds), *Conversational Agents and Natural Language Interaction: Techniques and Effective Practices* (pp. 1–22). IGI Global. doi:

Larsson, S. & Traum, D. (2000). Information state and dialogue management in the TRINDI dialogue move engine toolkit. *Natural Language Engineering*, 6, 323–340.

Mauldin, M. (1994). *CHATTERBOTS, TINYMUDS, and the Turing Test: Entering the Loebner Prize Competition*. AAAI, Palo Alto, CA. https://www.aaai.org/Papers/AAAI/1994/AAAI94-003.pdf

Ministry of Communications, Department of Posts. (2017). *All India Pincode Directory [CSV File]*. Retrieved from https://data.gov.in/resources/all-india-pincode-directory

Ministry of Health and Family Welfare, Government of India. (2020). *Helpline Numbers of States & Union Territories [PDF File]*. Retrieved from https://www.mohfw.gov.in/pdf/coronavirushelplinenumber.pdf

Mondal, A., Das, D., Cambria, E., & Bandyopadhyay, S. (2017). *Med-ConceptNet: an Affinity score based Medical Concept Network*. In the *Proceedings of 30th International FLAIRS Conference*, Florida, USA, pp. 335–340

Nuruzzaman, M. & Hussain, O. (2018). *A Survey on Chatbot Implementation in Customer Service Industry through Deep Neural Networks*. 54–61. doi:10.1109/ICEBE.2018.00019.

Peltason, J., & Wrede, B. (2010). *Pamini: A framework for assembling mixed-initiative human-robot interaction from generic interaction patterns*. Proceedings of the SIGDIAL 2010 Conference, The 11th Annual Meeting of the Special Interest Group on Discourse and Dialogue, 14–15 September 2010, Tokyo, Japan. https://www.researchgate.net/publication/220794459_Pamini_A_framework_for_assembling_mixed-initiative_human-robot_interaction_from_generic_interaction_patterns

PTI. (2011, July 15). *About 70 per cent Indians live in rural areas: Census report*. Retrieved December 26, 2020, from https://www.thehindu.com/news/national/About-70-per-cent-Indians-live-in-rural-areas-Census-report/article13744351.ece

Purohit, H., Dong, G., Shalin, V., Thirunarayan, K., & Sheth, A. (2015). *Intent Classification of Short-Text on Social Media*. 2015 IEEE International Conference on Smart City/SocialCom/SustainCom (SmartCity), 222–228.

Russo, A., D'Onofrio, G., Gangemi, A., Giuliani, F., Mongiovi, M., Ricciardi, F., Greco, F., Cavallo, F., Dario, P., Sancarlo, D., Presutti, V., & Greco, A. (2019). Dialogue Systems and Conversational Agents for Patients with Dementia: The Human-Robot Interaction. *Rejuvenation Research*, 22(2), 109–120. doi: 10.1089/rej.2018.2075

Soares, M.A., & Parreiras, F.S. (2020). A literature review on question answering techniques, paradigms and systems. *The Journal of King Saud University Computer and Information Sciences*, 32, 635–646.

Taylor, J. (2020, January 31). *Bat soup, dodgy cures and 'diseasology': The spread of coronavirus misinformation*. Retrieved December 26, 2020, from https://www.theguardian.com/world/2020/jan/31/bat-soup-dodgy-cures-and-diseasology-the-spread-of-coronavirus-bunkum

Traum, D.R., & Larsson, S. (2003) The Information State Approach to Dialogue Management. In: van Kuppevelt J., Smith R.W. (eds) *Current and New Directions in Discourse and Dialogue. Text, Speech and Language Technology*, vol 22. Springer, Dordrecht. doi: 10.1007/978-94-010-0019-2_15

Van der Kooij, J., Goossen, W. T., Goossen-Baremans, A. T., de Jong-Fintelman, M., & van Beek, L. (2006). Using SNOMED CT codes for coding information in electronic health records for stroke patients. *Studies in Health Technology and Informatics*, 124, 815–823.

Weizenbaum, J. (1966). ELIZA—a computer program for the study of natural language communication between man and machine. *Commun. ACM*, 9, 36–45.

8 COVID-19 Outbreak Prediction After Lockdown, Based on Current Data Analytics

Muhammad Kashif, Noor Ayesha, Tariq Sadad, and Zahid Mehmood

CONTENTS

8.1   BACKGROUND

The betacoronavirus SARS-CoV-2 occurred in Wuhan, China, and quickly spread worldwide in early December 2019. In late January, WHO declared a public health emergency, announcing the COVID-19 epidemic on March 11, 2020 (Khan et al., 2021). Confirmed cases had been reported among workers or those who lived near the Huanan Seafood Wholesale Market, even though the initial cases had no contact with this market (Harapan et al., 2020). The main COVID-19 indications reported at hospital admission are sore throat, cough, fever, shortness of breath, and abdominal pain. WHO published 79,394 COVID-19 confirmed cases and 2,838 deaths on February 29, 2020, the initial report of a COVID-19 case having been made on December 19, 2019 (Wang et al., 2020a). Later, WHO reported 9,472,473 COVID-19 confirmed cases and deaths of 484,236 individuals in 216 countries on June 26, 2020 (see Figure 8.1).

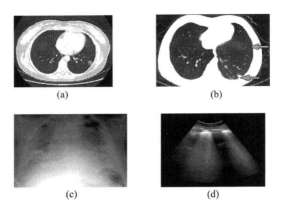

FIGURE 8.1   Lung evaluation of COVID-19 patients.

The number of COVID-19 cases and increase in the mortality rate is reported as approximately 3.4%, compared to 9.6% for SARS and 34.4% for Middle East respiratory syndrome (MERS) (Sohrabi et al., 2020). The mortality rate ranged from 11% to 15% among patients on hospital admission. This article evaluates the genomic structure, infection source, transmission, clinical features, diagnosis, treatments, and prevention strategies, with possible future challenges and solutions, of SARS-CoV-2.

8.2   COVID-19 CLASSIFICATION

CoV are RNA viruses, a subfamily of coronavirinae, belonging to the coronaviridae family and the order nidovirales (composed of coronaviridae, arteriviridae, mesoviri-didae, and roniviridae families). Coronavirinae and torovirinae are two subfamilies of the coronaviridae family. The coronavirinae subfamily consists of alpha-CoV, beta-CoV, gamma-CoV, and delta-CoV, based on their genomic structure. Alpha- and betacoronaviruses can cause respiratory illnesses in humans, and gastrointestinal ill-nesses in animals. by contrast, gamma- and deltacoronaviruses cause bird infections and may infect mammals, but have not been reported as causing any illnesses in humans. Genomic analysis found that SARS-CoV-2, SARS-CoV, and MERS-CoV belong to the betacoronavirus group (Rehman et al., 2021b; Khan et al., 2019a, 2020b; Marie-Sainte et al., 2019a; Rehman, 2021).

The study investigated the origin of the COVID-19 virus, using unsupervised clustering methods ("hierarchical clustering algorithm and density-based spatial clustering of applications with noise (DBSCAN)") and DNA sequences of 334 SARS-CoV-2 were collected from the NCBI GenBank database. Artificial intelligence (AI methods) were used to perform experiments on various clusters collected in Guangdong and Guangxi provinces in China, resulting in SARS-CoV-2 genomes that produce a cluster through the genome of bat "CoV RaTG13" pangolin CoV genomes (Nguyen et al., 2020). Bats are considered as COVID-19 origination, due to 96.2% genome similarity, affecting other mammal hosts such as

SARS-CoV in Himalayan palm civets and MERS-CoV in dromedary camels, before spreading to humans. However, the origin of SARS-Cov-2 is still unknown (Rehman et al., 2021d).

8.3 TRANSMISSION AND SPREAD

The outbreak of an unknown acute respiratory virus was reported in Wuhan China on December 12, 2019, possibly linked to a seafood market. Based on comparison, the virus was recognized as COVID-19 due to SARS-CoV similarity and bat coronavirus homology of more than 70% and 95%, respectively, on January 7, 2020. The positive cases reported from Huanan seafood market ecological samples indicate that the virus was initiated from the market. However, the evidence provided by a genomic study suggested that the virus had entered the Wuhan seafood market from a still unknown origin and had spread fast, and that its spread among humans may have happened earlier. Transmission of the COVID-19 virus occurs between humans through close interaction with infected individuals, via sneezing, coughing, respiratory aerosols, or droplets penetrating the lungs (body) through the mouth or nose via inhalation (Marie-Sainte et al., 2019b; Mughal et al., 2017, 2018).

The average incubation period estimation was 5.1 days (4.5–5.8 days, 95% CI) and 11.5 days (8.2–15.6 days, 97% CI) of those who develop symptoms of infection. At the same time, case definitions mainly depend on a 14-day gap. The study shows that the basic reproduction number R0 (varies by location) estimates the infection spread in individuals by assuming if R0 > 1, an epidemic occurs. Gradually, fadedness is revealed in the epidemic, if R0 < 1. The serial intervals method estimated the intrinsic growth rate (γ) of R0. COVID-19 reproductive number R0 ranging (2.24 to 3.58) estimated from mean significantly greater than 1 (Zhao et al., 2020). The reproductive number R0 may change constantly during a phase of an epidemic.

Pneumonia from COVID-19 was reported between January 20 and 31, 2020, in nine pregnant women at Wuhan University's Zhong nan Hospital. The results from these cases suggest that there is no indication of vertical transmission in women intrauterine in late pregnancy to develop COVID-19 pneumonia. In late January, healthcare workers were infected, 90% of them being from Hubei province. Wang et al. (2020) revealed the ratio (2.7%, 95% CI: 2.6e2.8) of healthcare workers with COVID-19 infection to those with SARS (21.1%, 95% CI: 20.2e22.0) infection was very low (Rehman et al., 2021b, 2021c; Saba, 2019, 2020, 2021).

Local epidemics have been established in several countries after the spread of COVID-19 to more than 100 countries/regions, including in the USA and Europe. The tremendous increase of reported cases in developed countries with high standards of health is higher than that in countries with a lower standard of health, apart from Iran.

8.4 RISK FACTORS

The median age was 56 years (IQR 46.0 to 67.0) of 191 patients between 18 and 87 years (Zhou et al., 2020; Saba et al., 2018, 2019, 2020). The patients had comorbidities such as cardiovascular disease, diabetes, hypertension, or chronic lung disease; the patients were most likely to be male. A study investigation showed that that death in severe COVID-19 male patients was due to cardiac injury, leukocytosis, hyperglycemia, or use of high-dose corticosteroid and a high lactate dehydrogenase level (Li et al., 2020). In severely affected COVID-19 patients, the coinfection rate of bacterial, fungal, and other viruses was significantly increased, compared with those who were not severely affected (Zhang et al., 2020; Iqbal et al., 2017, 2018, 2019).

The clinical results show that children's COVID-19 cases were less severe than those of adult patients while young children, particularly newborns, were at risk of COVID-19 infection (Dong et al., 2020). The data reported between February 12 and April 2, 2020, in the USA from 149,760 laboratories of confirmed coronavirus-infected patients were evaluated. The age was known for 99.6% (149,082) among the reported cases, and 1.7% (2,572) were children aged under 18 years. Another study analyzed 171 positive cases of children by presenting more detailed symptoms, mostly cough (48.5%), minimum fever of 37.5°C (41.5%), and pharyngeal erythema (46.2%), from the Wuhan Children's hospital (Ludvigsson, 2020). The COVID-19 infection seemed milder in children than in adults, with sporadic deaths and better prognosis (Ludvigsson 2020; Saba et al., 2020b).

8.5 DIAGNOSIS OF COVID-19 INFECTION

The common clinical manifestations of COVID-19 are cough, fever, and sputum. Normal body temperature may be manifested, due to low response of the patient's immune system. Dyspnea or shortness of breath is observed from a lack of oxygen and deprived lung function. COVID-19-related pneumonia has initiated by infecting host cells with SARS-CoV-2 through angiotensin-converting enzyme 2 (ACE2) receptors, causing chronic damage and acute myocardial injury to the cardiac system (Zheng et al., 2020; Saba et al., 2019b, 2019c).

A multivariable binary logistic model was used to evaluate severe COVID-19 potential risk factors, while severe patients' survival was analyzed by the Cox proportional hazard regression model (Li et al., 2020). The results recognized that 347 (68.7%) of 505 were positive cases for SARS-CoV-2 nucleic acid test preadmission, severe cases in 269 (49.1%) of 548 patients on admission, and 279 (50.9%) of 548 had no severe case on admission (Li et al., 2020). The risk factors were explored by univariable and multivariable logistic regression methods to evaluate the in-hospital deaths. The study of 191 patients (135 and 56 from Jinyintan and Wuhan Pulmonary Hospitals, respectively), of whom 54 died in hospital and 137 were discharged. 48% (91) patients had comorbidity, 30% (58) patients had hypertension, 19% (36) patients had diabetes, and 8% (15) patients had coronary heart disease (Zhou et al., 2020). The distinctive, abnormal laboratory radiographic findings and variables significantly related with COVID-19 were decreased lymphocytes (31%), leucopenia (19%), high body temperature and high level of creatine kinase MB (31%), and procalcitonin (17%) (Qiu et al., 2020).

Reverse transcription polymerase chain reaction (RT-PCR) method sensitivity was reported for 30% to 60% of throat swab samples because of limitations of sample collection and detection methods. The study revealed that positive chest CT results in certain individuals might have negative RT-PCR results. The patients must go through RT-PCR tests to detect COVID-19 after chest CT has revealed pneumonia. Chest CT diagnostic along with negative RT-PCR reports showed that the patients' epidemiological history had respiratory or fever indications. Wang et al. (2020) reported 97% and 90% for sensitivity and bilateral chest CT results, respectively. The sensitivity for CT scans is 88% and specificity is 25%, with a positive probability ratio of 1.17 and negative of 0.48 (Siordia, 2020; Saba et al., 2021).

Among 240 patients, the negative rate of chest radiographs was 36.7% (18/49) in group A, 28% (21/75) in group B, 18.8% (16/85) in group C, and 16.11% (5/31) in group D. Alterations of 73.3% of patients (132/180) and 26.6% of patients (48/180) were bilateral and unilateral, respectively (Vancheri et al., 2020). A convolutional neural network (CNN) POCOVID-Net was proposed by Born et al. (2020) for COVID-19 detection from a lung point-of-care ultrasound (POCUS) dataset, containing COVID-19, pneumonia, and healthy images. Out of 654 images COVID-19 images, 628 were classified accurately, 13 were misclassified as pneumonia, and 75 were misclassified as healthy. Lung images showed irregularities, due to the classification of 14 pneumonia patients as COVID-19. 75 patients' images are classified as healthy lungs because the lung POCUS dataset contains maximum images of COVID-19 as compared to healthy patients.

Loey et al. (2020) evaluated deep learning model ResNet-50 to detect COVID-19 from a chest CT dataset containing 742 images for two types of labels. 21 patients with non-COVID-19 infection were misclassified as having COVID-19 disease. A deep CNN, COVID-Net, was presented by Wang et al. (2020a) to detect COVID-19 cases from chest X-ray (CXR) images by creating an open-source COVIDx dataset. The network achieved a sensitivity of 91.0% and a high positive predicted value of 98.9% for COVID-19 cases by signifying very few false-positive COVID-19 detections, i.e. only one patient with non-COVID-19 infection was misclassified as COVID-19.

A deep CNN CoroNet model has been proposed that automatically detects COVID-19 infection from CXR images of two datasets (Rehman, 2020). ImageNet and the other dataset are prepared from two public datasets for normal, bacterial pneumonia, and viral pneumonia cases. The CoroNet model achieved an average accuracy of 89.6%. In contrast, for the COVID-19 class, the average accuracy, precision, recall, and F-measure (F1-score) of 96.6%, 93.17%, 98.25%, and 95.6%, respectively, were reported for the first dataset (Khan et al., 2020a). Table 8.1 shows the different performance measures on the same dataset. Therefore, false negatives and a comparatively long testing time may result from low sensitivity and high specificity from nucleic acid detection of SARS-CoV-2 (Wang et al., 2020). The analysis reported that imaging features and their variations for COVID-19 diagnosing have an essential use (Dong et al., 2020).

TABLE 8.1

Performance Measures of Different Studies and Techniques on the Same Dataset

Ref.	Dataset	Technique	Sensitivity	Specificity	Precision	Recall	F1 Score	Accuracy	AUC
Ko et al., 2020	CT (COVID-19 pneumonia)	FCONet model (ResNet-50)	97.39%	99.64%	–	–	–	98.67%	99.00%
He et al., 2020	CT	CAD approach	–	–	–	–	–	99.63%	99.40%
Mobiny et al., 2020	CT	Inception-v3	–	85.30%	84.40%	74.00%	78.10%	81.90%	89.40%
Mobiny et al., 2020	CT	DenseNet121	–	83.90%	81.50%	79.40%	80.10%	82.50%	90.30%
Ko et al., 2020	CT (COVID-19 pneumonia)	FCONet model (Xception)	90.50%	94.82%	–	–	–	92.97%	98.00%
He et al., 2020	CT	3D deep CNN to detect COVID-19 (DeCoVNet)	–	–	–	–	–	90.10%	97.50%
Mobiny et al., 2020	CT	ResNet-50	–	76.90%	75.39%	84.90%	79.50%	80.80%	88.00%

8.6   PREVENTION OF COVID-19

There is no precise treatment for the coronavirus, which exceeded SARS by the number of positive cases and the death rate (Khan et al., 2021). COVID-19 is transmitted mostly through respiratory droplets or close contact with someone infected with respiratory symptoms (such as sneezing, coughing, high temperature, etc.).

Two control SEIR-type models were formulated, followed by applying Pontryagin's maximum principle for establishing optimal controls of COVID-19. The results suggested that the epidemic could be prevented without vaccination, by quarantine measures only. By keeping the strongest quarantine for a planned time period, the infection level decreased to a certain low level (Grigorieva et al., 2020). The Chinese government encouraged people to take precautions against COVID-19 disease, such as staying at home, limiting mass gatherings and social contacts, canceling or postponing large public events, closing schools, and wearing face masks, to limit further spreading (Khan et al., 2019b, 2019c).

COVID-19 transmission can be controlled through proper diagnosis of patients having any abnormality or asymptomatic appearance (Dong et al., 2020). The suspected case should first be transferred for a screening for SARS-CoV-2 disease and nucleic acid tests should also be performed twice for confirmation, and a positive patient should be isolated immediately. The healthcare workers, nurses, surgeons, and other relevant personnel need to take protection during the operations, such as wearing protective face shields (medical protective inner masks and medical-surgical outer masks), double medical caps, medical protective (inner) clothing and (outer) isolation gowns, twofold latex gloves and shoe covers (Wu et al., 2020). The prevention measures are demonstrated in Figure 8.2.

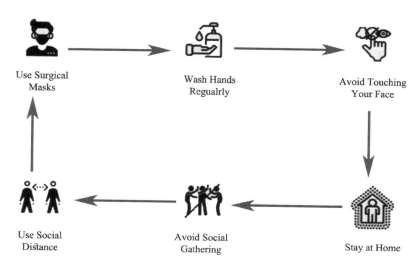

FIGURE 8.2   Preventative measures against COVID-19.

8.7 TREATMENTS

There is currently no effective antiviral vaccine or treatment available for COVID-19 (Sohrabi et al., 2020). A study of Western medicines, traditional Chinese medicines, and natural products revealed more than 30 antivirals that may have potential effectiveness against COVID-19, such as lopinavir/ritonavir, interferon α (IFN-α), ribavirin, chloroquine phosphate, and arbidol, that have been tested in clinical studies. The National Health Commission of China also included these antivirals against COVID-19 in diagnostic, treatment and prevention guidelines (Dong et al., 2020; Khan et al., 2020b). Effective antiviral drugs against COVID-19 investigated in different studies are shown in Table 8.2.

A study of the data of 135 cases of hospitalized patients with COVID-19 in northeast Chongqing was collected, with a median age of 47 years, who received antiviral therapy, antibacterial therapy, and corticosteroids. Several patients also received traditional Chinese medicine (TCM). The research recommended that patients should use Kaletra early. There were some limitations in that the sample size was small; the patients at study completion were still hospitalized (Wan et al., 2020). The clinical results presented through blood sample detection that 54 novel coronavirus pneumonia (NCP) patients boosted their immune system after TCM treatment for COVID-19. Multiple linear regression analysis shows improvement in the patients' recovery for the TCM category of removing dampness and invigorating the spleen. Intestinal

TABLE 8.2
Summary of Effective and Common Antiviral Drugs in the Treatment of COVID-19

Author	Drug Names	Therapy Types	Status
Gautret et al. (2020), Gao et al. (2020), Yao et al. (2020)	Chloroquine phosphate/ hydroxychloroquine	Antiviral anti-malaria anti-inflammatory	FDA approved (use in an emergency)
Martinez (2020)	Remdesivir	Antiviral	investigational (antiviral and clinical trials are in progress)
Dong et al. (2020)	Darunavir	HIV protease inhibitor	FDA approved
Lim et al., 2020	lopinavir/ritonavir	HIV protease inhibitor	FDA approved
Cai et al. (2020)	Favipiravir	Nucleoside analog	investigational
Hoffmann et al. (2020)	Camostat Mesylat	Transmembrane protease, serine 2 (TMPRSS2) inhibitor	Japan approved
Caly et al. (2020)	Ivermectin	Anti-parasite	FDA approved
Fan et al. (2020)	Cepharanthie, Selamectin, and mefloquine hydrochloride	Antiviral anti-inflammatory activities	Investigational

function regulation vitality and keeping microenvironmental balance are suggested in the TCM treatment (Luo et al., 2020).

Convalescent plasma (CP) therapy has been applied in the SARS, MERS, and 2009 H1N1 epidemic treatment with reasonable effectiveness and in the treatment of many infectious diseases (Duan et al., 2020). Acute respiratory distress syndrome (ARDS) and COVID-19 treated in five critically infected patients through CP transfusion resulted in a significant improvement in patients' survival (Shen et al., 2020). Data were collected from the clinical records of ten patients from three participating hospitals, recovered from COVID-19 infection, based on WHO guidelines. CP therapy's clinical outcomes show a satisfying effect and low risk in severe COVID-19 patients' treatment, using one dose with a high medication of neutralizing antibodies (Duan et al., 2020; Liaqat et al., 2020; Lung et al., 2014).

8.8 COVID-19 DATASETS

The main objective of this section is to alert researchers to a broad collection of available open-source datasets, including collections of case reports and medical images concerning COVID-19 (Rad et al., 2013, 2016; Saba et al., 2020b, 2019c).

Lopez et al. (2020) collected multilingual tweets from all countries from January 22, 2020, when reported COVID-19 cases were less than 600 worldwide. The dataset is continuously collected and updated using the Twitter API by using viruses, coronavirus, ncov19, ncov2019, and covid. Link for dataset: https://github.com/lopezbec/COVID19_Tweets_Dataset

"NAIST COVID: Multilingual COVID-19 Twitter and Weibo Dataset" (Gao et al., 2020) was published, comprising 20+ million COVID-19 related microblogs in Japanese, Chinese, and English from Weibo and Twitter, from January 20 to March 24, 2020. It contains the microblog IDs list by two fields—timestamps and query keywords. Dataset link: https://github.com/sociocom/covid19_dataset

The first lung ultrasound (POCUS) dataset was published by sampling 64 videos containing 1,103 images (277 bacterial pneumonia, 654 COVID-19, and healthy controls of 172) at present (Born et al., 2020). The videos are consistently labeled, and the dataset can be generated by dividing them into 1,000 images. Link of open-source dataset and available code: https://github.com/jannisborn/covid19_pocus_ultrasound

Zarei et al. (2020) gathered data uninterrupted between January 5 and March 30, 2020 by covering 18,500 comments and 329,000 likes from 5,300 posts (mainly English language), distributed by 2,500 publishers. The Instagram dataset data is collected by inquiring about certain hashtags such as virus, covid19, corona, quarantine, covid 19, stay at home, and covid. It is split into four parts (features, comment metrics, post content and publisher information) to organize the data. Dataset link: https://github.com/kooshazarei/COVID-19-InstaPostIDs

Cui and Lee (2020) evaluated CoAID (COVID-19 heAlthcare mIsinformation Dataset) which contains fact-checked confirmed true and fake news on articles, reliable websites, and social platform posts. The dataset comprises 1,896 COVID-19-associated news stories, 516 posts from social webpages, user engagements of

183,564 and their ground truth labels. Open-source dataset link: https://github.com/cuilimeng/CoAID

Wang et al. (2020) compiled the COVIDx dataset, including 13,870 patient cases with 13,975 CXR images, by creating it from the amalgamation and variation of five different open access data repositories (COVID-19 Image Data Collection, COVID-19 Chest X-ray Dataset Initiative, ActualMed COVID-19 Chest X-ray Dataset Initiative, RSNA Pneumonia Detection Challenge dataset and COVID-19 radiography database) comprising chest radiography images. Dataset link: https://github.com/lindawangg/COVID-Net

Sethy et al. (2020) prepared an X-ray images dataset of COVID-19, pneumonia, and healthy patients. The dataset is the collection of three datasets—Cohen et al. (2020), Kaggle (2020) (https://www.Kaggle.com/andrewmvd/convid19-X-rays), and Kermany et al. (2018)—and consists of 127 pneumonia, 127 COVID-19, and 127 healthy cases.

The dataset prepared, merging the Cohen et al. (2020) dataset with 50 images and Kaggle (2020) (https://www.kaggle.com/paultimothymooney/chest-xray-pneumonia) dataset with 50 images, equally divided into 50 positive COVID-19 and 50 normal cases (Narin et al., 2020).

Shuja et al. (2020) revealed the COVID-19 CT Lung and Infection Segmentation Dataset containing 20 labeled COVID-19 CT scans. The two radiologists labeled them left lung, right lung, and infections, followed by verification of an experienced radiologist. Dataset link: https://zenodo.org/record/3757476#.XxKOkigzbIU

A novel large-scale COVID-19 Weibo social media dataset was released, named the Weibo-COV dataset. From December 1, 2019, to April 30, 2020, it covers 40+ million tweets, including sufficient tweets, interaction, retweets, and regions. Dataset link: https://github.com/nghuyong/weibo-public-opinion-datasets

Zhou (2020) revealed the evacuee dataset, containing 2,666 overseas citizens evacuated after quarantine in Wuhan, China (between January 29 and February 2, 2020), by closely monitoring their health. COVID-19 tests result from them show positively that 12 persons became infected. Dataset link: https://figshare.com/articles/Evacuees_from_Wuhan/11859207/1

Wei et al. (2020) presented a small COVID-Q dataset of 1,690 COVID questions from 13 online sources (Quora, Google, Yahoo, and Bing search, etc.) annotated by classifying questions into 15 category labels and 207 classes. Dataset link: https://github.com/JerryWei03/COVID-Q

8.9   ANALYSIS & FINDINGS

Results of COVID-19 detection have been analyzed and reported using different performance measures, such as areas under the curve (AUC), receiver operating characteristic (ROC), and precision-recall curves, in several research studies. Therefore, the analysis evaluation reported the effectiveness of chest CT in COVID-19 diagnosing. The main infection of COVID-19 is pneumonia or lung, and for disease diagnosing and treatment evaluation, CT has been sufficiently used as aa medical imaging procedure (Dong et al., 2020). Islam and Islam (2020) analyzed the operational use of drones, robotics, AI, big data, and IoT in several countries in COVID-19 prediction and diagnosis, such as delivering medicines, blood samples, and warnings for staying at home, wearing masks, and temperature monitoring. New AI and data science methods have been developed for COVID-19 virus detection, tracking, treating, and antiviral vaccine development.

The combination of segmentation proposed by COVID-SDNet, data augmentation and data transformation of CXR images had better results in severe and moderate levels. By contrast, few mild and normal-PCR+ cases are most problematic to classify as they comprise insufficient or no visual features (Tabik et al., 2020; Mashood Nasir et al., 2020; Majid et al., 2020). The pooled positive rate of the 2,738 participants of 13 studies with CT imaging was 89.76% and 90.35%, using single-arm analysis while only including chest CT thin-section among patients suspected of COVID-19 (Bao et al., 2020). Zhou et al. (2020) found that with white blood cell counts below 10.0×109 per L or procalcitonin below 0.25 ng/mL in more than 70% of patients on hospital admission no findings of bacteriological infection were found. The pathogenic coinfections rate analyzed by Zhang et al. (2020) with bacteria and fungus were higher in severely infected patients (14 [25.5%] vs 3 [1.8%]; P<0.001) and (6 [10.9%] vs 1 [0.6%]; P = 0.001, respectively) than in non-severely infected patients.

The analysis of 61,742 confirmed COVID-19-infected patients of different 80 studies, including 69.5% patients who had an origin history to Wuhan. A rise of 91% in platelets, along with a 62.5% decrease in lymphocytes, was reported by laboratory analysis among a population of 4.5 % (2361/52251) patients. 76.8% chest radiography association by chest CT and X-ray bilaterally resulted in abnormality, 75.5% unification and 71% ground-glass opacity (GGO). The study has several limitations of disease severity overestimation, due to a screening deficiency of slightly symptomatic or asymptomatic individuals. The rates of hospitalization, critical condition and mortality were incorrectly elevated, due to various infected individuals not having been detected (Mittal et al., 2020; Pormohammad et al., 2020). Liver hypodensity and pericholecystic fat stranding regularly or irregularly were revealed by analyzing COVID-19 patients' upper abdominal CT scan. SARS-CoV-2 infection may affect various organ dysfunction syndromes, including liver damage. In COVID-19 patients, a GGO pattern of 66.76% and 35.15% consolidation were detected by chest CT findings from nine studies (Park et al., 2020).

The clinical findings mostly determined 38% fatigue, 60% cough, and 77% fever by evaluating 24 different research studies, including 11,950 confirmed cases of COVID-19. The study has some limitations of accessing full articles, varying laboratory findings, different disease-stage patients being involved, and clinical results not being described due to the ineffectual treatment period. Rodriguez-Morales et al. (2020) reported that the frequency rate of fever in adults was high compared to that in children (92.8%, 95%CI 89.4–96.2%, versus 43.9%, 95% CI 28.2–59.6%).

Wong et al. (2020) investigated 64 patients, including 26 men with a mean age of 56 years, with RT-PCR initial positive results of 58 patients (91%) and abnormal findings of 44 patients (69%) at baseline chest radiography, and also both for 38 patients (59%). Earlier COVID-19 tested positive abnormalities were exposed in six patients (9%) through chest radiography RT-PCR. The study with 69% sensitivity had several limitations reported, such as the disease course being shortened, with some patients not being followed to their final results. The precision of analysis was possibly affected because the tests between RT-PCR and serial chest radiograph breaks were not uniform. Control group deficiency limits chest radiography sensitivity and specificity estimation.

8.10 CONCLUSIONS AND FUTURE CHALLENGES

The COVID-19 (SARS-CoV-2) pandemic is obviously an international public health emergency that spread rapidly worldwide. Early indications of SARS-CoV-2 are cough, fever, and fatigue that are most similar to those of SARS. A recent history of travel to COVID-19 outbreak areas or contact with a positive case individual requires rapid clinical precautions, such as isolation and laboratory confirmation. Controlling the epidemic spread and the mortality rate is an issue, yet no precise antiviral drugs have been developed to protect the population's safety. The COVID-19 pandemic potential needs investigation to precisely track and predict future host evolution. Mathematical, statistical, deep learning, and AI models have been developed to investigate the pandemic efficiently by providing the best results. This chapter has reviewed the COVID-19 classification background, transmission, spread, risk factors, clinical manifestation and diagnosing strategies, treatments, control and prevention strategies, and future challenges in detail.

With the increasing number of cases of COVID-19 infection, the world is under huge pressure. Several studies have investigated possible future challenges and solutions relating to COVID-19. Some of them are as follows. Social and travel restrictions resulted in a lockdown to stop the spread of infection and save humanity. This lockdown situation led to global economic and GDP losses, and employment crises. There are plans and control strategies against the rise in air pollution, due to comorbidity in COVID-19-infected patients. COVID-19-related medical waste may become a major hazard if it is not properly handled. During patients' hospital treatment, the bacterial confrontation and healthcare-associated infections are more significant issues and challenges. Orthopedic patients have to face unique challenges during face-to-face communication with doctors to avoid coronavirus infection.

Possible solutions for these challenges are to increase industrial production, support small-scale industries, recover infrastructure-associated projects, and speed up essential strategies of increasing investment in the public health sector—approaches required for adapting to working from home.

REFERENCES

Bao, C., Liu, X., Zhang, H., Li, Y. and Liu, J., 2020. Coronavirus disease 2019 (COVID-19) CT findings: A systematic review and meta-analysis. *Journal of the American College of Radiology*, 17(6), p. 701.

Born, J., Brändle, G., Cossio, M., Disdier, M., Goulet, J., Roulin, J. and Wiedemann, N., 2020. POCOVID-Net: automatic detection of COVID-19 from a new lung ultrasound imaging dataset (POCUS). arXiv preprint arXiv:2004.12084.

Cai, Q., Yang, M., Liu, D., Chen, J., Shu, D., Xia, J., Liao, X., Gu, Y., Cai, Q., Yang, Y. and Shen, C., 2020. Experimental treatment with favipiravir for COVID-19: An open-label control study. *Engineering*, 6(10), 1192–1198.

Caly, L., Druce, J.D., Catton, M.G., Jans, D.A. and Wagstaff, K.M., 2020. The FDA-approved drug ivermectin inhibits the replication of SARS-CoV-2 in vitro. *Antiviral Research*, 178, 104787.

Cohen, J.P., Morrison, P., Dao, L., Roth, K., Duong, T.Q. and Ghassemi, M., 2020. COVID-19 image data collection: Prospective predictions are the future. arXiv preprint arXiv:2006.11988.

Cui, L. and Lee, D., 2020. CoAID: COVID-19 healthcare misinformation dataset. arXiv preprint arXiv:2006.00885.

Dong, L., Hu, S. and Gao, J., 2020. Discovering drugs to treat coronavirus disease 2019 (COVID-19). *Drug Discoveries & Therapeutics*, 14(1), 58–60.

Duan, K., Liu, B., Li, C., Zhang, H., Yu, T., Qu, J., Zhou, M., Chen, L., Meng, S., Hu, Y. and Peng, C., 2020. Effectiveness of convalescent plasma therapy in severe COVID-19 patients. *Proceedings of the National Academy of Sciences*, 117(17), 9490–9496.

Fan, H.H., Wang, L.Q., Liu, W.L., An, X.P., Liu, Z.D., He, X.Q., Song, L.H. and Tong, Y.G., 2020. Repurposing of clinically approved drugs for treatment of coronavirus disease 2019 in a 2019-novel coronavirus-related coronavirus model. *Chinese Medical Journal*, 133(9), 1051–1056.

Gao, Z., Yada, S., Wakamiya, S. and Aramaki, E., 2020. Naist covid: Multilingual covid-19 twitter and weibo dataset. arXiv preprint arXiv:2004.08145.

Gautret, P., Lagier, J.C., Parola, P., Meddeb, L., Mailhe, M., Doudier, B., Courjon, J., Giordanengo, V., Vieira, V.E., Dupont, H.T. and Honoré, S., 2020. Hydroxychloroquine and azithromycin as a treatment of COVID-19: Results of an open-label non-randomized clinical trial. *International Journal of Antimicrobial Agents*, 56(1)105949.

Grigorieva, E., Khailov, E. and Korobeinikov, A., 2020. Optimal quarantine strategies for covid-19 control models. arXiv preprint arXiv:2004.10614.

Harapan, H., Itoh, N., Yufika, A., Winardi, W., Keam, S., Te, H., Megawati, D., Hayati, Z., Wagner, A.L. and Mudatsir, M. 2020. Coronavirus disease 2019 (COVID-19): A literature review. *Journal of Infection and Public Health*, 13(5), 667–673.

He, X., Wang, S., Shi, S., Chu, X., Tang, J., Liu, X., Yan, C., Zhang, J. and Ding, G., 2020. Benchmarking deep learning models and automated model design for COVID-19 detection with chest CT scans. *medRxiv*.

Hoffmann, M., Kleine-Weber, H., Krüger, N., Mueller, M.A., Drosten, C. and Pöhlmann, S., 2020. The novel coronavirus 2019 (2019-nCoV) uses the SARS-coronavirus receptor ACE2 and the cellular protease TMPRSS2 for entry into target cells. *BioRxiv*, 181(2), 271–280.

Iqbal, S. Ghani, M.U. Saba, T. and Rehman, A. (2018). Brain tumor segmentation in multispectral MRI using convolutional neural networks (CNN). *Microscopy Research and Technique*, 81(4), 419–427. doi: 10.1002/jemt.22994.

Iqbal, S., Khan, M.U.G., Saba, T. Mehmood, Z. Javaid, N., Rehman, A., Abbasi, R. (2019) Deep learning model integrating features and novel classifiers fusion for brain tumor segmentation, *Microscopy Research and Technique*, 82(8), 1302–1315. doi:10.1002/jemt.23281

Iqbal, S., Khan, M. U. G., Saba, T., Rehman, A. (2017). Computer assisted brain tumor type discrimination using magnetic resonance imaging features. *Biomedical Engineering Letters*, 8(1), 5–28, doi. 10.1007/s13534-017-0050-3.

Islam, M.N. and Islam, A.N., 2020. A systematic review of the digital interventions for fighting COVID-19: the Bangladesh perspective. *IEEE Access*, 8, 114078–114087.

Kermany, D.S., Goldbaum, M., Cai, W., Valentim, C.C., Liang, H., Baxter, S.L., McKeown, A., Yang, G., Wu, X., Yan, F. and Dong, J., 2018. Identifying medical diagnoses and treatable diseases by image-based deep learning. *Cell*, 172(5), 1122–1131.

Khan, A.I., Shah, J.L. and Bhat, M.M., 2020a. Coronet: A deep neural network for detection and diagnosis of COVID-19 from chest x-ray images. *Computer Methods and Programs in Biomedicine*, 196, 105581.

Khan, M.A.; Akram, T. Sharif, M., Saba, T., Javed, K., Lali, I.U., Tanik, U.J. and Rehman, A. (2019a). Construction of saliency map and hybrid set of features for efficient segmentation and classification of skin lesion. *Microscopy Research and Technique*, 82(5), 741–763, doi: 10.1002/jemt.23220

Khan, M.A., Ashraf, I., Alhaisoni, M., Damaševičius, R., Scherer, R., Rehman, A. and Bukhari, S.A.C. (2020b) Multimodal brain tumor classification using deep learning and robust feature selection: A machine learning application for radiologists. *Diagnostics*, 10, 565.

Khan, M.A. Kadry, S., Zhang, Y.D., Akram, T., Sharif, M., Rehman, A. and Saba, T. (2021) Prediction of COVID-19- pneumonia based on selected deep features and one class kernel extreme learning machine, *Computers & Electrical Engineering*, 90, 106960.

Khan, M.A., Lali, I.U. Rehman, A. Ishaq, M. Sharif, M. Saba, T., Zahoor, S. and Akram, T. (2019b) Brain tumor detection and classification: A framework of marker-based watershed algorithm and multilevel priority features selection. *Microscopy Research and Technique*, 82(6), 909–922. doi:10.1002/jemt.23238.

Khan, M.A., Sharif, M. Akram, T., Raza, M., Saba, T. and Rehman, A. (2020c) Hand-crafted and deep convolutional neural network features fusion and selection strategy: An application to intelligent human action recognition. *Applied Soft Computing*, 87, 105986

Khan, S. A., Nazir, M., Khan, M. A., Saba, T., Javed, K., Rehman, A., … & Awais, M. (2019c). Lungs nodule detection framework from computed tomography images using support vector machine. *Microscopy Research and Technique*, 82(8), 1256–1266.

Ko, H., Chung, H., Kang, W.S., Kim, K.W., Shin, Y., Kang, S.J., Lee, J.H., Kim, Y.J., Kim, N.Y., Jung, H. and Lee, J., 2020. COVID-19 pneumonia diagnosis using a simple 2D deep learning framework with a single chest CT image: Model development and validation. *Journal of Medical Internet Research*, 22(6), e19569.

Li, X., Xu, S., Yu, M., Wang, K., Tao, Y., Zhou, Y., Shi, J., Zhou, M., Wu, B., Yang, Z. and Zhang, C., 2020. Risk factors for severity and mortality in adult COVID-19 inpatients in Wuhan. *Journal of Allergy and Clinical Immunology*, 146(1), 110–118.

Liaqat, A., Khan, M. A., Sharif, M., Mittal, M., Saba, T., Manic, K. S. and Al Attar, F. N. H. (2020). Gastric tract infections detection and classification from wireless capsule endoscopy using computer vision techniques: A review. *Current Medical Imaging*, 16(10), 1229–1242.

Lim, J., Jeon, S., Shin, H.Y., Kim, M.J., Seong, Y.M., Lee, W.J., Choe, K.W., Kang, Y.M., Lee, B. and Park, S.J., 2020. Case of the index patient who caused tertiary transmission of coronavirus disease 2019 in Korea: The application of lopinavir/ritonavir for the treatment of COVID-19 pneumonia monitored by quantitative RT-PCR. *Journal of Korean Medical Science*, 35(6), e79.

Loey, M., Manogaran, G. and Khalifa, N.E.M., 2020. A deep transfer learning model with classical data augmentation and cgan to detect covid-19 from chest ct radiography digital images. *Neural Computing and Applications*, 1–13. doi:10.1007/s00521-020-05437-x.

Lopez, C.E., Vasu, M. and Gallemore, C., 2020. Understanding the perception of COVID-19 policies by mining a multilanguage Twitter dataset. arXiv preprint arXiv:2003.10359.

Ludvigsson, J.F., 2020. Systematic review of COVID-19 in children shows milder cases and a better prognosis than adults. *Acta Paediatrica*, 109(6), pp. 1088–1095.

Lung, J.W.J., Salam, M.S.H., Rehman, A., Rahim, M.S.M. and Saba, T. (2014) Fuzzy phoneme classification using multi-speaker vocal tract length normalization. *IETE Technical Review*, 31 (2), pp. 128–136, doi. 10.1080/02564602.2014.892669.

Luo, E., Zhang, D., Luo, H., Liu, B., Zhao, K., Zhao, Y., Bian, Y. and Wang, Y., 2020. Treatment efficacy analysis of traditional Chinese medicine for novel coronavirus pneumonia (COVID-19): an empirical study from Wuhan, Hubei Province, China. *Chinese Medicine*, 15, pp. 1–13.

Majid, A., Khan, M. A., Yasmin, M., Rehman, A., Yousafzai, A. and Tariq, U. (2020). Classification of stomach infections: A paradigm of convolutional neural network along with classical features fusion and selection. *Microscopy Research and Technique*, 83(5), 562–576.

Marie-Sainte, S. L. Aburahmah, L., Almohaini, R. and Saba, T. (2019a). Current techniques for diabetes prediction: Review and case study. *Applied Sciences*, 9(21), 4604.

Marie-Sainte, S. L., Saba, T., Alsaleh, D., Alotaibi, A. and Bin, M. (2019b). An improved strategy for predicting diagnosis, survivability, and recurrence of breast cancer. *Journal of Computational and Theoretical Nanoscience*, 16(9), 3705–3711.

Martinez, M.A., 2020. Compounds with therapeutic potential against novel respiratory 2019 coronavirus. *Antimicrobial Agents and Chemotherapy*, 64(5), e00399–e00320.

Mashood Nasir, I., Attique Khan, M., Alhaisoni, M., Saba, T., Rehman, A. and Iqbal, T. (2020). A hybrid deep learning architecture for the classification of superhero fashion products: An application for medical-tech classification. *Computer Modeling in Engineering & Sciences*, 124(3), 1017–1033.

Mittal, A., Kumar, D., Mittal, M., Saba, T., Abunadi, I., Rehman, A. and Roy, S. (2020). Detecting pneumonia using convolutions and dynamic capsule routing for chest X-ray images. *Sensors*, 20(4), 1068.

Mobiny, A., Cicalese, P.A., Zare, S., Yuan, P., Abavisani, M., Wu, C.C., Ahuja, J., de Groot, P.M. and Van Nguyen, H., 2020. Radiologist-level COVID-19 detection using CT scans with detail-oriented capsule networks. arXiv preprint arXiv:2004.07407.

Mughal, B., Muhammad, N., Sharif, M., Rehman, A. and Saba, T. (2018). Removal of pectoral muscle based on topographic map and shape-shifting silhouette. *BMC Cancer*, 18(1), 1–14.

Mughal, B. Muhammad, N. Sharif, M. Saba, T. and Rehman, A. (2017) Extraction of breast border and removal of pectoral muscle in wavelet domain, *Biomedical Research*, 28(11), pp. 5041–5043.

Narin, A., Kaya, C. and Pamuk, Z., 2020. Automatic detection of coronavirus disease (covid-19) using x-ray images and deep convolutional neural networks. arXiv preprint arXiv:2003.10849.

Nguyen, T.T., Abdelrazek, M., Nguyen, D.T., Aryal, S., Nguyen, D.T. and Khatami, A., 2020. Origin of novel coronavirus (COVID-19): A computational biology study using artificial intelligence. *BioRxiv*.

Park, J.H., Jang, W., Kim, S.W., Lee, J., Lim, Y.S., Cho, C.G., Park, S.W. and Kim, B.H., 2020. The clinical manifestations and chest computed tomography findings of coronavirus disease 2019 (COVID-19) patients in China: A proportion meta-analysis. *Clinical and Experimental Otorhinolaryngology*, 13(2), p. 95.

Pormohammad, A., Ghorbani, S., Baradaran, B., Khatam, A., Turner, R., Mansournia, M.A., Kyriacou, D.N., Idrovo, J.P. and Bahr, N.C., 2020. Clinical Characteristics, laboratory findings, radiographic signs and outcomes of 52,251 patients with confirmed covid-19 infection: A systematic review and meta-analysis. *Microbial Pathogenesis*, 147, 104390.

Qiu, H., Wu, J., Hong, L., Luo, Y., Song, Q. and Chen, D., 2020. Clinical and epidemiological features of 36 children with coronavirus disease 2019 (COVID-19) in Zhejiang, China: An observational cohort study. *The Lancet Infectious Diseases*, 20(6), 689–696

Rad, A.E. Rahim, M.S.M., Rehman, A. Altameem, A. and Saba, T. (2013) Evaluation of current dental radiographs segmentation approaches in computer-aided applications *IETE Technical Review*, 30(3), pp. 210–222

Rad, A.E., Rahim, M.S.M., Rehman, A. and Saba, T. (2016) Digital dental X-ray database for caries screening, *3D Research*, 7(2), pp. 1–5, doi. 10.1007/s13319-016-0096-5.

Rehman, A. (2020). *Ulcer Recognition based on 6-Layers Deep Convolutional Neural Network*. In *Proceedings of the 2020 9th International Conference on Software and Information Engineering (ICSIE)* (pp. 97–101). Cairo Egypt.

Rehman, A. (2021) Light microscopic iris classification using ensemble multi-class support vector machine, *Microscopic Research & Technique*; https://doi.org/10.1002/jemt.23659

Rehman, A., Khan, M. A., Saba, T., Mehmood, Z., Tariq, U. and Ayesha, N. (2021a). Microscopic brain tumor detection and classification using 3D CNN and feature selection architecture. *Microscopy Research and Technique*, 84(1), pp. 133–149. doi: 10.1002/jemt.23597.

Rehman, A. Saba, T., Ayesha N. and Tariq, U (2021b) Deep learning-based COVID-19 detection using CT and X-ray images: Current analytics and comparisons. *IEEE IT Professional*. doi:10.1109/MITP.2020.3036820

Rehman, A., Sadad, T. Saba, T., Hussain A. and Tariq, U (2021c) Real-time diagnosis system of COVID-19 using X-ray images and deep learning. *IEEE IT Professional*. doi. 10.1109/MITP.2020.3042379

Rodriguez-Morales, A.J., Cardona-Ospina, J.A., Gutiérrez-Ocampo, E., Villamizar-Peña, R., Holguin-Rivera, Y., Escalera-Antezana, J.P., Alvarado-Arnez, L.E., Bonilla-Aldana, D.K., Franco-Paredes, C., Henao-Martinez, A.F. and Paniz-Mondolfi, A., 2020. Clinical, laboratory and imaging features of COVID-19: A systematic review and meta-analysis. *Travel Medicine and Infectious Disease*, p. 101623.

Saba, T. (2019). Automated lung nodule detection and classification based on multiple classifiers voting. *Microscopy Research and Technique*, 82(9), pp. 1601–1609.

Saba, T. (2020). Recent advancement in cancer detection using machine learning: Systematic survey of decades, comparisons and challenges. *Journal of Infection and Public Health*, 13(9), 1274–1289.

Saba, T. 2021. Computer vision for microscopic skin cancer diagnosis using handcrafted and non-handcrafted features. *Microscopy Research and Technique*. doi:10.1002/jemt.23686

Saba, T., Bokhari, S. T. F., Sharif, M., Yasmin, M. and Raza, M. (2018). Fundus image classification methods for the detection of glaucoma: A review. *Microscopy Research and Technique*, 81(10), pp. 1105–1121.

Saba, T., Haseeb, K., Ahmed, I., & Rehman, A. (2020a). Secure and energy-efficient framework using Internet of Medical Things for e-healthcare. *Journal of Infection and Public Health*, 13(10), pp. 1567–1575.

Saba, T., Khan, M. A., Rehman, A. and Marie-Sainte, S. L. (2019a). Region extraction and classification of skin cancer: A heterogeneous framework of deep CNN features fusion and reduction. *Journal of Medical Systems*, 43(9), 289.

Saba, T., Khan, S.U., Islam, N., Abbas, N., Rehman, A., Javaid, N. and Anjum, A., (2019b). Cloud-based decision support system for the detection and classification of malignant cells in breast cancer using breast cytology images. *Microscopy Research and Technique*, 82(6), pp. 775–785.

Saba, T., Mohamed, A.S., El-Affendi, M. Amin, J. and Sharif, M. (2020b) Brain tumor detection using fusion of hand crafted and deep learning features *Cognitive Systems Research* 59, pp. 221–230

Saba, T., Sameh, A., Khan, F., Shad, S. A. and Sharif, M. (2019c). Lung nodule detection based on ensemble of hand crafted and deep features. *Journal of Medical Systems*, 43(12), p. 332.

Saba, T., Rehman, A., Mehmood, Z., Kolivand, H., & Sharif, M. (2018). Image enhancement and segmentation techniques for detection of knee joint diseases: A survey. *Current Medical Imaging*, 14(5), 704–715.

Saba, T., Sameh, A., Khan, F., Shad, S. A., & Sharif, M. (2019). Lung nodule detection based on ensemble of hand crafted and deep features. *Journal of medical systems*, 43(12), 1–12.

Sadad, T., Rehman, A., Munir, A., Saba, T., Tariq, U., Ayesha, N. and Abbasi, R. (2021). Brain tumor detection and multi-classification using advanced deep learning techniques. *Microscopy Research and Technique*. 2021. doi:10.1002/jemt.23688.

Sethy, P.K., Behera, S.K., Ratha, P.K. and Biswas, P. 2020. Detection of coronavirus disease (COVID-19) based on deep features and support vector machine. *International Journal of Mathematical, Engineering and Management Sciences*, 5(4), 643–651.

Shen, M., Peng, Z., Xiao, Y. and Zhang, L., 2020. Modelling the epidemic trend of the 2019 novel coronavirus outbreak in China. *BioRxiv. The Innovation*, 1(3), 100048.

Shuja, J., Alanazi, E., Alasmary, W. and Alashaikh, A., 2020. COVID-19 open source data sets: A comprehensive survey. *medRxiv. Applied Intelligence*, 51, 1296–1325.

Siordia Jr, J.A., 2020. Epidemiology and clinical features of COVID-19: A review of current literature. *Journal of Clinical Virology*, 127, p. 104357.

Sohrabi, C., Alsafi, Z., O'Neill, N., Khan, M., Kerwan, A., Al-Jabir, A., Iosifidis, C. and Agha, R., 2020. World Health Organization declares global emergency: A review of the 2019 novel coronavirus (COVID-19). *International Journal of Surgery*, 76, 71–76.

Tabik, S., Gómez-Ríos, A., Martín-Rodríguez, J.L., Sevillano-García, I., Rey-Area, M., Charte, D., Guirado, E., Suárez, J.L., Luengo, J., Valero-González, M.A. and García-Villanova, P., 2020. COVIDGR dataset and COVID-SDNet methodology for predicting COVID-19 based on Chest X-Ray images. arXiv preprint arXiv:2006.01409.

Vancheri, S.G., Savietto, G., Ballati, F., Maggi, A., Canino, C., Bortolotto, C., Valentini, A., Dore, R., Stella, G.M., Corsico, A.G. and Iotti, G.A., 2020. Radiographic findings in 240 patients with COVID-19 pneumonia: time-dependence after the onset of symptoms. *European Radiology*, 90, p. 1.

Wan, S., Xiang, Y., Fang, W., Zheng, Y., Li, B., Hu, Y., Lang, C., Huang, D., Sun, Q., Xiong, Y. and Huang, X., 2020. Clinical features and treatment of COVID-19 patients in north-east Chongqing. *Journal of Medical Virology*, 92(7), 797–806.

Wang, L.S., Wang, Y.R., Ye, D.W. and Liu, Q.Q., 2020a. A review of the 2019 Novel Coronavirus (COVID-19) based on current evidence. *International Journal of Antimicrobial Agents*, 55(6), p. 105948.

Wei, J., Huang, C., Vosoughi, S. and Wei, J., 2020. What are people asking about COVID-19? a question classification dataset. arXiv preprint arXiv:2005.12522.

Wong, H.Y.F., Lam, H.Y.S., Fong, A.H.T., Leung, S.T., Chin, T.W.Y., Lo, C.S.Y., Lui, M.M.S., Lee, J.C.Y., Chiu, K.W.H., Chung, T. and Lee, E.Y.P., 2020. Frequency and distribution of chest radiographic findings in COVID-19 positive patients. *Radiology*, p. 201160.

Wu, J.T., Leung, K. and Leung, G.M., 2020. Nowcasting and forecasting the potential domestic and international spread of the 2019-nCoV outbreak originating in Wuhan, China: a modelling study. *The Lancet*, 395 (10225), pp. 689–697.

Yao, X., Ye, F., Zhang, M., Cui, C., Huang, B., Niu, P., Liu, X., Zhao, L., Dong, E., Song, C. and Zhan, S., 2020. In vitro antiviral activity and projection of optimized dosing design of hydroxychloroquine for the treatment of severe acute respiratory syndrome coronavirus 2 (SARS-CoV-2). *Clinical Infectious Diseases*, 71(15), 732–739.

Zarei, K., Farahbakhsh, R., Crespi, N. and Tyson, G., 2020. A first Instagram dataset on COVID-19. arXiv preprint arXiv:2004.12226.

Zhang, G., Hu, C., Luo, L., Fang, F., Chen, Y., Li, J., Peng, Z. and Pan, H., 2020. Clinical features and short-term outcomes of 221 patients with COVID-19 in Wuhan, China. *Journal of Clinical Virology*, 127, p. 104364.

Zhao, S., Lin, Q., Ran, J., Musa, S.S., Yang, G., Wang, W., Lou, Y., Gao, D., Yang, L., He, D. and Wang, M.H., 2020. Preliminary estimation of the basic reproduction number of novel coronavirus (2019-nCoV) in China, from 2019 to 2020: A data-driven analysis in the early phase of the outbreak. *International Journal of Infectious Diseases*, 92, pp. 214–217.

Zheng, Y.Y., Ma, Y.T., Zhang, J.Y. and Xie, X., 2020. COVID-19 and the cardiovascular system. *Nature Reviews Cardiology*, 17(5), pp. 259–260.

Zhou, C., 2020. Evaluating new evidence in the early dynamics of the novel coronavirus COVID-19 outbreak in Wuhan, China with real time domestic traffic and potential asymptomatic transmissions. *medRxiv*.

Zhou, F., Yu, T., Du, R., Fan, G., Liu, Y., Liu, Z., Xiang, J., Wang, Y., Song, B., Gu, X. and Guan, L., 2020. Clinical course and risk factors for mortality of adult inpatients with COVID-19 in Wuhan, China: A retrospective cohort study. *The Lancet*, 395(10229), 1054–1062.

9 A Deep Learning CNN Model for Genome Sequence Classification

Hemalatha Gunasekaran, K. Ramalakshmi, Shalini Ramanathan, and R. Venkatesan

CONTENTS

9.1 INTRODUCTION

The world has seen many types of virus, such as Ebola, Nipah, SARS, MERS, influenza B, rotavirus, hepatitis, HIV, and coronavirus. Some newly discovered viruses have been recognized in the recent past because of the sudden appearance of a new disease. Therefore, the classification of a virus's genomic sequence is very important for virus subtyping and taxonomy classification problems. A virus has either a DNA or an RNA genome and is called a DNA virus or an RNA virus, respectively.

9.1.1 DNA AND ITS STRUCTURE

Deoxyribonucleic acid (DNA) is the blueprint of any living organism. DNA is made up of four nucleotides—namely, adenine (A), cytosine (C), guanine (G), and thymine

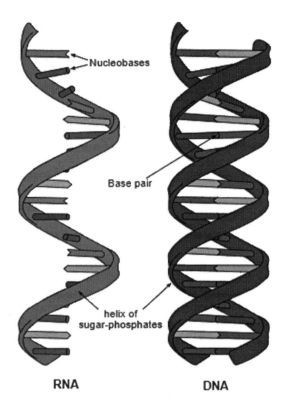

FIGURE 9.1   DNA and RNA Structure.

(T). These are known as the building blocks of DNA. DNA appears as single-stranded or double-stranded, as shown in Figure 9.1. In double-stranded, each type of nucleotide attaches with its complementary pair on the opposite strand. Adenine pairs with thymine, and cytosine pairs with guanine. Ribonucleic acid (RNA) can also be single- or double-stranded. In RNA, uracil (U) replaces thymine (T). Therefore, the genome is a sequence of nucleotides A, C, G, T for a DNA virus, and A, C, U, G for an RNA virus (Stephen and Kari, 2018).

Correct classification and typing of genomic sequences may assist in enhancing phylogenetic and functional studies of viruses,. The classification of genomic sequences assigns a sequence to a category of recognized sequences with similar properties, features, or characteristics. The more we are aware of the virus, the more we have the ability to create a vaccine to protect humankind.

The classification of viruses is difficult, however, as their genetic sequences may have little in common with viral genomes already known and available in reference databases. Various methods are available to classify the genomic sequence. The methods are broadly classified as (1) alignment and (2) alignment-free methods. The alignment process includes (1) general-purpose similarity search tools such as BLAST and USEARCH, and (2) organism-specific tools such as REGA and

SCUEAL for the classification of human immunodeficiency virus 1 (HIV-1). However, there are many limitations in using alignment-based methods; they use more memory and time. When used in a large database, they create a real bottleneck. Similarly, when the virus has higher mutation rates, it is difficult to classify using an alignment-based method. Alignment also requires the adaptation of a number of parameters (substitute matrices, gap penalties, statistical parameters, threshold values, etc.) which depend on a priori knowledge of the sequence to be compared. Therefore, changing these parameters is often arbitrary and requires a trial-and-error technique. Many studies have shown that the consistency of the alignment can be significantly influenced by minor variations in these parameters (Dylan et al., 2019). To overcome the above limitations, alignment-free methods are gaining more popularity in genomic sequence classification. In these methods, features are extracted from the DNA sequence and machine-learning models are used to classify the genomic sequence.

In this chapter, we discuss how machine-learning (ML) techniques can be used in the field of virology. ML and natural language processing (NLP) have recently gained much interest, mostly in the field of classification. We compare the accuracy of different supervised classification models, such as Bayes, random forest, support vector machine (SVM), and decision table, with an NLP classification model using k-mer. Deep learning (DL) is one type of advanced ML algorithm that learns features from the input and predicts the output using a deep artificial neural network (ANN). In this chapter, we develop a convolutional neural network (CNN) model to classify the genomic sequence. We find that NLP classification accuracy is higher than the state-of-the-art classification model, and that the other models have been outperformed by deep neural network models.

9.2 LITERATURE SURVEY

In pathogen identification, epidemiological surveys, and evolutionary studies, virus sequence classification plays a vital role. We will address the latest ML techniques used for the classification of genomes in this section.

Nurul and Abdullah (2017) proposed a CNN approach to classify the hepatitis B virus to identify the genetic marker for liver cancer. The author obtained an accuracy of more than 90% in the training dataset. The proposed model obtains the same accuracy with different sequence length. The author compared the proposed model with other types of classifiers, such as SVM, neural network, decision tree (DT), naïve Bayes (NB), nonlinear integral classifier, and rule learning.

Amine and Diallo (2019) used a linear classifier to classify the viruses. In his dissertation, the author updated many linear classifier parameters, such as the types of classifiers (generative and discriminatory), their hyper-parameters (smoothing value and regularization penalty function), the role of classification (genotyping and sub-typing), the sequence duration (partial and complete), and the length of k-mer words.

Riccardo et al. (2016) used the LeNet-5 network for DNA sequence classification, which was originally used for speech detection, character recognition, and image processing. In this work the author converted the DNA sequence into a fixed sequence length using the k-mer counting method. The author proposed a CNN architecture that is capable of identifying the feature specified in the k-mer and successfully classified the sequence. The accuracy of the classification outperformed all the previously defined models.

Fahad et al. (2019) used the DNA sequences of cancer patients and diagnosed the cancer easily. The author used different classifiers such as ANN, k-nearest neighbors, DT, fuzzy classifier, NB classifier, random forest, and SVM, and compared the models based on accuracy, recall, and other parameters.

Stephen and Kari (2018) proposed a stand-alone open source software KAMERIS, which operates on k-mer portions of DNA sequences. In this work, the DNA sequence is converted into a vector for every possible k-mer (for every length k). These vectors are called feature vectors and are used for DNA classification. The author used this software for HIV-1 virus subtyping and bacterial taxonomic classification. The author found that this method is much more efficient than other methods that use complex distance functions or correlation metrics, or methods that require profound biological knowledge.

Nguyen et al. (2016) used a CNN for the classification of a DNA sequence. In this work, the DNA sequence is converted into a sequence of word size 3 nucleotides. Through this, the author created a dictionary of size 64 different words. Each word is represented as a one-hot vector, which preserves the positional information of each nucleotide in the sequence. This matrix is given as an input to the convolutional layer for classification. This model contains four convolutional layers followed by a sub-sampling layer, three fully connected layers, and the output layer.

Amani and El Allali (2019) proposed a DL technique to classify a metagenome in a raw DNA sequence. The CNN-MGP is capable of extracting the characteristics of coding and non-coding regions with the help of GC content in the DNA sequence. In this work, the coding and non-coding regions are encoded as numerical values and given as an input to the CNN for prediction. The author evaluated this model by testing it with ten different CNN models with ten mutually exclusive datasets. The CNN-MGP model achieved an accuracy of around 91% when compared to other state-of-the-art classification algorithms.

Riccardo et al. (2016) proposed a DL neural network for DNA sequence classification based on spectral sequence representation. The author used some important k-mers and represented the sequence in a fixed dimension like spectral representation. The author tested the proposed model with two different sequence lengths, one with full sequence length and other with length less than 500 bp. The author compared the accuracy, recall, and precision of the proposed model with the general regression neural network algorithm with different distance functions. The author found that the sequence with lower length obtained good precision and recall values and also noted that the model with a large kernel value performed poorly.

9.3 PROPOSED METHOD

DNA sequence classification is a supervised ML problem. This classification problem is stated as the task of approximating a mapping function (f) from input variables (X) to discrete output variables (Y), where X is the DNA sequence of different viruses, and Y is the species label in which the input data X has to be classified

TABLE 9.1

Label Encoding

DNA letter sequence (TEXT)	DNA letter sequence (Numeric)
A	0
C	1
G	2
T	3

(Greiet and Jasper 2019). Supervised classification methods, such as Bayes net, random forest, random tree, logistic model tree (LMT), locally weighted learning (LWL), and DT, are used for DNA sequence classification. For all these algorithms, the input and the output variable should be of numerical type. Therefore, data preprocessing has to be done on the input dataset before applying the ML model. The preprocessing technique, such as one-hot encoding or label encoding, is used to convert the DNA dataset into numerical type.

Label Encoder

This is a quite simple and straightforward approach where each value in a column is converted into a number. In the case of a DNA dataset, it includes nucleotide sequences (A, C, G, and T). For each nucleotide, unique numerical values are assigned, as defined in Table 9.1.

The problem with this approach is that the numeric value assigned to each nucleotide makes the algorithm interpret that there is some ordering between each type, which is not true. One-hot encoding overcomes the limitation of this approach.

The Figure 9.2 shows the virus dataset in pandas DataFrame format. The column sequence contains the genomic sequence of different viruses such as SARS, COVID, and MERS. The column label contains the class of each genomic sequence. In this example, there are three classes: 0–COVID, 1–MERS, and 2–SARS.

ML models cannot be applied to data in string format. So the sequence given in Figure 9.2 is converted into individual nucleotides and then a label-encoding technique is applied to convert the sequence into numeric value, as shown in Figure 9.3.

One-Hot Encoder

In this method, each category value is converted into a vector of all zeros except one in a specific position. A is encoded as [1,0,0,0], T as [0,0,0,1], C as [0,1,0,0], and G as [0,0,1,0] as shown in Figures 9.4 and 9.5 representing one-hot encoding for a raw DNA sequence.

index		sequence	label	len
0	0	ATTAAAGGTTTATACCTTCCCAGGTAACAAACCAACCAACTTTCGA...	0	29903
1	26	TTTAAAGGTTTATACCTTCCCAGGTAACAAACCATTCAACTTTCGA...	0	29903
2	28	ATTAAAGGTTTATACCTTCCCAGGTAACAAACCAACCAACTTTCGA...	0	29903
3	29	TTGGTTGGTTTATACCTTCSCAGGTAACAAACCAACCAACTTTCGA...	0	29903
4	52	ATTAAAGGTTTATACCTTCCCAGGTAACAAACCAACCAACTTTCGA...	0	29903
...	...		...	...

FIGURE 9.2 Virus Dataset.

```
['C' 'G' 'C' 'G' 'A' 'T' 'C' 'A' 'A' 'A' 'A' 'C' 'A' 'A' 'C' 'G' 'T' 'C'
 'G' 'G' 'C' 'C' 'C' 'C' 'A' 'A' 'G' 'G' 'T' 'T' 'T' 'A' 'C' 'C' 'C' 'A'
 'A' 'T' 'A' 'A' 'T' 'A' 'C' 'T' 'G' 'C' 'G' 'T' 'C' 'T' 'T' 'G' 'G' 'T'
 'T' 'C' 'A' 'C' 'C' 'G' 'C' 'T' 'C' 'T' 'C' 'A' 'C' 'T' 'C' 'A' 'A' 'C'
 'A' 'T' 'G' 'G' 'C' 'A' 'A' 'G' 'G' 'A' 'A' 'G' 'A' 'C' 'C' 'T' 'T' 'A'
 'A' 'A' 'T' 'T' 'C' 'C' 'C' 'T' 'C' 'G' 'A' 'G' 'G' 'A' 'C' 'A' 'A' 'G'
 'G' 'C' 'G' 'T' 'T' 'C' 'C' 'A' 'A' 'T' 'T' 'A' 'A' 'C' 'A' 'C' 'C' 'A'
 'A' 'T' 'A' 'G' 'C' 'A' 'G' 'T' 'C' 'C' 'A' 'G' 'A' 'T' 'G' 'A' 'C' 'C'
 'A' 'A' 'A' 'T' 'T' 'G' 'G' 'C' 'T' 'A' 'C' 'T' 'A' 'C' 'C' 'G' 'A' 'A'
 'G' 'A' 'G' 'C' 'T' 'A' 'C' 'C' 'A' 'C' 'G' 'A' 'C' 'G' 'A' 'A' 'T' 'T' 'C'
 'G' 'T' 'G' 'G' 'T' 'G' 'G' 'T' 'G' 'A' 'C' 'G' 'G' 'T' 'A' 'A' 'A' 'A'
 'T' 'G' 'A' 'A' 'A' 'G' 'A' 'T' 'C' 'T' 'C' 'A' 'G' 'T' 'C' 'C' 'A' 'A'
 'G' 'A' 'T' 'G' 'G' 'T' 'A' 'T' 'T' 'T' 'C' 'T' 'A' 'C' 'T' 'A' 'C' 'C'
 'T' 'A' 'G' 'G' 'A' 'A' 'C' 'T' 'G' 'G' 'G' 'C' 'C' 'A' 'G' 'A' 'A' 'G'
 'C' 'T' 'G' 'G' 'A' 'C' 'T' 'T' 'C' 'C' 'C' 'T' 'A' 'T' 'G' 'G' 'T' 'G'
 'C' 'T' 'A' 'A' 'C' 'A' 'A' 'A' 'G' 'A' 'C' 'G' 'G' 'C' 'A' 'T' 'C' 'A'
 'T' 'A' 'T' 'G' 'G' 'G' 'T' 'T' 'G' 'C' 'A' 'A' 'C' 'T' 'G' 'A' 'G' 'G'
 'G' 'A' 'G' 'C' 'C' 'C' 'T' 'G' 'A' 'A' 'T' 'A' 'C' 'A' 'C' 'C' 'A' 'A'
 'A' 'A' 'G' 'A' 'T' 'C' 'A' 'C' 'A' 'T' 'T' 'G' 'C' 'A' 'C' 'C' 'C' 'C'
 'G' 'C' 'A' 'A' 'T' 'C' 'C' 'T' 'G' 'C' 'T' 'A' 'A' 'C' 'A' 'A' 'T' 'G'
 'C' 'T' 'G']
[1 2 1 2 0 3 1 0 0 0 0 1 0 0 1 2 3 1 2 2 1 1 1 1 0 0 2 2 3 3 3 0 1 1 1 0 0
 3 0 0 3 0 1 3 2 1 2 3 1 3 3 2 2 3 3 1 0 1 1 2 1 3 1 3 1 0 1 3 1 0 0 1 0 3
 2 2 1 0 0 2 2 0 0 2 0 1 1 3 3 0 0 0 3 3 1 1 1 3 1 2 0 2 2 0 1 0 0 2 2 1 2
 3 3 1 1 0 0 3 3 0 0 1 0 1 1 0 0 3 0 2 1 0 2 3 1 1 0 2 0 3 2 0 1 1 0 0 0 3
 3 2 2 1 3 0 1 3 0 1 1 2 0 0 2 0 2 1 3 0 1 1 0 2 0 1 2 0 0 3 3 1 2 3 2 2 3
 2 2 3 2 0 1 2 2 3 0 0 0 0 3 2 0 0 0 2 0 3 1 3 1 0 2 3 1 1 0 0 2 0 3 2 2 3
 0 3 3 3 1 3 0 1 3 0 1 1 3 0 2 2 0 0 1 3 2 2 2 1 1 0 2 0 0 2 1 3 2 2 0 1 3
 3 1 1 1 3 0 3 2 2 3 2 1 3 0 0 1 0 0 0 2 0 1 2 2 1 0 3 1 0 3 0 3 2 2 2 3 3
 2 1 0 0 1 3 2 0 2 2 2 0 2 1 1 1 3 2 0 0 3 0 1 0 1 1 0 0 0 0 2 0 3 1 0 1 0
 3 3 2 2 1 0 1 1 1 2 1 0 0 3 1 1 3 2 1 3 0 0 1 0 0 3 2 1 3 2]
```

FIGURE 9.3   Sequence Transformed into Vector.

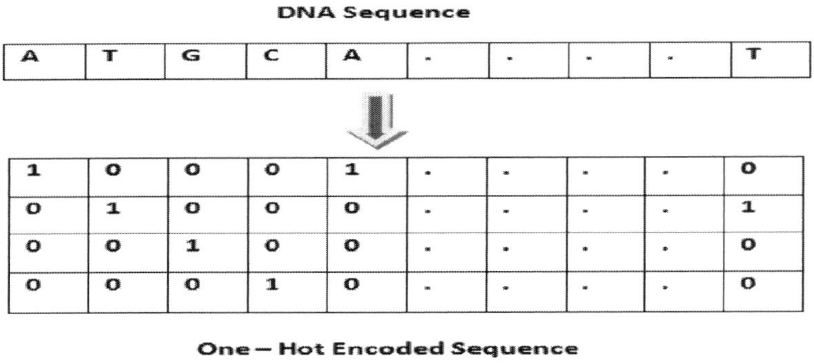

FIGURE 9.4   One-Hot Encoding.

9.3.1   ML CLASSIFICATION MODELS

Random forest

Random forest works with the concept of DT and contains multiple individual DT models. Row sampling and feature sampling are done on the dataset and fed to individual DTs. This process of sampling the data and feeding to the DT model is called a bootstrap. Each DT predicts the class output; the majority class prediction will be the output of the random forest. This process of selecting the majority vote is named as bagging. In the random forest, the base learners are DTs. The high variance

```
{0: array([[0., 1., 0., 0.],
           [0., 0., 1., 0.],
           [0., 1., 0., 0.],
           ...,
           [0., 1., 0., 0.],
           [0., 0., 0., 1.],
           [0., 0., 1., 0.]]), 1: array([[0., 1., 0., 0.],
           [0., 0., 1., 0.],
           [0., 1., 0., 0.],
           ...,
           [0., 1., 0., 0.],
           [0., 0., 0., 1.],
           [0., 0., 1., 0.]]), 2: array([[0., 1., 0., 0.],
           [0., 0., 1., 0.],
           [0., 1., 0., 0.],
           ...,
           [0., 1., 0., 0.],
           [0., 0., 0., 1.],
           [0., 0., 1., 0.]]), 3: array([[0., 1., 0., 0.],
           [0., 0., 1., 0.],
           [0., 1., 0., 0.],
           ...,
           [0., 1., 0., 0.],
           [0., 0., 0., 1.],
           [0., 0., 1., 0.]]), 4: array([[0., 1., 0., 0.],
           [0., 0., 1., 0.],
           [0., 1., 0., 0.],
           ...,
```

FIGURE 9.5   One-Hot Encoding of DNA Sequence.

problem in DT can be avoided by using random forest as only the majority of the vote is considered in random forest. The number of DTs is the hyper-parameter of the model.

Random tree

DT is restricted to decision making when compared with random forest. Each decision node is treated as a subtree and hence the name random forest. Random forest gives multiple possible decisions on a larger dataset with better accuracy. The independent decision nodes are involved in the collective decision-making process. If the classification problem has a continuous variable instead of a binary target variable then the random forest is called a regression tree. When a regression tree has categorical value then it is called a classification tree. Regression tree and classification tree both come under the classification and regression tree (CART) designation. A decision split in a regression tree is constructed by an attribute that returns the highest standard deviation reduction (i.e. the most homogeneous branches). By pruning the branches, the overfitting problem is controlled.

LMT

This is a type of instance-based learning called lazy learning. Linear regression and tree induction are popular models for solving supervised learning problems. The benefits of both models are combined to form "model trees." Logistic regression

(LR) is implemented at the leaf node and the attribute selection is done in a natural way. The LogitBoost algorithm is used to generate an LR model at the nodes of the tree (Landwehr et al., 2005). The algorithm runs iteratively to fine-tune the attributes to complete the tree structure.

LWL

For locally weighted learning, LWL is a general algorithm. Using an instance-based method, it assigns weights and creates a weighted instance classifier. Different classifiers can be selected, but for classification issues and linear regression for regression problems, NB is a good choice.

DT

DT is a supervised ML algorithm suitable for solving regression and classification problems (Breiman 2001). Entropy is the measure to understand the impurity or homogeneity of our dataset. Since the data has class labels, the DT is constructed through the training set to maximize the purity among the class at the leaf nodes. The impurity reduction from parent to child is maximum. The information gain is the measure of reduction in entropy after the split of nodes.

9.3.2 NLP Technique

K-mer counting

K-mer counting is an NLP technique that works by establishing a vocabulary of words. In the case of a DNA sequence, the vocabulary contains every possible k-mer of maximum length k, using the four letters of the DNA alphabet (A, C, T, G). The vocabulary also includes the shortest k-mer of size 1. The size of the vocabulary is $V = \Sigma_{i=1}^{k} 4^i$. After creating the vocabulary, every sample in the dataset is split into its respective k-mers but with a sliding window of size k moving over the DNA sequence, one step at a time. Figure 9.6 shows the k-mer of size 4 applied over the input DNA sequence.

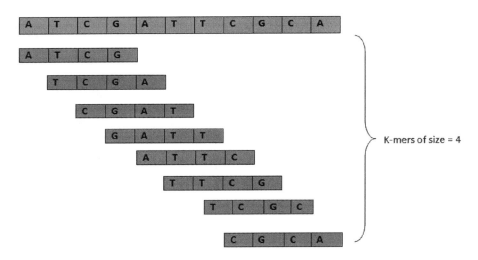

FIGURE 9.6 K-mers (size=4).

	label	words
0	MERS	[atttaa, tttaag, ttaagt, taagtg, aagtga, agtga...
1	MERS	[gattta, atttaa, tttaag, ttaagt, taagtg, aagtg...
2	MERS	[atgata, tgatac, gataca, atacac, tacact, acact...
3	MERS	[tatgcc, atgcct, tgccta, gcctaa, cctaac, ctaac...
4	MERS	[tatgcc, atgcct, tgccta, gcctaa, cctaac, ctaac...

FIGURE 9.7 K-mers (size=6) for DNA of MERS Virus.

For each sample in the dataset, it counts how many times each word in the vocabulary appears for the current sample and divides the count by the length of sample sequence. This way every sample in the dataset is transformed into a 1D numerical array with length equal to the size of the vocabulary. This method can be used to convert samples of different lengths to samples of equal length. Its use stems from the bag-of-words model used in NLP. Figure 9.7 shows the k-mers of size 6 for the dataset given in Figure 9.7.

9.3.3   DL MODEL FOR CLASSIFICATION

The CNN architecture as shown in Figure 9.8 contains an input layer, an output layer, sequences of one or more convolutions and the pooling layers, and finally a fully connected layer. Learning takes places in the convolutional layers by looking at the patterns in the data. The more complex the data, the more layers are required to learn the features of the data.

9.3.3.1   Convolutional Layer

The convolutional layer is responsible for extracting the important features of the data. In this layer, a kernel of size F slides across the entire width and height of the input

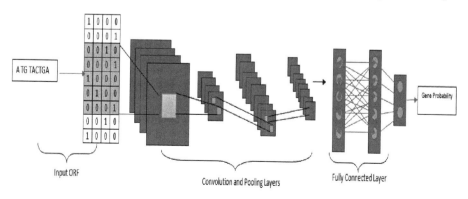

FIGURE 9.8 High-level Overview of a CNN for Use with One-hot Encoded DNA Data (Amani and El Allali, 2019).

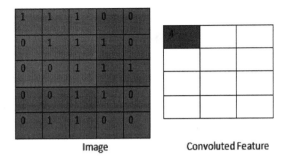

Image   Convoluted Feature

FIGURE 9.9   Kernel Size 3 × 3 Sliding over the Input.

volume V in steps of size S to extract the convolved feature. Figure 9.9 shows a filter/ kernel of size 3 × 3 matrix (marked in color yellow) that moves around the input in step size one, resulting in a dot product to produce the convolved feature. The kernel size F and the step size S are the hyper-parameter of CNN, that can be independently adjusted and becomes the weight of the layer. For every slide, the kernel performs on the input; it makes the dot product between the kernel and input and creates one value exactly for that particular position. As the kernel slides on the entire input, it produces a 2D feature map. Similarly, other filters are used to extract the different features from the input. Each kernel/filter produces a 2D feature map, the maps then being stacked along the depth dimension resulting in a 3D output of depth K, where K is equal to the number of kernels/filters applied to the input. This K is also the hyper-parameter of CNN. When the value of the step size S is larger than one, the filter will skip some neighboring values during the computation of the dot product. Sometimes the filters will not be applied to the edges or corners of the input volume. To overcome this problem, the zero-padding P is added to the input volume to ensure the filter is applied to the corner and edges. The zero-padding P is also the hyper-parameter of CNN.

The step size S, zero-padding P, filter/kernel size F, and input volume V determine the output volume W.

9.3.3.2   Pooling Layer

The main purpose of this layer is to reduce the size of the output volume and to avoid overfitting. In this layer a kernel/filter size F slides in step size F over the entire feature map and applies a function to the feature map. The function can be average or L2-norm or max where maximum is the most commonly applied function during which only the maximum value is retained over all inputs. In Figure 9.10, a filter size of 2 × 2 matrix is moved over the input and a function maximum is applied to retain the maximum value read along the entire input volume.

9.3.3.3   Fully Connected Layer

The sequence of convolutional layer and max-pooling layer is followed by the flatten layer. The main role of the flatten layer is to convert the 3D arrangement of the convolutional and pooling layer to a 1D vector. The layer takes the 3D output of the last convolutional layer and puts in a 1D vector by just stacking the feature maps one

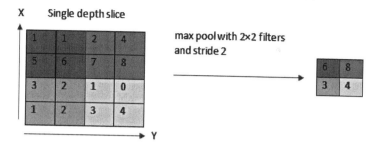

FIGURE 9.10 Max-Pooling Layer.

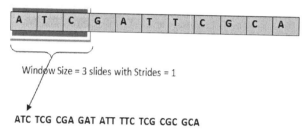

FIGURE 9.11 Translating DNA Sequence into Sequence of Words.

after the other. One or more fully connected layers are added to pass the output to the next fully connected layer. When the last fully connected layer is reached, a prediction is made by the network and outputted.

The input for the convolutional layer should be in the form of a numeric vector. Therefore, the DNA sequence is converted into a 1D numerical vector using a label encoder, or converted into a 2D numerical vector using a one-hot encoder. The DNA sequence is converted into a sequence of words by using the k-mer counting method. A window of fixed size strides across the DNA sequence to obtain the sequence of words, as shown in Figure 9.11.

The words obtained through k-mer counting are converted into one-hot encoding, as shown in Figure 9.12. After the DNA sequence has been converted into a 2D numerical vector, it is given as input to the convolutional layer to obtain the features of the dataset. This extracted feature is used to classify the unknown sequence.

The CNN is a layered architecture, and is built using the Keras library in Python. The model used in our method is sequential. In this work, we use a 1D convolutional model, as the DNA sequences are sequences of letters without spaces, unlike text data. Initially, the DNA sequence is converted into a numeric value, using one-hot encoding. To convert the text data into a numerical value, an embedding layer is added in Keras. This layer requires the input data to be integer encoded, so that each letter in the DNA sequence is represented by a unique number. The first convolutional layer is added with 128 filters, each of size 3 × 3 matrix with the ReLU

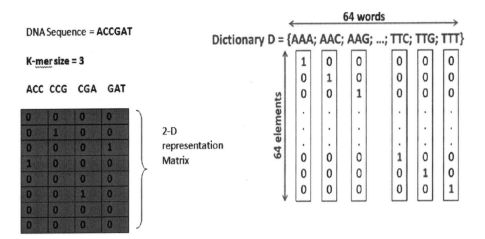

FIGURE 9.12   One-Hot Vector Representation of K-mer (Stephen and Kari, 2018).

```
Model: "sequential"
_____
Layer (type)                 Output Shape              Param #
=================================================================
embedding (Embedding)        (None, 32000, 8)          128
_____
conv1d (Conv1D)              (None, 32000, 128)        3200
_____
max_pooling1d (MaxPooling1D) (None, 16000, 128)        0
_____
conv1d_1 (Conv1D)            (None, 16000, 64)         24640
_____
max_pooling1d_1 (MaxPooling1 (None, 8000, 64)          0
_____
conv1d_2 (Conv1D)            (None, 8000, 32)          6176
_____
max_pooling1d_2 (MaxPooling1 (None, 4000, 32)          0
_____
flatten (Flatten)            (None, 128000)            0
_____
dense (Dense)                (None, 128)               16384128
_____
dense_1 (Dense)              (None, 64)                8256
_____
dense_2 (Dense)              (None, 3)                 195
=================================================================
Total params: 16,426,723
Trainable params: 16,426,723
Non-trainable params: 0
_____
None
```

FIGURE 9.13   Summary of CNN Architecture.

activation function. In the first layer, 128 different features are extracted from the dataset. The max-pooling layer of pool size 2 × 2 matrix is stacked between the two convolutional layers to reduce the size of the feature map. The second convolutional

```
Epoch 1/10
51/51 [==============================] - 32s 624ms/step - loss: 0.4179 - accuracy: 0.8494 - val_loss: 0.2299 - val_accuracy: 0.9089
Epoch 2/10
51/51 [==============================] - 32s 625ms/step - loss: 0.2035 - accuracy: 0.9351 - val_loss: 0.2016 - val_accuracy: 0.9176
Epoch 3/10
51/51 [==============================] - 32s 632ms/step - loss: 0.1635 - accuracy: 0.9515 - val_loss: 0.1592 - val_accuracy: 0.9413
Epoch 4/10
51/51 [==============================] - 33s 642ms/step - loss: 0.1368 - accuracy: 0.9605 - val_loss: 0.1601 - val_accuracy: 0.9476
Epoch 5/10
51/51 [==============================] - 33s 642ms/step - loss: 0.1209 - accuracy: 0.9654 - val_loss: 0.1644 - val_accuracy: 0.9476
Epoch 6/10
51/51 [==============================] - 33s 639ms/step - loss: 0.1087 - accuracy: 0.9679 - val_loss: 0.1570 - val_accuracy: 0.9587
Epoch 7/10
51/51 [==============================] - 33s 640ms/step - loss: 0.0981 - accuracy: 0.9725 - val_loss: 0.1245 - val_accuracy: 0.9663
Epoch 8/10
51/51 [==============================] - 33s 639ms/step - loss: 0.0902 - accuracy: 0.9730 - val_loss: 0.1338 - val_accuracy: 0.9625
Epoch 9/10
51/51 [==============================] - 33s 640ms/step - loss: 0.0859 - accuracy: 0.9750 - val_loss: 0.1263 - val_accuracy: 0.9644
Epoch 10/10
51/51 [==============================] - 33s 640ms/step - loss: 0.0795 - accuracy: 0.9763 - val_loss: 0.1259 - val_accuracy: 0.9663
```

FIGURE 9.14   Training of CNN.

layer has 64 filters of size 3×3 matrix with the ReLU activation function. The second max-pooling layer is added to reduce the dimensionality of the output. The last convolutional layer is added with 32 filters. The flatten layer is added to convert all the output into a 1D vector. Then two dense layers are added with 128 neurons and 64 neurons respectively, with the ReLU activation function. Finally, a dense layer is added with three neurons with the softmax activation function to predict the outcome. The CNN model summary is given in Figure 9.13 and the model is trained for ten epochs, as shown in Figure 9.14.

9.4   RESULTS AND DISCUSSIONS

9.4.1   INPUT

The virus dataset is downloaded from NCBI—a community portal for virus sequence datasets. The reference sequence genomes of MERS, SARS, and SARS-CoV-2 is downloaded in CSV format from the NCBI portal. The number of samples in MERS, SARS, and SARS-CoV-2 is around 5,000, 1,418, and 6,503 samples, respectively. Figures 9.15–9.17 show the reference sequence genomes of MERS, SARS, and SARS-CoV-2, respectively.

Figure 9.18 shows the length of the reference sequence genome. The maximum length of the sequence is 29,903.

All the ML models suffer from underfitting and overfitting issues. If the model cannot capture the structure available in the training dataset and cannot classify the unknown or unseen data then it is termed underfitting. When the model fails to generalize for the additional data, then it is termed overfitting. To overcome underfitting and overfitting, the training and test dataset has to be selected carefully. If the dataset is not very large, k-fold cross-validation is used to reduce the risk of overfitting.

	id	sequence	description	label	uid
0	NC_045512.2	ATTAAAGGTTTATACCTTCCCAGGTAACAAACCAACCAACTTTCGA...	NC_045512.2 |Severe acute respiratory syndrome...	NC_045512.2	AHemePPIw0
1	MT373156.1	CTCATACCACTTATGTACAAAGGACTTCCTTGGAATGTAGTGCGTA...	MT373156.1 |Severe acute respiratory syndrome ...	MT373156.1	Z770ftRV5B
2	MT373157.1	CTCATACCACTTATGTACAAAGGACTTCCTTGGAATGTAGTGCGTA...	MT373157.1 |Severe acute respiratory syndrome ...	MT373157.1	YiAZ60ZKD4
3	MT373158.1	CTCATACCACTTATGTACAAAGGACTTCCTTGGAATGTAGTGCGTA...	MT373158.1 |Severe acute respiratory syndrome ...	MT373158.1	2AkJsJiy88
4	MT373159.1	CTCATACCACTTATGTACAAAGGACTTCCTTGGAATGTAGTGCGTA...	MT373159.1 |Severe acute respiratory syndrome ...	MT373159.1	43XFAg3pNR

FIGURE 9.15   SARS-CoV-2 Reference Sequence Genome.

	id	sequence	description	label	uid
0	NC_038294.1	ATTTAAGTGAATAGCTTGGCTATCTCACTTCCCCTCGTTCTCTTGC...	NC_038294.1 |Betacoronavirus England 1| comple...	NC_038294.1	1UOMCsGUPv
1	NC_019843.3	GATTTAAGTGAATAGCTTGGCTATCTCACTTCCCCTCGTTCTCTTG...	NC_019843.3 |Middle East respiratory syndrome ...	NC_019843.3	nDYo11S2h8
2	MK910259.1	ATGATACACTCAGTGTTTCTACTGATGTTCTTGTTAACACCTACAG...	MK910259.1 |Middle East respiratory syndrome-r...	MK910259.1	swl4OZWflF
3	MN312749.1	TATGCCTAACATGTGTAGGATTTTCGCGTCTCTGATTTTGGCACGC...	MN312749.1 |Bat MERS-like coronavirus isolate ...	MN312749.1	PwDXWJiVce
4	MN312750.1	TATGCCTAACATGTGTAGGATTTTTGCATCTCTGATTTTGGCACGC...	MN312750.1 |Bat MERS-like coronavirus isolate ...	MN312750.1	Ba8lBjNSPe

FIGURE 9.16   MERS Reference Sequence Genome.

With this model, the dataset is divided into training and testing datasets with

	id	sequence	description	label	uid
0	NC_045512.2	ATTAAAGGTTTATACCTTCCCAGGTAACAAACCAACCAACTTTCGA...	NC_045512.2 |Severe acute respiratory syndrome...	NC_045512.2	a1dKA3H4LL
1	NC_004718.3	ATATTAGGTTTTTACCTACCCAGGAAAAGCCAACCAACCTCGATCT...	NC_004718.3 |SARS coronavirus| complete genome	NC_004718.3	k9GiTTjUGL
2	MT373156.1	CTCATACCACTTATGTACAAAGGACTTCCTTGGAATGTAGTGCGTA...	MT373156.1 |Severe acute respiratory syndrome ...	MT373156.1	Qurcolaih1
3	MT373157.1	CTCATACCACTTATGTACAAAGGACTTCCTTGGAATGTAGTGCGTA...	MT373157.1 |Severe acute respiratory syndrome ...	MT373157.1	ZV6WLmB8t2
4	MT373158.1	CTCATACCACTTATGTACAAAGGACTTCCTTGGAATGTAGTGCGTA...	MT373158.1 |Severe acute respiratory syndrome ...	MT373158.1	1m2jgZDzJ7

FIGURE 9.17   SARS Reference Sequence Genome.

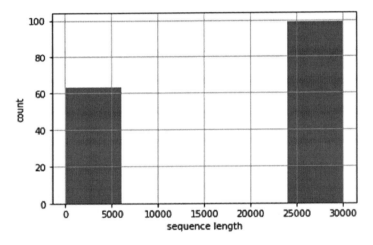

FIGURE 9.18   Length of the Reference Sequence Genome.

80/20% ratio. To obtain reliable classification results, the training dataset should contain an appropriate number of samples per species. Consequently, the testing dataset should contain only the DNA sequences of the same virus that is already presented in the training dataset.

9.4.2 Results

In this section, the proposed CNN model is compared with other classifiers, such as Bayes net, random forest, random tree, LMT, LWL, DT, and k-mer counting, to validate the improvement of classification. It clearly shows that CNN has the highest value of accuracy with 97.86%, followed by k-mer, random tree, random forest,

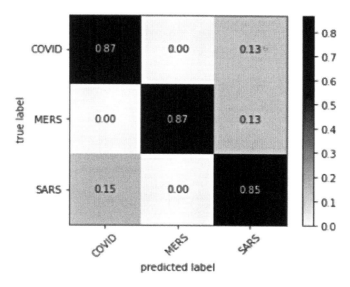

FIGURE 9.19 Confusion Matrix for K-mer Counting.

LWL, Bayes net, and decision table with 96.9%, 89.73%, 89.67%, 89.7%, 87.8%, and 82.5%, respectively.

The models are tested using precision, accuracy, and recall metrics. Precision is the ratio of the actual number of samples belonging to a certain class to the expected number of samples belonging to that class, showing how many of the results obtained are really relevant (Greiet and Jasper 2019). Recall demonstrates how the model correctly predicts several samples belonging to a certain class, and thus highlights how effective the model is to retrieve that class (Greiet and Jasper 2019).

$$\text{Accuracy} = \frac{TP + TN}{all} \quad \text{Precision} = \frac{TP}{TP + FP} \quad \text{Recall} = \frac{TP}{TP + FN}$$

where TP= True Positive, FP = False Positive, TN= True Negative, FN = False Negative.

K-mer Confusion Matrix
The confusion matrix for the classification of DNA sequences using the k-mer counting method is given in Figure 9.19.
CNN Confusion Matrix

The confusion matrix for the classification of DNA sequences using the CNN method is given in Figure 9.20.

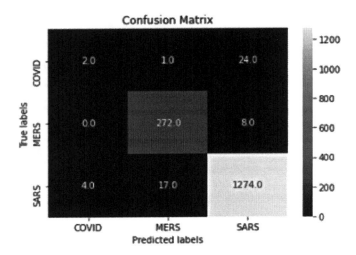

FIGURE 9.20 Confusion Matrix for CNN.

9.5 CONCLUSION AND FUTURE WORK

In our work, we compared the various ML models for DNA sequence classification. Since DNA is a sequence of character, it is considered as a text classification problem and the sequences are converted into numerical values using a one-hot or label encoder. Firstly, state-of-the-art ML models like random tree, random forest, LWL, Bayes net classifier, DT, and LMT are applied to classify the DNA sequence. We found that random tree provided the highest accuracy of around 89.73% when compared to other ML models, as shown in Table 9.2.

TABLE 9.2

Comparison of Different Classification Algorithms

Classifier	Correctly classified instances %	Incorrectly classified instances	Mean absolute error	Root mean squared error	Precision	Recall
Bayes net	87.8	12.1	0.21	0.281	0.894	0.878
Decision table	82.5	17.4	0.20	0.299	0.840	0.825
LWL	89.7	10.26	0.12	0.246	0.908	0.897
Random forest	89.67	10.32	0.13	0.24	0.905	0.897
Random tree	89.73	10.26	0.12	0.24	0.908	0.897
Multinomial NB Classifier (k-mer)	96.9	3.1	0.03	0.04	0.96	0.97

Secondly, we applied the NLP technique called k-mer counting to classify the DNA sequence. The k-mer counting technique also produced a better accuracy, of around 96.6%, when compared to state-of-the-art classification models.

Finally, we created a CNN to classify the DNA sequence and found that DL CNN outperforms the other ML methods with an accuracy of 97.86%. However, there are some limitations in our model; it requires the sequence to be of a uniform length. The other limitation is that the numbers of samples in each class are not uniformly distributed.

To improve the accuracy in future, we can use an ensemble model. An ensemble model is the process of combining two or more learning models to obtain a better prediction accuracy. In the ensemble model, the CNN can be used for feature extraction and the extracted features can be classified using recurrent neural networks.

REFERENCES

Amani, A.-A. and A. El Allali, CNN-MGP: Convolutional Neural Networks for Metagenomics Gene Prediction, *Interdisciplinary Sciences: Computational Life Sciences*, 11, 628–635, 2019.

Amine, M. R. and A. B. Diallo, *Statistical Linear Models in Virus Genomic Alignment-free Classification: Application to Hepatitis C Viruses*, IEEE International Conference on Bioinformatics and Biomedicine (BIBM) San Diego, CA, USA, 2019.

Breiman, L. Random Forests, *Machine Learning*, 45, 5–32 (2001). doi:10.102 3/A:1010933404324.

Dylan, L., A. M. Remita and A. B. Diallo, Toward an Alignment-Free Method for Feature Extraction and Accurate Classification of Viral Sequences, *Journal of Computational Biology*, 26, 6, 2019.

Fahad, H., U. Saeed, G. Muhammad, N. Islam and G. S. Sheikh, Classifying cancer patients based on DNA sequences using machine learning, *Journal of Medical Imaging and Health Informatics*, 9, 3, 436–443, March 2019.

Greiet, D. C. and Z. Jasper, Deep Learning for classification of DNA functional sequence, Master's thesis submitted to Ghent University in partial fulfilment of the requirements for the degree of Master of Science in Bioinformatics: Bioscience Engineering in the academic year 2018–2019)2019.

Hemalatha, G., R. M. Arokiaraj and K. Ramalakshmi, CNN Deep-Learning Technique to Detect Covid-19 using Chest X-Ray, *Journal of Mechanics of Continua and Mathematical Sciences*, 15, 9, 368–379, September (2020).

Landwehr, N., M. Hall and E. Frank, Logistic Model Trees, *Internation Conference on Machine Learning*, 59, 161–205 (2005).

Nguyen, N. G., V. A. Tran, D. L. Ngo, D. Phan, F. R. Lumbanraja, M. R. Faisal, B. Abapihi, M. Kubo and K. Satou, DNA Sequence Classification by Convolutional Neural Network, *Journal of Biomedical Science and Engineering*, 95), 280–286, 2016.

Nurul, A. K. and A. Abdullah, Classification of DNA Sequences Using Convolutional Neural Network Approach, *UTM Computing Proceedings Innovations in Computing Technology and Applications*, 2, 1–6, 2017.

Riccardo, R., A. Fiannaca, M. La Rosa and A. Urso, A Deep Learning Approach to DNA Sequence Classification, 9874, 129–140. doi:10.1007/978-3-319-44332-4_102016.

Stephen, S.-R. and L. Kari, DNA Sequence Classification: It's Easier Than You Think: An opensource k-mer based machine learning tool for fast and accurate classification of a variety of genomic datasets, Master's thesis submitted to The University of Western Ontario, 2018.

10 The Impact of Lockdown Strategies on COVID-19 Cases with a Confined Sentiment Analysis of COVID-19 Tweets

Tanzila Saba, Hind Alaskar, Dalyah Ajmal, and Erum Afzal

CONTENTS

10.1   INTRODUCTION

On December 31, 2019, the World Health Organization's (WHO) Regional Office in China was informed of pneumonia cases caused by an unknown disease discovered in Wuhan, Hubei Province, China. The new coronavirus was announced as the virus that had caused these cases by the Chinese authorities on January 7, 2020 (En.wikipedia, 2020). Within a relatively short period of time, a new coronavirus, originated in Wuhan, Hubei Province, in China, known as SARS-CoV-2, appeared and became a global health issue. WHO has declared it to be a global pandemic[1]. It is an unusual disease, due to its ability to infect people rapidly. It is highly transmissible between humans. It had infected over 79.2 million people worldwide, with over 1.7 million cases of death, by December 24, 2020 (Worldometers.info. 2021).

There are great efforts being made by countries to slow down the rapid spread of the disease. Due to the shortage of emergency rooms, effective vaccines, and other therapeutic medicines, countries' health systems have been affected. Precautions have been taken, and lockdowns in many countries worldwide have already been carried out to deal with the rapid spread. Other precautionary steps, such as curfews, working from home, and online schooling, have been enforced in different countries and cities to ensure social distancing. Despite the advanced health systems in many countries, it is difficult and cumbersome to accommodate many patients requiring intensive care. The situation has affected and strained the health system. Without vaccines and a proper cure, countries have adapted different strategies to flatten the pandemic curve and brace themselves for the epidemic.

This chapter analyzes the impact of lockdown strategies imposed by countries in the three categories (complete lockdown, partial lockdown, and herd immunity) by taking one country as an example for each category. For a complete lockdown strategy, the Kingdom of Saudi Arabia's (KSA) lockdown policies and their impact are analyzed and discussed. For a partial lockdown, the United Kingdom (UK) is considered, to understand the validity of their lockdown policies on the number of COVID-19 cases. Finally, Sweden is considered for the herd immunity strategy, to assess its effect on the number of COVID-19 cases. Table 10.1 summarizes the major lockdown dates for the specified countries.

The main contributions of the chapter are detailed below:

- Impact analysis of lockdown strategies—complete lockdown, partial lockdown, and herd immunity—on COVID-19
- A confined sentiment analysis of COVID-19 tweets from the perspective of these strategies.

This research's analytical approach is a comparative analysis based on COVID-19 graphs/charts generated using the JHU dataset (Anon, 2020a) as the data source. The charts in this chapter are generated using the Power BI tool. The method followed is to randomly choose one country from the three categories (complete lockdown, partial lockdown, and no lockdown) and analyze its statistics, using the percentage of change during and after the lockdown period.

TABLE 10.1
Lockdown Dates in Saudi Arabia, UK, and Sweden

Country	Lockdown		
	City/Region	Start Date	End Date
	Jeddah	29/03/2020	21/06/2020
	Mecca	26/03/2020	21/06/2020
Saudi Arabia	Medina	26/03/2020	21/06/2020
	Qatif	09/03/2020	21/06/2020
	Riyadh	26/03/2020	21/06/2020
	England	23/03/2020	04/07/2020
	Scotland	23/03/2020	29/06/2020
United Kingdom	Northern	23/03/2020	03/07/2020
	Wales	23/03/2020	13/07/2020
	Leicester	30/06/2020	24/07/2020
Sweden	NA	NA	NA

The further sections present analyses of COVID-19 control strategies for KSA, the UK, and Sweden respectively.

10.2 ANALYSIS OF KINGDOM OF SAUDI ARABIA (KSA)

10.2.1 LOCKDOWN POLICIES

Compared to other countries, KSA is particularly renowned for its religious tourism that makes the kingdom even more vulnerable to contagions from tourists. At various times of the year, KSA receives many people from all over the world to perform Hajj and Umrah and to visit religious places. The kingdom instituted strict blocking as early as possible. It has suspended many events and activities, including sports, religious tourism, and in-campus schools.

On February 2, 2020, Saudi Airlines suspended all flights between Riyadh, Jeddah, and Guangzhou, China. A few days later, on February 6, Saudi Arabia suspended citizens and residents' travel to the Republic of China (Awsat, 2020). Later, on February 27, KSA announced a temporary suspension of entry for individuals wishing to perform Umrah rituals in Makkah in Medina, as well as tourists. The order has also been expanded to include visitors coming from countries where the virus is spreading. On February 28, the Minister of Foreign Affairs of Saudi Arabia announced a temporary suspension of Gulf countries' citizens' permission to travel to Mecca and Medina. There were no reported cases of confirmed COVID-19 during February (Staff, 2020). On March 2, 2020, the Ministry of Health reported the first incident of COVID-19 in the kingdom, of a Saudi citizen returning from Iran via Bahrain (one of the Gulf countries) (Arab News, 2020).

In the following months, the kingdom had the maximum number of reported cases in the Arab countries of the Persian Gulf. On March 3, Saudi Arabia instigated precautionary health measures. Table 10.2 summarizes the timeline of the kingdom's precautionary health measures and lockdown dates.

As seen in Table 10.2, on March 21, the kingdom announced that all domestic and foreign travel would be suspended. Following the administration of curfews and lockdowns, the number of reported cases decreased significantly. All curfews had been lifted by the tri-phased program across the country except in the city of Mecca by June 21. The kingdom was seeing more regular recoveries than cases by mid-July. The Hajj was conducted with a small number of participants in the final week of July, and in first week of August with 10,000 socially distanced pilgrims.

TABLE 10.2

Measures Imposed by Saudi Arabian Government from February to September 2020

March (Arab News, 2020) (Saudigazette, 2020) (Nasrallah, 2020) (Al Arabiya English, 2020a) (Khudair, Deemah, 2020)

Mar. 4	Umrah is temporarily suspended for citizens and residents of the kingdom.
Mar. 5	Temporary daily closure of the Grand Mosque for sterilization purposes.
Mar. 8	Temporarily suspending entry and exit from Qatif Governorate, Saudi Arabia.
Mar. 9	The Ministry of Education announced the temporary suspension of studies in all regions and governorates of the kingdom.
Mar. 14	■ Temporarily suspending the travel of citizens and residents to United Arab Emirates, Kuwait, Bahrain, Lebanon, Syria, South Korea, Egypt, Italy, and Iraq, as well as suspending the entry of those coming from those countries, or the entry of those who were there during the 14 days before arrival.
	■ Stopping air and sea flights between the kingdom and the abovementioned countries, excluding evacuation, shipping, and trade trips.
Mar. 16	■ Suspending physical attendance at the workplace in all government agencies (public sector) for 16 days, except for some sectors.
	■ Closing open-air and covered markets and commercial complexes, except pharmacies, supermarkets, and food-supply activities.
	■ Preventing gathering in public places.
Mar. 17	■ Suspending Friday and congregational prayers for all obligatory prayers in mosques, except the Two Holy Mosques.
	■ Suspending physical attendance at the workplace in all private sectors for 15 days, except for some sectors.
Mar. 20	Suspension of all domestic flights, buses, taxis, and trains for 14 days.
Mar.21	Closure of shops from 8:00 p.m. until 6:00 a.m. except for some shops.
Mar. 24	The kingdom ordered a curfew at specific hours of the day. From 7:00 p.m. to 6:00 a.m., people cannot move outside their house unless a special case is declared.
Mar. 29	Extending the suspension of domestic and international flights and physical attendance at workplaces in private and public sectors until further notice.
April 2020 (Al Arabiya English, 2020b)	
Apr. 1	A 24-hour curfew in Makkah and Madinah's cities and prevention of entry into or exit from them until further notice.
Apr. 6	A 24-hour curfew in Riyadh, Tabuk, Dammam, Hafuf, Dhahran, Qatif, Jeddah, and Khubar until further notice.
Apr. 8	Curfew in all cities starts from 3:00 p.m. to 6:00 a.m.
Apr. 15	Children under the age of 15 are not allowed to enter any shopping stores except in some cases.
Apr. 26	During Ramadhan, curfew hours start from 5:00 p.m. until 9:00 a.m. except for Makkah and Madinah, where there is a 24-hour curfew.

(Continued)

TABLE 10.2
(Continued)

March (Arab News, 2020) (Saudigazette, 2020) (Nasrallah, 2020) (Al Arabiya English, 2020a)
(Khudair, Deemah, 2020)

May, 2020 (Tuqa, 2020) (Tommy, 2020)	
May 28	Changing the permissible times for movement for three days only in all cities of the kingdom to be from 6:00 a.m. until 3:00 p.m., except Makkah.
May 31	■ Changing the permissible times for movement for three weeks in all cities of the kingdom to be from 6:00 a.m. until 8:00 p.m., except Makkah.
	■ Allowing the conducting of Friday prayers and all obligatory prayers in the kingdom's mosques, except Makkah.
	■ Lifting the suspension of attendance at private and public agencies.
	■ Lifting the suspension of domestic flights.
June 2020 (WSJ, 2020)	
Jun. 21	Completely lifting the curfew and allowing activities to return, provided that gatherings do not exceed 50 persons.
Jun. 22	Hajj of 1411 AH year will be performed by a limited number of people from inside the kingdom.
September 2020 (WSJ, 2020)	
Sept. 13	The Ministry of the Interior announced the complete lifting of restrictions on citizens leaving and returning to the kingdom, only in special cases.

10.2.2 UNDERSTANDING THE STATISTICS OF CASES

On March 8, 2020, the Saudi Arabian government declared that it would temporarily stop transportation to and from the Qatif regional capital, although residents could enter the area. The Ministry of the Interior reported that everyone in the country with confirmed cases came from Qatif. On March 24, a national curfew with limited movement between 7:00 p.m. and 6:00 a.m. was implemented. On March 30, the provincial capital of Jeddah was placed under curfew by the Ministry of the Interior, and all travel to and from the city was suspended. Starting on April 2, Makkah and Madinah were subject to a 24-hour curfew. On April 6, 24-hour curfews were declared with only movement of vital importance being permitted between 6:00 a.m. and 3:00 p.m.

In addition, from June 6 to 20, Saudi Arabia reimposed regulations and a curfew in Jeddah. There was suspension of prayers for a few weeks in all the mosques of the city.

Figure 10.1 shows the total confirmed cases throughout the period in KSA. The figure clearly shows that, despite the lockdown measures, the number of confirmed cases was rising. However, as observed in Figure 10.2, it can be seen that, with the curfew measure, the number of recoveries was on the rise too.

According to cases reported by the Ministry of Health in Saudi Arabia, the daily number of cases peaked during June and July. It started to decrease slowly afterward; this may have resulted from Saudi Arabia's precautionary measures to deal with the spread of COVID-19. The daily reported cases are shown in detail in Figure 10.3. Saudi Arabia has been very successful in treating COVID-19 patients. Through the

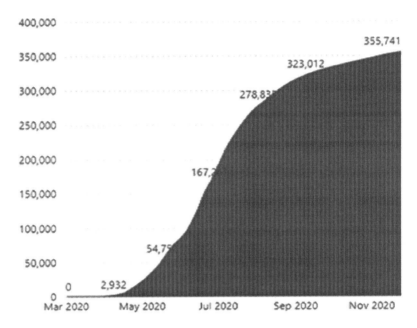

FIGURE 10.1 Total confirmed cases in Saudi Arabia.

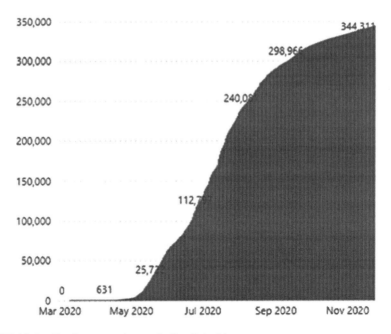

FIGURE 10.2 Total recovered cases in Saudi Arabia.

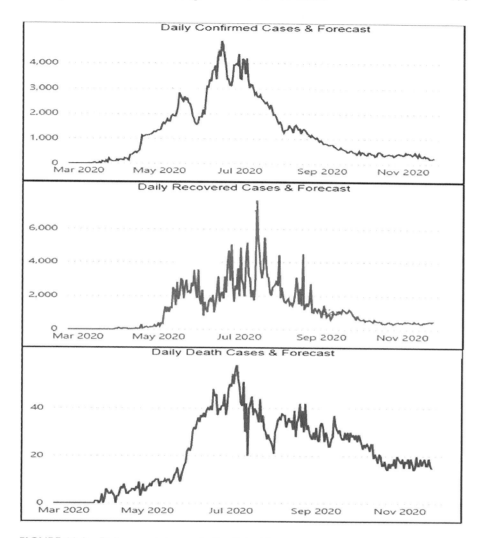

FIGURE 10.3   Daily reported cases in Saudi Arabia.

use of lockdown measures, an effort was made to lessen the number of people admitted to hospital. This resulted in a manageable number of people being admitted to hospital for special care due to COVID-19. The number of daily deaths was low compared to other countries.

10.2.3   VALIDATION OF THE COMPLETE LOCKDOWN STRATEGY IN SAUDI ARABIA

The government in the kingdom has been taking steps to limit and eradicate diffusion by closing schools and offices and enforcing day-long curfews in major cities. Alrashed (2020) predicted the following forecasts after lockdown, as shown in

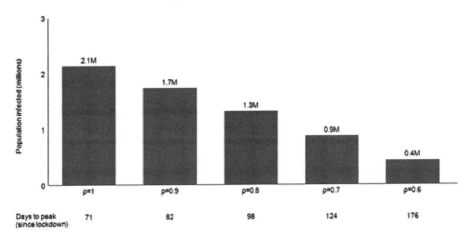

FIGURE 10.4   Predicted number of COVID-19 cases after the lockdown in Saudi Arabia.

TABLE 10.3
COVID-19 Summary Statistics for Saudi Arabia

Calculations as of November 24, 2020
Data source JHU

Recovery Rate	**96.79%**
Mortality Rate	1.63%
Total Confirmed Cases	355,741
Total Recovered Cases	344,311
Total Death Cases	5,811
Active Cases	5,619

Figure 10.4. The number of cases after lockdown is predicted to decrease over time. This is actually validated in the total number of real cases reported by the Ministry of Health daily.

Table 10.3 summarizes the mortality and recovery rate in the kingdom and summary statistics of COVID-19 results as of November 24, 2020. It can be clearly seen that Saudi Arabia has a high rate of recovery compared to other countries, and a low death rate.

Up to June 21, the total confirmed cases were 161,000, which then rose to 255,000 in a month, showing an increase of 58%. The deaths per day during lockdown peaked at 48, rising to 60 after lockdown; however, it declined later. The average new cases per day during lockdown were 2,853, rising to 3,142 during the first month after lockdown, showing a 10% increase; however, it declined later. Our research indicates that the peak would have arrived much sooner without the lockdown and that the health system around the peak would have been under immense pressure.

During lockdown, the exposed population can be protected. It should be noted that the essence of COVID-19 is more complicated, and current prediction trends must be broadened by taking variables, such as older population, birth/death rates, and herd immunity, into account in order to give more precise forecasts.

10.3 ANALYSIS OF THE UNITED KINGDOM (UK)

10.3.1 LOCKDOWN POLICIES

The UK did not impose a complete lockdown within the country but initiated some partial lockdowns by identifying those areas and regions where cases were rising. According to Anon (2020a, 2020b), on January 20, 2020, the first confirmed cases outside China were in Japan, South Korea, and Thailand, and a day later in the United States. However, Public Health England announced the risk level from COVID-19 to be "Low," though not "Very Low". The first two cases occurred in China on January 29, 2020, the people affected experiencing COVID-19-like symptoms and testing positive for the virus (BBC News, 2020a). The UK government advised its citizens to leave China as soon as possible. On March 13, some sporting events were postponed, including the London Marathon (BBC News, 2020b). Prime Minister Boris Johnson advised people to work from home, wear masks, ensure social distance, and avoid public places. However, the death cases rose to 55 and confirmed cases to 1,543 (Stewart, 2020).

The UK government also ordered the cessation of all non-essential travel and connections with people outside of their own homes, promoting the slogan "Stay Home, Protect the NHS, Save Lives" and shutting down almost every school, company, location, amenity, and place of worship (Conservative Party, 2020). People with COVID-19-like symptoms were advised to isolate themselves, while those with certain conditions were told to protect themselves by observing health measures. People were ordered to stay apart in public.

According to Aspinall (2020), the UK government ordered a lockdown for all restaurants, gyms, pubs, and other stores by March 20. Long-lasting restrictions were expected to significantly harm the UK's economy, exacerbate mental health issues and suicide rates, and cause more deaths due to loneliness, delays, and deteriorating living standards. Table 10.4 summarizes the timeline of the UK's precautionary health measures and lockdown dates.

10.3.2 UNDERSTANDING THE STATISTICS OF CASES

In late April, Prime Minister Boris Johnson said that the UK had overcome the height of its outbreak (UK ONS, 2020). Regular cases and deaths steadily decreased in May and June, and continued in July and August at a reasonably low pace. The estimated number of deaths in the UK was just over 65,000 from the beginning of the outbreak up to mid-June, as shown in Figure 10.5.

Cases increased considerably from late August, and a "tiered" lockdown was imposed in England. These restrictions evolved in late October, as more COVID-19-impacted countries in Europe responded to the substantial strain on their health systems, whether real or possible, by announcing national lockdowns or circuit breakers.

TABLE 10.4

Measures imposed by the UK from February to September 2020

March (Conservative Party, 2020) (The Guardian, 2020)

Mar. 20	**Cafés, pubs, and restaurants will be closed until further notice.**
Mar. 23	Partial Lockdown:
	■ People in the UK are permitted to leave their homes for restricted purposes, such as food-shopping, once-daily exercise, or medical needs.
	■ Traveling is only allowed for work.
	■ Shops for non-essential goods are closed.
	■ Gatherings of more than two persons are prohibited.
Mar. 24	The Church of England announced its closure.

May (Aspinall, 2020) (BBC News, 2020c)

May 10	There will be a phased plan for reopening schools in England.
	On June 1, it is planned to reopen for early years.
	On June 15, it is planned to reopen for Years 10 and 12.
May 22	Plan to impose a 14-day quarantine period from June 8 on individuals coming from abroad.
May 25	Outdoor markets, car showrooms, and sports events will open from June 1.
	All other non-essential stores will open from June 15.

June (Aspinall, 2020) (BBC News, 2020d)

Jun. 1	Schools reopen for children.
Jun. 15	■ People will be allowed to visit each other's houses, as part of showing support.
	■ It is compulsory to wear face masks on public transport.
	■ Non-essential shops, parks, zoos, and shopping malls are opened for the first time since their closure in March.
Jun. 25	The government announces that restaurants and pubs will be allowed to use outdoor spaces such as terraces.
Jun. 29	The government advise people in Leicester to stay at home as much as they can.

July (Aspinall, 2020) (Anon, 2020a) (Legislation UK Gov., 2020)

Jul. 4	■ Each local authority will have the power to declare further restrictions.
	■ Pubs, barbers, restaurants, and shops are reopened.
Jul. 13	The UK government orders a lockdown in some areas of England.
Jul. 14	The UK government announces a mask policy; it is compulsory to wear a mask in public starting from July 24.

September (Stretton, 2020)

Sept. 9	UK government restricts group gatherings to no more than six people.

10.3.3 VALIDATION OF PARTIAL LOCKDOWN STRATEGY IN THE UK

The new cases reported during the lockdown period fell considerably from 4,352 to 2,780 cases a day. Before the lockdown, the death rate rose around 1,000 per day. During the lockdown period, a second wave was seen, with new cases rising to more than 30,000 a day with a rising death rate during November. Coronavirus infections in England fell by around 30%, according to early findings from a spontaneous swab testing survey of analysts at Imperial College London after three weeks of a second nationwide lockdown (UK ONS, 2020).

The reduction in the number of infections is not even. "There is a country-wide mixed picture. In the north, but not so rapidly in the midlands, south of London, infections are dropping." This possibly represents the fact that large parts of the north of England were under restrictions until they were locked down. Table 10.5

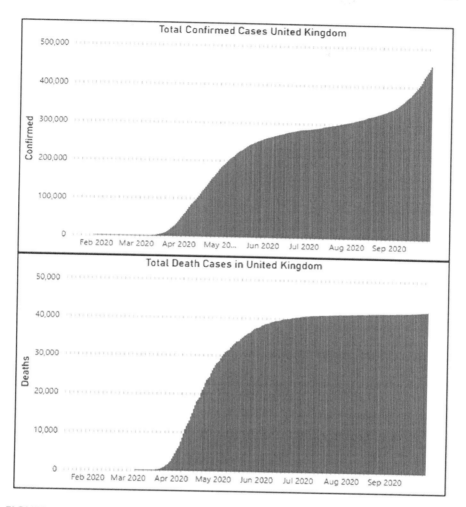

FIGURE 10.5 Total confirmed and death cases in the UK.

TABLE 10.5

COVID-19 Summary Statistics for the UK

Calculations as of November 24, 2020

Recovery Rate	NA
Mortality Rate	3.63%
Total Confirmed Cases	1,542,611
Total Recovered Cases	3,495 reported
Total Death Cases	55,953
Active Cases	1,483,271

summarizes the statistics for UK cases. The mortality rate is high compared to that of Saudi Arabia. Recovered cases were not reported. Thus, it is not possible to compare these statistics with rates of recovery elsewhere.

10.4 ANALYSIS OF SWEDEN

10.4.1 LOCKDOWN POLICIES

Unlike several other nations, Sweden has not instituted a lockdown and has kept large parts of its community free. The Swedish Constitution guarantees people's freedom of movement, and consequently a lockdown was not considered constitutional. Instead, a series of guidelines from the government agency responsible for this sector, the Swedish Public Health Agency, must be adopted by the Swedish public (BBC News, 2020e).

Following the Health Agency's recommendations, the government enacted laws restricting freedom of speech, barring gatherings of more than 50 people, prohibiting visits to nursing homes, and physically closing high schools and universities. Primary schools were left open, partly to discourage healthcare staff from staying at home with their children (Ludvigsson, 2020).

The Public Health Agency has made recommendations: to work from home, stop excessive country travel, engage in social distancing, and stay at home as much as possible for people over 70. Those with even mild symptoms that could be triggered by COVID-19 should remain at home. The government has abolished the initial day without paid sick leave, and the length of time you can stay at home with pay without a note has been increased from 7 to 21 days (Ludvigsson, 2020).

10.4.2 UNDERSTANDING THE STATISTICS OF CASES

The number of confirmed cases is increasing in Sweden. Figure 10.7 represents the daily confirmed cases. It can be seen that the number of cases reported has been increasing since late May. Due to the time limitation of this study, further months were not evaluated.

In Figures 10.6 and 10.7, it can be clearly seen that the number of cases has been increasing at a steeper rate as the curve is more vertical. This shows that no lockdown has increased the total number of cases.

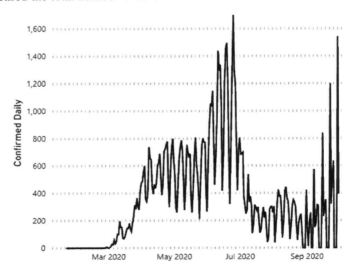

FIGURE 10.6   Daily confirmed COVID-19 cases in Sweden.

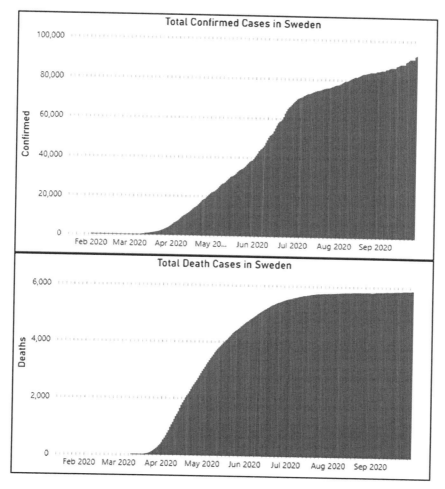

FIGURE 10.7   Total confirmed and death cases in Sweden.

10.4.3   VALIDATION OF NO LOCKDOWN STRATEGY IN SWEDEN

The pandemic has put the Swedish health system under serious pressure and has delayed tens of thousands of operations. Swedish hospitals and other facilities initially reported a lack of personal protective equipment. At the beginning of the pandemic, there were fears that Swedish hospitals would not be able to treat all those who might get ill, particularly those in need of intensive care. In a couple of weeks, the number of intensive care beds was finally doubled by Swedish hospitals, and full capacity was never surpassed (Ludvigsson, 2020).

Table 10.6 summarizes the mortality and recovery rate in Sweden and summary statistics of COVID-19 results as of November 24, 2020. It can be clearly seen that Sweden has a high death rate for the number of reported cases.

Throughout the period, the average number of cases rose by a maximum of 1,698 each day, which lowered eventually. During November, daily cases have risen to a

TABLE 10.6
COVID-19 summary statistics for Sweden

Calculations as of November 24, 2020
Data source JHU

Recovery Rate	NA
Mortality Rate	2.1%
Total Confirmed Cases	225,560
Total Recovered Cases	NA
Total Death Cases	6,500
Active Cases	219,060

maximum of 7,629 per day, showing that the herd immunity concept has not worked well in the second wave. The deaths per day initially rose to 70 and eventually come down close to 0. However, the cases are again rising in November, with an average death rate of 25 each day.

While Sweden was thought to have managed to ensure the hospitals kept pace, it did not protect its older people because most of its deaths have taken place in nursing homes or among people receiving home care.

10.5   A CONFINED SENTIMENT ANALYSIS OF COVID-19 TWEETS

The world is currently facing many changes and challenges due to the COVID-19 pandemic. Since the pandemic, countries have started enforcing many new rules to limit the spread of the virus. These rules affect people's lives, in such areas as how they interact with one another and how they carry on their daily activities. The pandemic enforced or proposed a new working and studying model in which people connect and work remotely. Governments have imposed many limitations on travel, tourism, and entertainment. People reacted differently to the new norm that they were forced to live with and adapt to. To understand how the public perceives the new rules enforced, a sentiment analysis study was applied on the Twitter platform. This is a widely used media platform for sharing opinions and reacting to changes.

Sentiment analysis classifies social data into positive, negative, and neutral, based on the sentiment/meaning they carry. For English tweets, the wordlist-based approach, Affin (Farha, 2019), was used for sentiment analysis. Mazajak (Nielsen, 2011), an online Arabic sentiment analyzer, was used to analyze Arabic tweets. Sentiments help understand the impact that COVID-19 has on the different aspects of people's lives.

The study's primary goal is to understand how people from different countries have reacted to the new way of living during the COVID-19 era. The countries selected for this study applied different policies to contain the spread of the virus. However, some shared rules are addressed by all. Therefore, this study focused on understanding people's opinions on the new way of living across three countries: Saudi Arabia, the United States (USA), and India. The study analyzed people's reactions to three main aspects introduced during the COVID-19 era: homeschooling, working from home, and social constraints such as wearing masks and social distancing.

10.5.1 Tools and Languages

A Twitter developer account was set up to get authorization to use Twitter APIs. The premium search API (free sandbox) was used to retrieve tweets for up to 30 days. The programming language used to conduct the analysis is Python, which is widely used in data science. On top of the language, the following Python packages were used to read tweets, apply sentiment analysis and visualize results: search tweets (PyPI, 2020), seaborn (Waskom et al., 2020), wordcloud (Anon, 2020b), Mazajak (Nielsen, 2011), and Affin (Farha, 2019).

10.5.2 Analysis Approach

10.5.2.1 Data Pulling

Around 2,000 tweets were pulled per each country and study aspect for a month-long period. Arabic tweets were retrieved to analyze KSA data, while English tweets were extracted for USA and India data. Tweets were scraped based on specific keywords and hashtags for each country and objective. Table 10.7 shows the keywords considered for each analysis group.

TABLE 10.7
Keywords Considered for Each Analysis Group

objective/ country	KSA	USA	India
Home schooling	بعد_عن# (عن بعد) المنزل_من_الدراسة# المنزل_من_االتعليم# (المنصة) (منصةعين) (لتعليميعن بعد) (لتعليممنالمنزل) (الدراسةفيالبيت)	#online #homeschooling #study_from_home #remote_leaning #virtual_learning #study_online #home_schooling	#online #homeschooling #study_from_home #remote_leaning #virtual_learning #study_online #home_schooling
Work from home	بعد_عن_العمل #العمل_من_المنزل#	#work_from_home #work_from_anywhere (work from home) #WFH (WFH) (wfh) (work online) (work remotely) (remote work) (remote job)	#work_from_home, #work_from_anywhere (work from home) #WFH (WFH) (wfh) (work online) (work remotely) (remote work) (remote job)
Social constraints	لبساللكمامة الكمامة #نسبنااها_ال_الكمامة #الاجتماعي_التباعد ضروري_لتعقيم# (التباعدالاجتماعي)	masks #wearmask #washyourhands #socialDistance #wearamask	masks #wearmask #washyourhands #socialDistance #wearamask

10.5.2.2 Data Preprocessing and Cleansing

Twitter data was preprocessed before applying the sentiment analysis for better accuracy. Entries with null tweets were excluded, then tweets were classified into Arabic and English to use the suitable analysis technique based on language. Then, tweets' texts were cleaned using NLP techniques: punctuation was removed, texts were tokenized, stop words were removed, and lemmatization was used to get the roots of words.

10.5.3 RESULTS & DISCUSSION

One of the main changes that most of the countries have enforced is homeschooling and studying remotely. Homeschooling impacts students and their families, so it was interesting to study the public's opinion about studying remotely and the different perceptions of it in each country. Sentiment analysis was performed on tweets tweeted between September 25 and October 20, 2020, filtered by homeschooling keywords and limited to 2,000 tweets per country.

In Saudi Arabia, most of the tweets were neutral. Only a small portion was negative, indicating that most people who tweeted are comfortable with homeschooling and agree with the rule applied to all schools in the country. Figure 10.8 shows the sentiment distribution of homeschooling tweets in KSA.

In the USA, sentiment distribution is somewhat similar to KSA's distribution. However, US tweets have more negative tweets than KSA ones. Figure 10.9 shows the sentiment distribution of homeschooling tweets in the USA.

In India, the sentiment distribution is very similar to the US distribution. The exact distributions of the sentiments are clear in Figure 10.10.

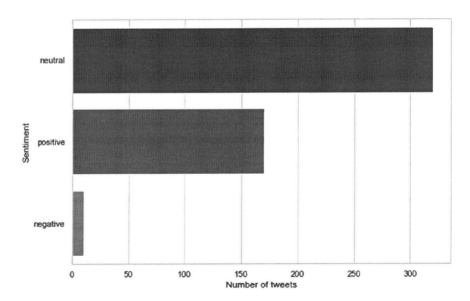

FIGURE 10.8 Sentiment distribution across homeschooling tweets in KSA.

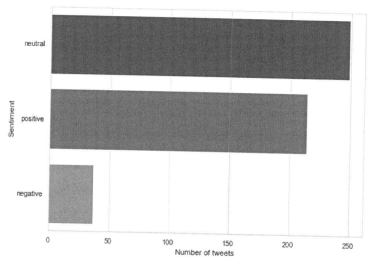

FIGURE 10.9 Sentiment distribution across homeschooling tweets in the USA.

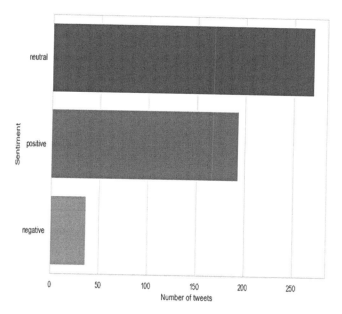

FIGURE 10.10 Sentiment distribution across homeschooling tweets in India.

As the neutral sentiment dominates in all three countries, we can say that there is a general acceptance of homeschooling by most people in the three countries.

Considering that the tweets were pulled at the end of September and the start of October, when people were accepting or adapting to the new changes, it is very

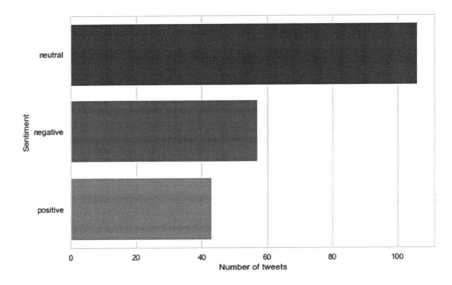

FIGURE 10.11   Sentiment distribution across homeschooling tweets.

normal to see many neutral sentiments and fewer negative ones. To better understand people's opinions when the virtual schools rule was enforced in KSA and see how people's opinions can change with time, related tweets at the time of the rule announcement were analyzed. Tweets were extracted using the trending hashtags at that time: '#العلم_الدراسي_الجديد,#', 'بعد_عن_المعلمات_المعلمين#' and 'العملية_المقررات#' from August 15 to 22, 2020.

Sentiments distribution was different in this period, with more negative sentiments appearing. Figure 10.11 shows the detailed sentiment distribution at this period.

One of the new models proposed in this era that has had a huge impact on people is working from home. Many governments and companies adopted remote working as a new way of working. People perceived and reacted to this change differently. Some enjoyed it, while others felt it left them disconnected and not motivated. Studying the sentiments on remote working tweets for KSA, the USA and India helped get an overview of opinions on the topic. Sentiment analysis was performed on tweets tweeted between September 25- and October 20, 2020, filtered by work from home keywords and limited to 2,000 tweets per country.

Analyzing KSA remote working tweets, we can see that most tweets are neutral or positive, meaning general acceptance and satisfaction about remote working. The huge volume of neutral tweets might also indicate that they are announcement tweets, or tweets with work from home tips. Figure 10.12 shows the exact sentiment distribution.

For the US and Indian results, we can notice a similar sentiment distribution to homeschooling tweets. The majority of tweets are neutral. However, there are some

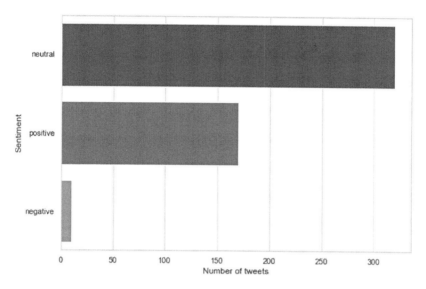

FIGURE 10.12 Sentiment distribution across work from home tweets in KSA.

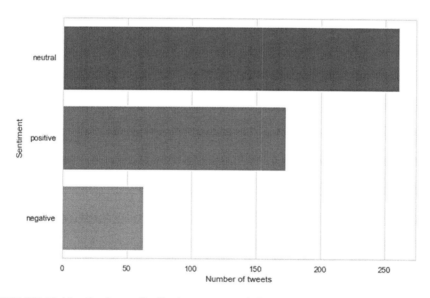

FIGURE 10.13 Sentiment distribution across work from home tweets in the USA.

with negative labels. This distribution reflects how people are reacting to the new model of working differently. Figures 10.13 and 10.14 give more insights about the sentiment distribution of US and Indian tweets.

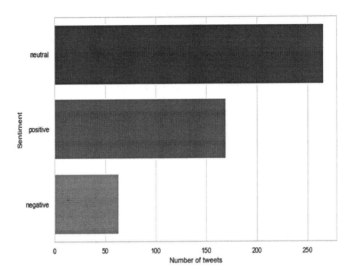

FIGURE 10.14   Sentiment distribution across work from home tweets in India.

New constraints on social interactions were introduced after the start of the COVID-19 pandemic. Social distancing and wearing masks are the common rules enforced to limit social interactions. To understand how the public reacted to these newly proposed rules, tweets with related keywords (refer to Table 10.7) were analyzed. KSA, the USA, and India were considered to study this objective. Around 2,000 tweets were extracted from October 16 to November 15, 2020, for each country.

The results obtained after applying sentiment analysis on KSA tweets were interesting. The majority of the sentiments were negative, which means people find it hard to practice social distancing and wear masks. However, a good percentage of neutral and positive tweets reflect another population that believes in the importance of the social constraints applied to limit the spread of the virus. Figure 10.15 shows the exact distribution of the sentiments for KSA social tweets.

Tweets related to COVID-19 social constraints in the USA and India have a very similar sentiment distribution. The majority of sentiments are neutral and positive. However, the number of negative tweets for these two countries is significantly higher than for homeschooling and work from home tweets. Hence, people are more accepting of homeschooling and work from home as a new norm. However, they face difficulties adapting to social distancing and wearing masks in public. Figures 10.16 and 10.17 give more insights about the sentiment distribution of US and Indian tweets.

10.5.4   Limitations & Constraints

The sentiment analysis conducted contributes useful insights into how people feel and react to the three main topics discussed during the COVID-19 era. However, the

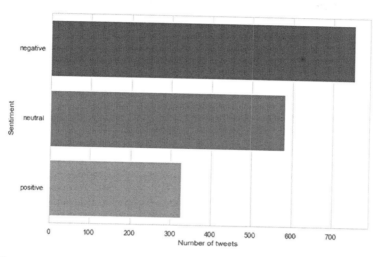

FIGURE 10.15 Sentiment distribution across Covid-19 social constraints-related tweets in KSA.

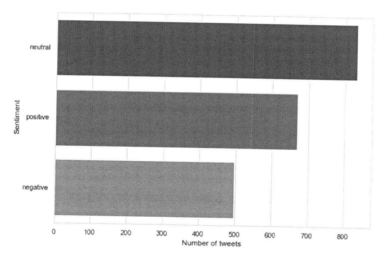

FIGURE 10.16 Sentiment distribution across COVID-19 social constraints-related tweets in the USA.

analysis was conducted under specific constraints and limitations. The analysis constraints are:

- Tweets extracted are filtered based on the location being studied. However, retweets do not carry geo/location data, so they are usually returned no matter where they are extracted. Hence, we can see huge similarities between US and Indian tweets.

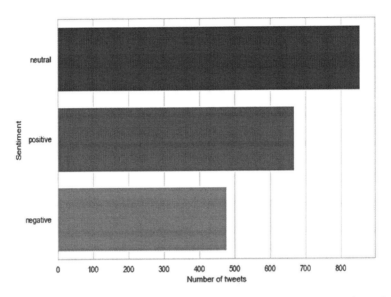

FIGURE 10.17   Sentiment distribution across COVID-19 social constraints-related tweets in India.

- This analysis was conducted using the Twitter premium API free tier, which only returns up to 30 days' tweets. So, the results presented reflect people's opinions during the month being studied.
- Arabic tweets were considered to study KSA public opinion and English tweets to analyze tweets in the USA and India. However, that does not reflect all the tweets available for each particular scope.
- We cannot generalize the results presented to how people feel about each topic studied, as people may change their opinions over time.
- Results do not reflect all populations within each country being studied, but only a sample of active people on Twitter.

10.6 CONCLUSION

The analysis of the different lockdown strategies of the three countries in question to control the spread of COVID-19 shows that no system can accurately measure the impact of lockdown as it depends on multiple factors, such as age, environment, and density. However, it is observed that lockdown strategies came out with reduced numbers of infected and death cases. In most countries, a declining case fatality rate has been observed. A variety of direct factors could have contributed to this decline, including the ability of the health service to cope with COVID-19 patients, increased and enhanced viral testing and tracing, the effectiveness of lockdown techniques, the growth of herd immunity, age effect on the affected population, variations in viral infection and deadliness. It is suggested that further periods be studied to understand the long-term impact of the lockdown strategy. Finally, sentiment analysis of

COVID-19 tweets was carried out to understand how people from different countries reacted to the new way of living during the COVID-19 era. The countries selected for this study applied different policies to control the spread of the virus. However, some shared rules are addressed by all. In the overall analysis, people easily accepted homeschooling and working from home as a new norm. However, there was a reluctance to adopt social distancing and the wearing of masks in public.

ABBREVIATIONS AND ACRONYMS

- SARS-CoV-2: Is the scientific name of the novel coronavirus named by the World Health Organization (WHO) after understanding the virus's DNA composition.
- COVID-19: Is the most common name used to refer to the novel coronavirus.
- Lockdown: Lockdown is a strategy adopted by local or national governments to impose restrictions on people to ensure social distancing and prevent virus spread.
- Pandemic: Pandemic is the categorization by WHO of any global disease impacting a vast population.

NOTE

1 https://www.snia.org/events/storage-developer

REFERENCES

Al Arabiya English, 2020a. Coronavirus: Saudi Arabia records 1,453 cases. *Al Arabiya English.* [online] Available at: https://english.alarabiya.net/en/News/gulf/2020/03/30/Coronavirus-Saudi-Arabia-records-1-453-cases.html [Accessed January 11, 2021].

Al Arabiya English, 2020b. Coronavirus: Saudi Arabia records 3,531 recoveries, 1,644 new cases. *Al Arabiya English.* [online] Available at: https://english.alarabiya.net/en/coronavirus/2020/05/28/Coronavirus-Saudi-Arabia-records-3-531-recoveries-1-644-new-cases.html [Accessed January 11, 2021].

Alrashed, S. et al., 2020. Impact of lockdowns on the spread of COVID-19 in Saudi Arabia. *Informatics in Medicine Unlocked.* [online] Available at: https://www.sciencedirect.com/science/article/pii/S2352914820305700. [Accessed January 12, 2021].

Anon, 2020a. Local lockdown ordered in English city of Leicester over high Covid-19 infection rate. *CNBC.* Available at: https://www.cnbc.com/2020/06/29/local-lockdown-ordered-in-english-city-of-leicester-over-high-covid-19-infection-rate.html [Accessed January 12, 2021].

Anon, 2020b. wordcloud. *PyPI.* Available at: https://pypi.org/project/wordcloud/ [Accessed January 12, 2021].

Arab News. 2020. *Saudi Arabia Announces First Case of Coronavirus.* [online] Available at: https://www.arabnews.com/node/1635781/saudi-arabia [Accessed January 11, 2021].

Aspinall, E., 2020. *COVID-19 Timeline - British Foreign Policy Group.* [online] British Foreign Policy Group. Available at: https://bfpg.co.uk/2020/04/covid-19-timeline/ [Accessed January 12, 2021].

BBC News, 2020a. *Coronavirus: Latest Patient Was First To Be Infected In UK.* [online] Available at: https://www.bbc.com/news/uk-51683428 [Accessed January 12, 2021].

BBC News, 2020b. *Coronavirus: Follow Virus Advice Or 'Tougher Measures' Likely, Says PM.* [online] Available at: https://www.bbc.com/news/uk-51998559 [Accessed January 12, 2021].

BBC News, 2020c. *Coronavirus: Schools In England Reopening On 1 June Confirmed, PM Says.* [online] Available at: https://www.bbc.com/news/education-52792769 [Accessed January 12, 2021].

BBC News 2020d. *Coronavirus: New Laws To Ease Outdoor Drinking And Dining Rules.* [online] Available at: https://www.bbc.com/news/business-53176005 [Accessed January 12, 2021].

BBC News 2020e. *Did Sweden's Coronavirus Strategy Succeed Or Fail?.* [online] Available at: https://www.bbc.com/news/world-europe-53498133 [Accessed January 12, 2021].

Conservative Party, 2020. *New National Restrictions In England - Stay At Home. Protect The NHS. Save Lives.* [online] Available at: https://www.conservatives.com/news/stay-at-home-protect-the-nhs-save-lives [Accessed January 12, 2021].

Farha, 2019. An Online Arabic Sentiment Analyser. *ACL Anthology.* Available at: https://www.aclweb.org/anthology/W19-4621/ [Accessed January 12, 2021].

Khudair, Deemah. 2020. *COVID-19 Cases In Saudi Arabia Surpass 10,000.* [online] Available at: https://www.arabnews.com/node/1662136/saudi-arabia [Accessed January 11, 2021].

Legislation UK Gov. 2020. *The Health Protection (Coronavirus, Restrictions) (No. 2) (England). Regulations 2020.* [online]

Ludvigsson, J.F., 2020. The first eight months of Sweden's COVID-19 strategy and the key actions and actors that were involved. *Wiley Online Library.* Available at: https://onlinelibrary.wiley.com/doi/10.1111/apa.15582 [Accessed January 12, 2021].

Nielsen, F.Å., 2011. A new ANEW: Evaluation of a word list for sentiment analysis in micro-blogs. *arXiv.org.* Available at: http://arxiv.org/abs/1103.2903 [Accessed January 12, 2021].

PyPI. 2020. *Searchtweets.* [online] Available at: https://pypi.org/project/searchtweets/ [Accessed January 12, 2021].

Saudigazette. 2020. *Saudi Arabia Reports 51 New Cases, Total Now 562.* [online] Available at: https://saudigazette.com.sa/article/591172/SAUDI-ARABIA/Saudi-Arabia-reports-51new-cases-total-now-562 [Accessed January 11, 2021].

Staff, R. 2020. *Saudi Arabia Temporarily Suspends Entry Of GCC Citizens To Mecca And Medina: Foreign Ministry.* [online] U.S. Available at: https://www.reuters.com/article/us-health-china-saudi-idUSKCN20M31T [Accessed January 11, 2021].

Stewart, H. 2020. *Boris Johnson Orders UK Lockdown To Be Enforced By Police.* [online] the Guardian. Available at: https://www.theguardian.com/world/2020/mar/23/boris-john-son-orders-uk-lockdown-to-be-enforced-by-police [Accessed January 12, 2021].

Stretton, R., 2020. *Solihull Hit By New Coronavirus Lockdown Rules As Cases Rise Across Town.* [online] *BirminghamLive.* Available at: https://www.birminghammail.co.uk/news/midlands-news/solihull-hit-new-coronavirus-lockdown-18905437 [Accessed January 12, 2021].

Nasrallah, T., 2020. *Coronavirus: Saudi Arabia Reports First Death, 205 New Cases Of COVID-19.* [online] Gulfnews.com. Available at: https://gulfnews.com/world/gulf/saudi/coronavirus-saudi-arabia-reports-first-death-205-new-cases-of-co-vid-19-1.1585054457152 [Accessed January 11, 2021].

The Guardian, 2020. *The New Corona Restrictions.* [online] the Guardian. Available at: https://www.theguardian.com/world/2020/mar/23/uk-lockdown-what-are-new-coronavirus-restrictions [Accessed January 12, 2021].

Tommy, H., 2020. Coronavirus: Saudi Arabia now has over 100,000 COVID-19 cases. *Al Arabiya English*. [online] Available at: https://english.alarabiya.net/en/coronavirus/2020/06/07/Coronavirus-Saudi-Arabia-now-has-more-than-100-000-COVID-19-cases-.html [Accessed January 12, 2021].

Tuqa, K., 2020. Coronavirus: Saudi Arabia sets guidelines for malls, industrial sector, as it reopens. *Al Arabiya English*. [online] Available at: https://english.alarabiya.net/en/coronavirus/2020/05/29/Coronavirus-Saudi-Arabia-sets-guidelines-for-malls-industrial-sector-as-it-reopens.html [Accessed January 11, 2021].

UK ONS, 2020. *Coronavirus (COVID-19) Infection Survey Pilot - Office For National Statistics*. [online] Available at: https://www.ons.gov.uk/peoplepopulationandcommunity/healthandsocialcare/conditionsanddiseases/bulletins/coronaviruscovid19infectionsurveypilot/england14may2020 [Accessed January 12, 2021].

Waskom, M., Gelbart, M., Botvinnik, O., Ostblom, J., Hobson, P., Lukauskas, S., Gemperline, D., Augspurger, T., Halchenko, Y., Warmenhoven, J., Cole, J., Ruiter, J., Vanderplas, J., Hoyer, S., Pye, C., Miles, A., Swain, C., Meyer, K., Martin, M., Bachant, P., Quintero, E., Kunter, G., Villalba, S., Fitzgerald, C., Evans, C., Williams, M., O'Kane, D., Yarkoni, T., & Brunner, T., 2020. *Mwaskom/Seaborn: V0.11.1 (December 2020)*. [online] Zenodo. Available at: https://doi.org/10.5281/zenodo.592845 [Accessed January 12, 2021].

Worldometers.info. 2021. *Coronavirus Update (Live): 91,167,632 Cases And 1,950,186 Deaths From COVID-19 Virus Pandemic - Worldometer*. [online] Available at: https://www.worldometers.info/coronavirus/ [Accessed January 11, 2021].

WSJ. 2020. *Dozens Of U.S. Diplomats To Leave Saudi Arabia As Coronavirus Outbreak Worsens*. [online] Available at: https://www.wsj.com/articles/dozens-of-u-s-diplomats-to-leave-saudi-arabia-as-coronavirus-outbreak-worsens-11593812097 [Accessed January 12, 2021].

11 A Mathematical Model and Forecasting of COVID-19 Outbreak in India

G. Maria Jones, S. Godfrey Winster,
A. George Maria Selvam, and D. Vignesh

CONTENTS

11.1 INTRODUCTION

The world has faced various types of pandemic problems over many centuries. The Black Death was one of the highly destructive epidemics in human history; it occurred during the thirteenth century in Europe and millions of people were affected, mainly elderly adults, with many individuals being exposed to physiological stressors (Dewitte, 2014). The next epidemic was smallpox in the late fifteenth century which killed around 8 million people in Mexico alone. The country became catastrophic due to the pandemic which resulted in a cocoliztli epidemic and led to the deaths of about 15 million native Mexican people. Again in 1576, cocoliztli killed 2.5 million people (Shastri et al., 2020). Another deadly pandemic which affected humankind in the early twentieth century, killing around 10 million people, was called Spanish Flu. These kinds of pandemic occurred throughout the centuries and had a devasting effect on humankind, and then another pandemic was identified in December 2019.

In that month, a novel coronavirus termed as coronavirus disease 2019 (COVID-19) appeared in Wuhan, China. Initially the reason behind the epidemic was unknown and later many scientists discovered a similar pattern to the SARS virus was found in

213

the coronavirus. The transmission of COVID-19 was observed and it was confirmed that the spread occurred through human to human contact with close or indirect contact with an affected person. By this time, the spread of the virus was uncontrollable in Wuhan city and it began to spread all over the world. On March 11, 2020, the World Health Organization (WHO) declared COVID-19 as a pandemic, with the infection reaching about 118,000 over 118 countries. The disease rapidly spread and created a huge impact in the healthcare systems of many nations, such as Spain, France, the United States, and Italy. According to a Johns Hopkins University report, there were 54,387,570 confirmed cases of COVID-19 across the globe as of November 16, 2020. Almost every nation has been implementing safety measures, such as social distancing, travel restrictions, lockdowns, quarantine, and testing. Based on infections in each country, lockdowns have been implemented to reduce human contact in the hope of decreasing the transmission rate.

Four phases of lockdown were implemented in India. The first phase, lockdown 1.0, ran from March 25 to April 14, 2020, with relaxation of restrictions for medical shops, pharmacies, and grocery shops, while the activities of nearly all services, factories, educational institutions, and IT companies were suspended. The second phase, lockdown 2.0, ran from April 15 to May 3, with some relaxation in non-affected areas. The third phase, lockdown 3.0, was implemented from May 4 to May 17 with the nation being divided into red, orange, and yellow zones. Based on the zone, transportation was started with limited passenger numbers. The red zone remained under full lockdown. Finally, the fourth phase of lockdown ran from May 18 to May 31, with the additional relaxation of dividing the red zone into containment and buffer zones. The number of cases of infection decreased from 11.8% to 6.3% on a daily basis, due to the lockdown (Arora, Kumar, & Ketan, 2020). The unlocking phases 1.0 to 6.0 lasted from June to November. In each phase, the government declared a relaxation with guidelines to be followed.

Advanced technologies like artificial intelligence (AI), Internet of things, machine learning (ML), deep learning (DL), and blockchain are being used in many fields and also in the healthcare system to predict outcomes and to understand human health patterns. There are some AI-based CT and X-ray scanners for detecting and monitoring the coronavirus infection in the human body. By September 2020, there was still no vaccination or treatment for coronavirus. Medical teams have been trying hard to provide a vaccination and, according to WHO, in December 2020 there were more than 100 people working on a vaccination, while WHO is collaborating with other global health organizations and scientists to speed up progress. In the meantime, social distancing, travel restrictions, using hand sanitizer, and wearing masks can help to slow down transmission.

Forecasting models like long short-term memory (LSTM), autoregressive integrated moving average (ARIMA), SARIMAX, Prophet and many more algorithms are being used to forecast the real-time series rates with seasonal and non-seasonal time series. Many researchers have analyzed the forecasting model and mathematical models for COVID-19 with various parameters and methods. In this chapter, we introduce a SEIRD (Susceptible, Exposed, Infected, Recovered, and Deceased), LSTM, Prophet, and ARIMA model to analyze the behavior of COVID-19. The work is classified into two segments. The mathematical model which includes SEIRD is analyzed in the first segment of the chapter, while the second part works with the LSTM and ARIMA model which is based on a prediction model. The following aspects are the main contribution of the chapter:

1. To propose a novel SEIRD model for India and Brazil.
2. To employ an LSTM and Prophet time series prediction model to predict the spread of COVID-19 for India, with respect to confirmed, recovered, and death cases.
3. The ARIMA model is employed to forecast the infected COVID-19 cases for the world, with respect to total confirmed cases.
4. The Prophet model is also employed to predict the confirmed cases of India.
5. In addition to this forecasting, the error rate is also calculated. They are mean absolute error (MAE), root mean squared error (RMSE), mean squared error (MSE), and R-squared.

The rest of the article is organized as follows: section 11.2 describes the various methodologies proposed by researchers to model coronavirus; section 11.3 provides the SEIRD mathematical model with the data fitting for some countries; section 11.4 gives the forecasting model based on DL to find the propagation of coronavirus; finally, Section 11.5 sets out the conclusion of the article.

11.2 RELATED WORK

Parul Arora and Himanshu Kumar (2020) used DL to predict COVID-19 positive cases in India. Deep, convolutional, and bi-directional LSTM models were used, based on the error function. The final results showed that the bi-directional method produced a good result compared with other methods, whereas convolutional LSTM gives a low prediction rate. Sarkar et al. (2020) proposed a mathematical model that predicts the dynamic of coronavirus, which was comprised of six compartments such as $SARII_qS_q$. The final result evaluated that there is a reduction in contact rate between infected and uninfected which reduces the reproduction number and they also demonstrated the model with the effects of social distancing and contact tracing. Amit et al. (2020) developed and validated the coronavirus data to analyze and predict the total infected population for India, USA, Canada, Iran, Germany, Japan, China, and Italy. The model was found to be the best fit for the actual data of all the countries except Iran.

Pai et al. (2020) analyzed the COVID-19 cases for India using the chi-square method for fitting the data and a SEIR model. The analysis resulted in the peak of cases being attained in May with the end being reached by August, if the lockdown continued. Yadav et al. (2020) proposed a novel support vector regression method for analyzing the five tasks of predicting the spread, analyzing the growth rates, predicting the recovered patients, analyzing the transmission, and correlating the virus and weather conditions. The authors have used support vector regression (SVR) for better classification accuracy instead of simple regression. In five tasks, the accuracy of SVR is 37.9%, 92.1%, 91.5%, 43.7%, and 91.1%, which is comparatively higher than other models. Kaxiras (2020) presented a model called FSIR (forced-SIR) and also fitted the model to real-time data for a number of countries. The author also compared the findings of the model with reported results. Maleki et al. (2020) proposed a time series model for a COVID-19 dataset and it was fitted to a historical dataset. ARIMA modeling based on two-piece distribution was used to predict the model, which works better than Gaussian time series.

Meehan et al. (2020) reviewed the mathematical role in modeling COVID-19 and also discussed the challenges posed over data availability. Nour et al. (2020) proposed a model based on a convolutional neural network (CNN) for analyzing chest X-ray (CXR) images. The proposed CNN was carried out with five layers, and for feature extraction ML algorithms like K-nearest neighbors (KNN), support vector machine (SVM), and decision tree (DT) were used. The final result worked well for SVM which had good accuracy compared with the other two algorithms. Smita Rath and Alakananda Tripathy (2020) predicted the infected rate by analyzing the daily cases of COVID-19. Linear and multiple regression models were performed with rates of 0.99 and 1.0 which show a strong prediction model. Salehi et al. (2020) summarized the AI-based approach for spreading of coronavirus. The analysis resulted in CNN being a significant method for detecting coronavirus in X-ray images. Olaide and Absalom (Olaide N. Oyelade, 2020) proposed a model which supports medical teams to diagnose coronavirus based on their medical records without analyzing the sample.

Nemati et al. (2020) used ML and statistical methods to predict the discharge of patients from hospitals and the final findings showed that male and older people's discharge rate has lower probability. Khan and Gupta (2020) predicted coronavirus based on daily infected cases, using ARIMA and NAR models. The ARIMA model worked on a 1, 1, 0 model which reached R^2 values of 0.95 whereas NAR reached R^2 values of 0.97. Peng and Yang (n.d.) proposed an SEIR model to analyze the COVID-19 virus for China, and their prediction model stated that the propagation of the virus was expected to end in the middle of March, and the worm propagation in wireless sensor networks (Selvam et al., 2020) was also carried out using a SIR model to analyze the spread of worms in the network. Bhardwaj (2020) predicted the evolution of coronavirus using a logistic model for Germany, Italy, USA, Brazil, Spain, and India. The predictions for selected countries are analyzed and discussed.

11.3 MATHEMATICAL FORMULATION OF COVID-19 OUTBREAK

COVID-19 has preyed on people with non-communicable diseases like diabetes, cardiovascular diseases, cancer, and respiratory diseases. The spread of the novel coronavirus between humans is by direct contact with surfaces that are contaminated, and inhalation of droplets from infected people. Though there are various efforts that were put forth in discovery of a vaccine or antiviral treatments, there is no approved vaccine for the prevention of the outbreak of the virus. This failure in discovery of a vaccine is due to little knowledge about the pathology, ecology, and epidemiology of the virus. One of the major issues in infection by coronavirus is that it does not trigger permanent immunity. The immunity triggered in humans begins to fade away over the course of time, resulting in reinfection of individuals by coronavirus.

Various strict measures have been imposed by governments of various countries to bring the outbreak under control, but the individual's response to those measures will be key in reducing the spread. The outbreak of COVID-19 cannot be stopped with biological and medical research alone, but with interdisciplinary research constructing mathematical models that will be very useful in predicting the spread in future. Here a mathematical model is considered which takes into account the possibility of reinfection of recovered individuals. The human population (N(t)) is split into compartments of Susceptible (S(t)), Exposed (E(t)), Infected (I(t)), and Recovered (R(t)).

$$N(t) = S(t) + E(t) + I(t) + R(t) \qquad (11.1)$$

The schematic representation of the model is presented in Figure 11.1.
The model of COVID-19 is given by

$$
\begin{aligned}
\frac{dS}{dt} &= \frac{-\beta SI}{N} + mR(t) \\
\frac{dE}{dt} &= \frac{CbSI}{N} - gEI \\
\frac{dI}{dt} &= \frac{(1-C)\beta SI}{N} - (q+\omega)I + \gamma EI \\
\frac{dR}{dt} &= \omega I - mR \\
\frac{dD}{dt} &= qI
\end{aligned}
\qquad (11.2)
$$

A. Susceptible Population (S(t))

The population is a fraction of individuals who are able to contract the disease. The decrease in the population at the rate β occurs when a susceptible individual interacts with the infected population. The population that recovered from infection and is subjected to infection again tends to increase the population.

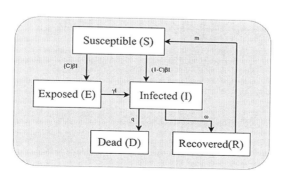

FIGURE 11.1 SEIRD Model.

$$\frac{dS}{dt} = \frac{-\beta SI}{N} + mR \qquad (11.3)$$

B. Exposed Population (E(t))

This compartment considers the fraction of the population who are around infected people but are not yet infected themselves. The population increases at rate 'C' by interaction between the susceptible and the infected population. The interaction with infectious individuals decreases the population at the rate γ.

$$\frac{dE}{dt} = \frac{C\beta SI}{N} - \gamma EI \qquad (11.4)$$

C. Infected Population (I(t))

The population who test positive for infection by coronavirus belong to this compartment. The increase in infections is from fraction '(1-C)' of the susceptible population and the exposed population who test positive at the rate γ. A part of infected individuals tend to develop immunity against the virus and individuals after suitable treatment recover at the rate ω. The failure to develop immunity and the presence of respiratory diseases, cancer, and so on can lead to death of the individuals at the rate q.

$$\frac{dI}{dt} = \frac{(1-C)\beta SI}{N} - (q+\omega)I + \gamma EI \qquad (11.5)$$

D. Recovered Population (R(t))

The population that are infected and are assumed to recover at the rate ω belong to this compartment and it is also assumed that the fading away of immunity in these individuals at the rate of m increases the probability of their getting infected again which results in adecrease in the population (Table 11.1).

$$\frac{dR}{dt} = \omega I - mR \qquad (11.6)$$

TABLE 11.1
Parameter Description

Parameter	Description
β	Effective contact rate
C	Fraction of population who are surrounded by infected people but are not infected
γ	Rate of progression from exposed to infected compartment
q	Recovery rate of individuals in infected compartment
ω	Disease-induced mortality rate
m	Rate of reinfection of recovered individuals

11.3.1 Data Fitting

Data fitting plays a vital role in expressing our views about a particular mechanism quantitatively and precisely. Comparing our model with data is also very important in understanding and predicting possible events in the future. Let us now fit the model to real-time data of two countries—namely, India and Brazil.

1. Data Fitting of India

   India, being the second most populated country in the world, has witnessed many outbreaks, starting from cholera in 1992 to COVID-19 in 2020. Though the death rate is low in comparison with other countries with a similar infection rate, increasing infections and reinfections are a great concern. With a population of about 138 crores in 2020, the first infection of COVID-19 was recorded in Kerala on January 30, 2020. In over 8 months the number of infections has risen to more than 7.5 million. The government of India has taken steps to control the spread by implementation of lockdowns and necessary medical facilities for the treatment of the infected. Here we fit data to our model with initial populations of compartments considered as $E(0) = 105$, $I(0) = 1$, $R(0) = 0$, $D(0) = 0$.

   Figure 11.2 illustrates the relation between the real-time data of infections, recovery and death of people in India and the time plot of the mathematical model in (2). The data were collected from datahub.io. An increasing number of recover and decreasing infections are observed from Figure 11.2. It is clear that the number of recovered has been increasing over the number of infected after 250 days. From Figure 11.3 it is observed that the peak of infection was attained after 200 days and the infections begin to decrease gradually. The number of people recovered from the coronavirus infection are plotted against the real-time data in Figure 11.4.

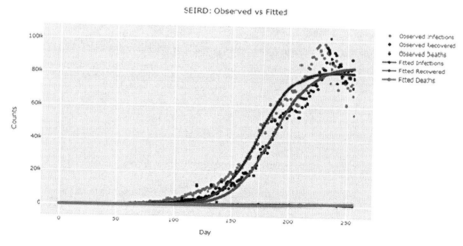

FIGURE 11.2 Fitting observed data and model for infected, recovered, and death compartments.

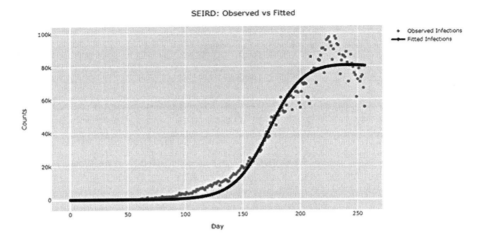

FIGURE 11.3   Fitting observed data and model for infected compartment.

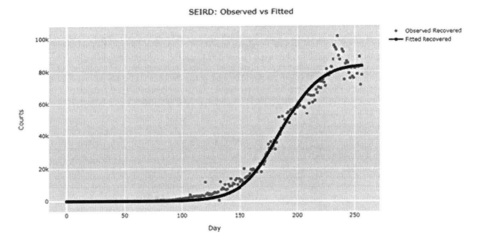

FIGURE 11.4   Fitting observed data and model for recovered compartment.

2. Data Fitting for Brazil

Here we consider a South American country, Brazil, that has recorded about 5.2 million infections from coronavirus. The population of Brazil is about 21 crores and the first infection was recorded on February 26, 2020. The country is reportedly the worst affected, due to the contradictory measures proposed by local leaders and higher government officials. Mayors and state governors in Brazil had implemented measures to restrict the movement of people to control the spread, but the higher government officials have been prioritizing political battles and growth of the economy over the pandemic. This has created widespread confusion among the people about the pandemic. Lockdown

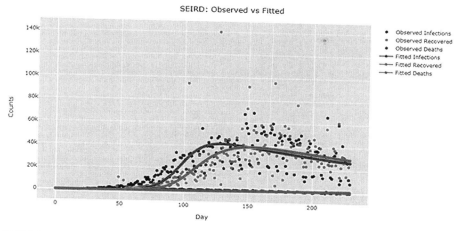

FIGURE 11.5   Fitting observed data and model for infected, recovered, and death compartments.

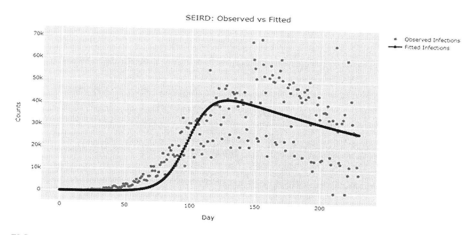

FIGURE 11.6   Fitting observed data and model for infected compartment.

systems were partially implemented and the infections were so many that the price of basic safety requisites like masks, sanitizers etc. grew rapidly and the increasing demand could not be met. Here the model is fitted to data with the initial populations of the compartments being E(0) = 1000, I(0) = 1, R(0) = 0, D(0) = 0.

The scattered plot performed in Figures 11.5–11.7 represents the real-time data of the pandemic in Brazil. From Figure 11.5, it can be seen that the number of infections recovered on each day has great variations. Figures 11.6 and 11.7 compare the observed number of infections and recovered due to the pandemic and model (2) with fitted values.

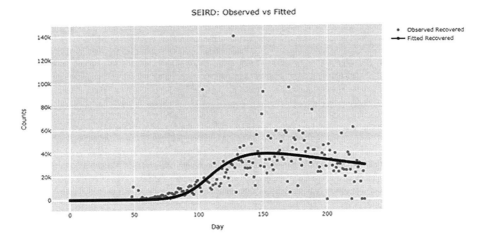

FIGURE 11.7   Fitting observed data and model for recovered compartment.

11.4 TIME SERIES MODELING

Time series is a series of time points collected in chronological order, mostly collected/gathered at regular time intervals. Analysis of time series can be applied to any variable which changes over a time period. Trend, seasonality, observed, noise, curve, and level are common components in a time series analysis.

The data were collected from ourworldindata.org for the number of cases reported in India up to November 23, 2020. These data were initially used for time series analysis to attain a basic understanding of the top 5 countries' infected and dead ratio among all other countries. From Figure 11.8, it can be seen that India is in second place, while from Figure 11.9 it can be seen that the death count in India is in third place.

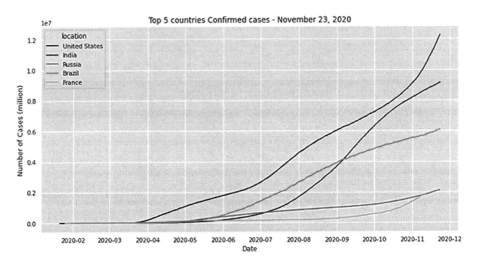

FIGURE 11.8   Top 5 countries affected by COVID-19.

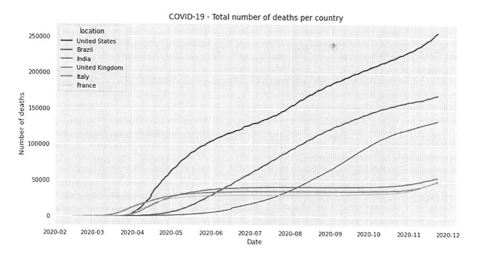

FIGURE 11.9   Top 6 countries' death rate.

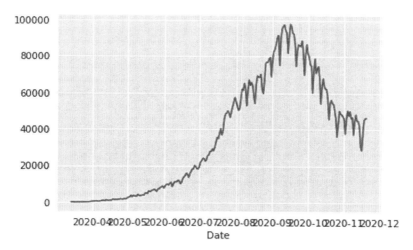

FIGURE 11.10   Time series model of affected cases in India.

The time series for the spread of coronavirus as it affected India is shown in Figure 11.10 as the infection reached its peak and started decreasing, while the proposed forecast model is shown in Figure 11.11.

11.4.1 LSTM MODEL FOR INDIA

LSTM, or long short-term memory, is a variant of a recurrent neural network (RNN). LSTM is mainly designed to avoid long-term dependency and the vanishing gradient problem in which the network stops learning, due to updates of the weights within a network.

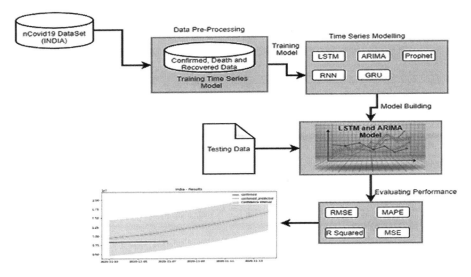

FIGURE 11.11   Proposed working model.

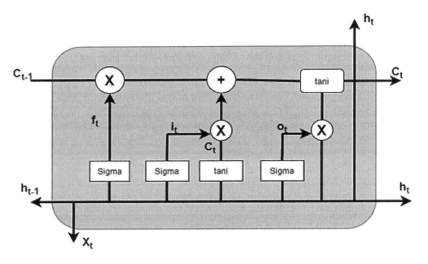

FIGURE 11.12   Structure of LSTM.

In RNN, the repeating mode of tan$_i$ takes places in a simple structure, whereas in LSTM the repeating mode takes place in various structures; the representation is presented in Figure 11.12. The initial step in LSTM is about deciding the information which is made by the sigmoid function termed "forgot gate" as equated in (11.7). It can be represented as ht-$_1$ and x$_t$. The output in cell state Ct-$_1$ is represented by 0 and 1. A 1 represents "to keep" whereas 0 represents "to get rid of."

$$f_t = s\left(W_f.\left[h_{t-1},x_t\right]+b_f\right)$$
$$i_t = s\left(W_i.\left[h_{t-1},x_t\right]+b_i\right) \tag{11.7}$$
$$C'_t = \tan_i\left(w_c.\left[h_{t-1},x_t\right]+b_c\right)$$

The next step is to decide what new information needs to be stored in the cell state. In this step, the sigmoid layer called "input gate" decides about the values to update. The next $\tan_i$ layer creates a vector of new state Ct to be added to the state. Now, the old state called previous cell C_{t-1} needs to be updated with new state C_t as equated in (11.8).

$$C_t = f_t.C_{t-1} + i_t.C'_t \tag{11.8}$$

Finally, the output gate is used to produce the result based on the cell state. The sigmoid later decides which cell state is going to produce the output. Then the $\tan_i$ is multiplied to the output of the sigmoid layer and formulated as (11.9).

$$O_t = \sigma\left(W_o\left[h_{t-1},x_t\right]+b_o\right) \tag{11.9}$$

The experiment was carried out in Google Colab using Python notebook 3.0 with libraries such as Pandas, Numpy, Keras, and TensorFlow under the 64 bit Windows 10 operating system. The forecasting dataset was modeled with two models—LSTM and ARIMA—based on the number of infected cases in India to predict the future based on historical data from the dataset. For modeling LSTM, a sequential model is used with the ReLU activation function and the Adam optimizer for forecasting. In Figures 11.13 and 11.14, the forecast for 7 days is shown, i.e. from November 16 to November 22, 2020.

Figure 11.15 represents the confirmed, dead, and recovered cases in India. This provides the understanding that the infected and recovered rates are proceeding similarly whereas the death rate is increasing.

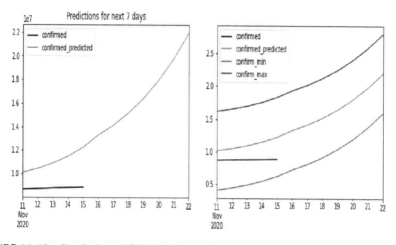

FIGURE 11.13 Prediction of COVID-19 cases for next 7 days.

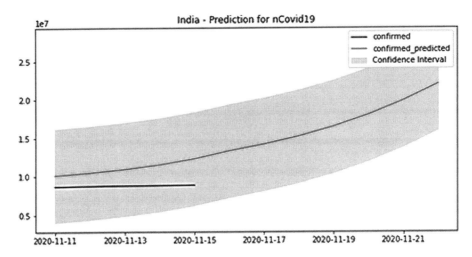

FIGURE 11.14   COVID-19 prediction with confidence level.

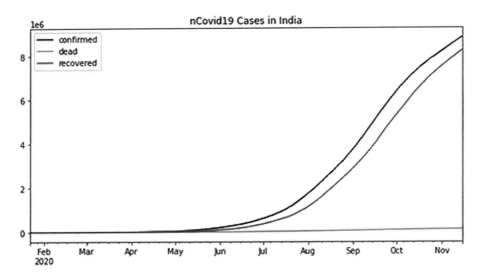

FIGURE 11.15   Analysis of Confirmed, Dead, and Recovered cases.

The Table 11.2 provides the forecasting result of COVID-19 where the prediction accuracy is 74%.

TABLE 11.2
LSTM Prediction

	confirmed	confirmed_predicted	confirm_min	confirm_max	Model accuracy	Country	Execution date
2020-11-11	8683916.0	1.011207e+07	4.069933e+06	1.615420e+07	0.74	India	2020-11-16
2020-11-12	8728795.0	1.045851e+07	4.416375e+06	1.650065e+07	0.74	India	2020-11-16
2020-11-13	8773479.0	1.091649e+07	4.874357e+06	1.695863e+07	0.74	India	2020-11-16
2020-11-14	8814579.0	1.149701e+07	5.454878e+06	1.753915e+07	0.74	India	2020-11-16
2020-11-15	8845127.0	1.226420e+07	6.222060e+06	1.830633e+07	0.74	India	2020-11-16
2020-11-16	NaN	1.330243e+07	7.260292e+06	1.934456e+07	0.74	India	2020-11-16
2020-11-17	NaN	1.414524e+07	8.103109e+06	2.018738e+07	0.74	India	2020-11-16
2020-11-18	NaN	1.517400e+07	9.131861e+06	2.121613e+07	0.74	India	2020-11-16
2020-11-19	NaN	1.642849e+07	1.038635e+07	2.247062e+07	0.74	India	2020-11-16
2020-11-20	NaN	1.796025e+07	1.191811e+07	2.400239e+07	0.74	India	2020-11-16
2020-11-21	NaN	1.980500e+07	1.376286e+07	2.584713e+07	0.74	India	2020-11-16
2020-11-22	NaN	2.198007e+07	1.593793e+07	2.802221e+07	0.74	India	2020-11-16

11.5 ARIMA MODEL FOR WORLD

ARIMA is used to model the time series data and also to predict the future forecasting. The ARIMA model has three order parameters, p, q, and d:

p is an autoregressive model
d is a difference order
q is a moving average model.

When dealing with seasonal ARIMA (SARIMA), the parameters are different from those of ARIMA and are:

P—seasonal autoregressive model

D—seasonal difference order

Q—seasonal moving model

M—a seasonal time period.

The ARIMA forecasting model is a univariate linear function used to predict the future points based on historical periods. It combines the past values to determine the future points versus a linear regression model. There are two parameters, which are:

I. Autoregressive (AR) model
   An AR model is a type of regression model where dependent values depend on the previous variables. The present variables are correlated with the previous time value which is termed partial autocorrelation (PAC) and the AR model can be represented as the equation below (11.10)

$$X_t = b_1 + a_1 X_{t-1} + a_2 X_{t-1} + \ldots a_n X_{t-n} \tag{11.10}$$

   where t is the lag order which represented the previous lag value presented in the model.

II. Moving average (MA)
   MA factors the error rate from lag observations. The effect of lagged observation on the current observation depends on autocorrelation which is similar to PAC and the error rate of MA is described in the equation below (11.11):

$$X_t = \beta_2 + \gamma_1 \varepsilon_{t-1} + \gamma_2 \varepsilon_{t-2} + \ldots + \gamma_p \varepsilon_{t-p} - \varepsilon_n \tag{11.11}$$

The term ε represents the error rate observed at lag and the weights are calculated based on the correlation function. The term p is the number of observed lag errors which has an impact on the current observation. Finally, the combined ARMA model is equated as below ((11.12),

$$X_t = \beta 1 + \beta 2 + \left(\alpha_1 X_{t-1} + \ldots + \alpha_n X_{t-n} \right) + \left(\gamma_1 \varepsilon_{t-1} + \ldots + \gamma_p \varepsilon_{t-p} + \varepsilon_n \right) \tag{11.12}$$

The AR and MA order can only be applied to a univariate stationary series. If the time series is not stationary, the series needs to be differentiated by d i.e d = 1,2,3,... then the model is termed an ARIMA (p, d, q) forecasting model.

The ARIMA model (p,d,q) is used to forecast the data with 95% lower and upper bound and also to analyze the goodness of fit statistics as shown in Tables 11.3 and 11.4 representing the ARIMA value parameters. The model can be checked by residual plot of the proposed model using ACF and PACF of the different orders. The forecasting model using ARIMA for world infected cases is depicted in Figure 11.16. The residual model is proportional to the difference between measured values with

TABLE 11.3
ARIMA Statistics

Goodness of fit statistics:

Observations	303
DF	292
SSE	42240474729
MSE	139407507.4
RMSE	11807.09564
WN Variance	139407507.4
MAPE(Diff)	255.1229978
MAPE	3.816095189
-2Log(Like)	6592.123373
FPE	154951201.6
AIC	6614.123373
AICC	6615.03059
SBC	6654.974434
Iterations	72

TABLE 11.4
ARIMA Model Parameters

Model parameters

Parameter	Value	Hessian standard error	Lower bound (95%)	Upper bound (95%)
Constant	181.432	38.900	105.189	257.675
Parameter	Value	Hessian standard error	Lower bound (95%)	Upper bound (95%)
AR(1)	0.705	0.068	0.571	0.838
AR(2)	−0.623	0.074	−0.769	−0.478
AR(3)	−0.163	0.081	−0.322	−0.004
AR(4)	−0.303	0.081	−0.461	−0.146
SAR(1)	−0.137	0.072	−0.278	0.005
MA(1)	−1.149	0.051	−1.249	−1.049
MA(2)	0.832	0.030	0.772	0.891
SMA(1)	−1.000	0.059	−1.116	−0.884

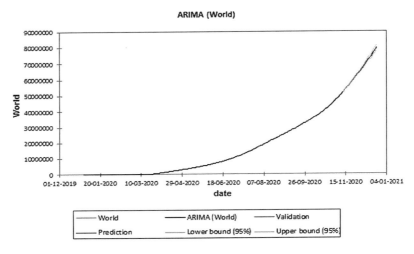

FIGURE 11.16 Model forecasting for world infected cases.

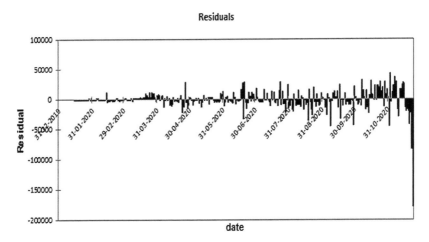

FIGURE 11.17 Residual plot.

respect to fit series (Khan & Gupta, 2020). The residual values of COVID-19 are represented in Figure 11.17.

ACF and PACF stand for autocorrelation function and partial autocorrelation function. The correlation between the observed time plots separated by t time units is termed an autocorrelation where ACF represents the correlation including the lag unit. The ACF representation for infected cases around the world is depicted in Figures 11.18 and 11.19. The correlation is represented in the Y axis and lag units are represented in the X axis.

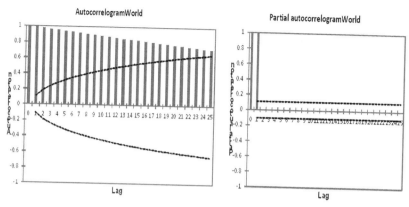

FIGURE 11.18 ACF and PACF plot.

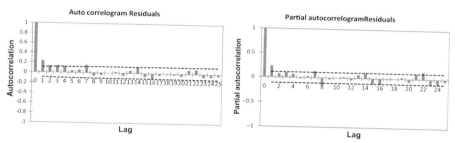

FIGURE 11.19 ACF and PACF plot.

11.6 PROPHET MODEL FOR INDIA

The Prophet forecasting model is a decomposable time series which is comprised of three main components—in this case, trend, seasonality, and holiday. They are represented as the following equation:

$$X(t) = h(t) + s(t) + q(t) + et \qquad (11.13)$$

where,

- h(t) is a linear or nonlinear curve for non-periodic changes which refers to a trend value
- s(t) is a periodic change which refers to a seasonality like yearly, monthly, weekly
- q(t) is a holidays effect which had irregular schedules
- $\varepsilon(t)$ is called error term.

The Prophet model is used to forecast the total infected cases of coronavirus in India. The model predicts the spread of the virus for next 20 days with respect to

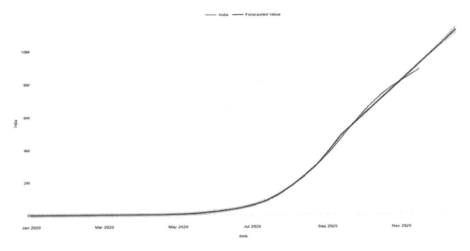

FIGURE 11.20   Forecasting using the Prophet model.

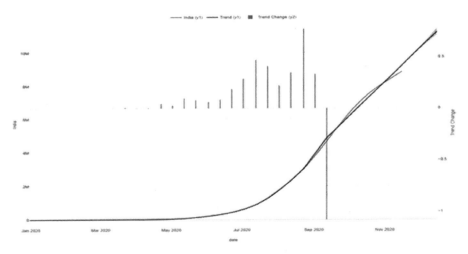

FIGURE 11.21   Trend change for COVID-19.

historical data. Figure 11.20 represents the Prophet forecast which implies the spread of the virus will increase.

The model is forecast for next 20 days and also the trend change for India is ana-lyzed as represented in Figure 11.21. The Prophet model's RMSE and MAE are 87,729.8522074289 and 42,899.9586785236 respectively.

11.7   CONCLUSION

In this chapter, we have proposed a SEIRD mathematical model and DL models for data fitting and forecasting the number of confirmed cases of COVID-19 in India. Initially, the SEIRD model was analyzed for two countries—India and Brazil—with infected, recovered, and death rate. An RNN-based LSTM was used for prediction which was tested on confirmed cases in India and forecast the prediction for next 7 days, i.e from November 16 to November 22, 2020. The results predicted with 74% accuracy. The next forecasting algorithm used was ARIMA with the effect of trend, seasonal, observed and residual parts for infected rate of the world with error ratio, and finally the Prophet forecasting model was implemented with RMSE and MAE. These prediction models will be useful for government authorities, researchers, and to medical teams for preparing themselves for arranging extra needs and medical equipment for the people.

REFERENCES

Amit, K., Rani, P., Kumar, R. and Vasudha Sharma, S. R. P. (2020) 'Data-driven modelling and prediction of COVID-19 infection in India and correlation analysis of the virus transmission with socio-economic factors,' *Diabetes & Metabolic Syndrome: Clinical Research & Reviews* 14(5), 1231–1240. doi: 10.1016/j.dsx.2020.07.008.

Arora, P., Kumar, H. and Ketan, B. (2020) 'Prediction and analysis of COVID-19 positive cases using deep learning models : A descriptive case study of India,' 139. doi: 10.1016/j.chaos.2020.110017.

Bhardwaj, R. (2020) 'A Predictive Model for the Evolution of COVID-19,' *Transactions of the Indian National Academy of Engineering. Springer Singapore*, 5(2), pp. 133–140. doi: 10.1007/s41403-020-00130-w.

Dewitte, S. N. (2014) 'Mortality risk and survival in the aftermath of the medieval black death,' 9(5). doi: 10.1371/journal.pone.0096513.

Kaxiras, E., Neofotistos, G. and Angelaki, E. (2020) 'The first 100 days: Modeling the evolution of the COVID-19 pandemic,' *Chaos, Solitons and Fractals: The Interdisciplinary Journal of Nonlinear Science, and Nonequilibrium and Complex Phenomena*, 138, 110114. doi: 10.1016/j.chaos.2020.110114.

Khan, F. M. and Gupta, R. (2020) 'ARIMA and NAR based prediction model for time series analysis of COVID-19 cases in India.' 1(April), 12–18. doi: 10.1016/j.jnlssr.2020.06.007.

Maleki, M., Mahmoudi, M.R., Heydari, M.H. and Pho, K.H. (2020) 'Modeling and forecasting the spread and death rate of coronavirus (COVID-19) in the world using time series models,' *Chaos, Solitons and Fractals: The Interdisciplinary Journal of Nonlinear Science, and Nonequilibrium and Complex Phenomena*. 140, 110151. doi: 10.1016/j.chaos.2020.110151.

Meehan, M. T. et al. (2020) 'Modelling insights into the COVID-19 pandemic,' *Paediatric Respiratory Reviews*. 35, 64–69. doi: 10.1016/j.prrv.2020.06.014.

Nemati, M., Ansary, J. and Nemati, N. (2020) 'Machine-learning approaches in COVID-19 survival analysis and discharge-time likelihood prediction using clinical data,' *Patterns* 1(5), 100074. doi: 10.1016/j.patter.2020.100074.

Nour, M., Cömert, Z. and Polat, K. (2020) 'A novel medical diagnosis model for COVID-19 infection detection based on Deep Features and Bayesian Optimization.' *Applied Soft Computing Journal*. 106580. doi: 10.1016/j.asoc.2020.106580.

Olaide, N. Oyelade, A. E. E. (2020) 'A case-based reasoning framework for early detection and diagnosis of novel coronavirus,' *Informatics in Medicine Unlocked*. 20, 100395. doi: 10.1016/j.imu.2020.100395.

Pai, C., Bhaskar, A. and Rawoot, V. (2020) 'Investigating the dynamics of COVID-19 pandemic in India under lockdown.' *Chaos, Solitons and Fractals: The Interdisciplinary Journal of Nonlinear Science, and Nonequilibrium and Complex Phenomena*, 138, 109988. doi: 10.1016/j.chaos.2020.109988.

Parul, A., Himanshu Kumar, B. K. P. (2020) 'Prediction and analysis of COVID-19 positive cases using deep learning models: A descriptive case study of India.' 139. doi: 10.1016/j. chaos.2020.110017.

Peng, L. and Yang, W. (n.d.) 'Epidemic analysis of COVID-19 in China by dynamical modeling.,' medRxiv, Feb 2020,22–24. doi: 10.1101/2020.02.16.20023465.

Salehi, A.W., Baglat, P. and Gupta, G. (2020) 'Review on machine and deep learning models for the detection and prediction of Coronavirus,' *Materials Today: Proceedings*, doi: 10.1016/j.matpr.2020.06.245.

Sarkar, K., Khajanchi, S. and Nieto, J.J. (2020) 'Chaos, Solitons and Fractals Modeling and forecasting the COVID-19 pandemic in India,' 139, 1–16. doi: 10.1016/j. chaos.2020.110049.

Selvam, A. G. M. et al. (2020) 'Modeling worm proliferation in wireless sensor networks with discrete fractional order system,' (5), 1815–1820. doi: 10.35940/ijrte.E4594.018520.

Shastri, S. et al. (2020) 'Time series forecasting of Covid-19 using deep learning models: India-USA comparative case study,' *Chaos, Solitons and Fractals: the interdisciplinary journal of Nonlinear Science, and Nonequilibrium and Complex Phenomena*, 140, 110227. doi: 10.1016/j.chaos.2020.110227.

Smita, R., Alakananda Tripathy, A. R. T. (2020) 'Prediction of new active cases of coronavirus disease (COVID-19) pandemic using multiple linear regression model.' *Diabetes & Metabolic Syndrome: Clinical Research & Reviews*, 14(5), 1467–1474. doi: 10.1016/j. dsx.2020.07.045.

Yadav, M., Perumal, M. and Srinivas, M. (2020) 'Analysis on novel coronavirus (COVID-19) using machine learning methods,' *Chaos, Solitons and Fractals: The Interdisciplinary Journal of Nonlinear Science, and Nonequilibrium and Complex Phenomena*. 139, 110050. doi: 10.1016/j.chaos.2020.110050.

12 Automatic Lung Infection Segmentation of COVID-19 in CT Scan Images

Mohsen Karimi, Majid Harouni, Afrooz Nasr, and Nakisa Tavakoli

CONTENTS

12.1 INTRODUCTION

The unknown coronavirus spread from Wuhan, China, in 2019. The disease caused by the coronavirus called COVID-19 has been identified as a major new concern by the World Health Organization (WHO) (Wang et al., 2020). The number of people infected with the virus has now exceeded 50,000,000, and the death toll has exceeded 1,000,000 (Benvenuto et al., 2020). The number of infected and victims of this virus is increasing every day. Although the coronavirus comes in different forms in people, symptoms such as fever, headache, cough, difficulty in breathing, body aches, sneezing and runny nose are common symptoms (Khan et al., 2021; Abbas et al., 2018). However, having these symptoms is not enough to diagnose a person with COVID-19 (Reusken et al., 2020). There are several methods for the definitive diagnosis of the virus in the human body, including stomach tests (Khan et al., 2019a), blood tests (Abbas et al., 2018, 2019a), lung imaging (Saba, 2019), and blood cancer (Abbas et al., 2019a, 2019b, 2019c). Because the coronavirus also has respiratory symptoms, it also affects the infected person's lungs (Shi et al., 2020). In other words, if the lungs of a person suspected of having the coronavirus can be examined, the presence of the virus in the person's body can likely be predicted. Lung imaging techniques such as computed tomography (CT) images, radiology and magnetic resonance imaging (MRI) can be used for this purpose (Fahad et al., 2018; Ejaz et al., 2019, 2020). Lung experts and doctors interpret the images. Since the outbreak of COVID-19, many doctors, experts, and nurses have worked hard to combat the virus. These people's enormous efforts have caused them much fatigue and exhaustion. This extreme fatigue has sometimes led to errors in the correct and timely diagnosis of the virus (Xu et al., 2020b). Although human error is inevitable, it is even more important in dealing with the virus due to the volume and pressure of work and the high levels of stress of medical staff (Adeel et al., 2020; Amin et al., 2018; Wu et al., 2020). There are several ways to reduce human error, such as methods based on image processing, machine vision and machine learning (ML) (Sharif et al., 2017; Saba et al., 2018). The use of ML and image processing methods based on artificial intelligence has been used successfully in many fields such as industry, medical image segmentation, security (Amin et al., 2019a, 2020; Saba et al., 2020b), handwriting recognition (Harouni et al., 2010, 2014; Rehman et al., 2020a), identity and gender classification (Meethongjan et al., 2013; Yaseen et al., 2018; Rehman et al., 2018a, 2018b, 2018c). Doctors use medical imaging techniques such as CT, MRI and radiology to diagnose coronavirus (Norbash et al., 2020). These imaging techniques are used to identify a variety of tumors such as brain tumors (Nazir et al., 2019; Rehman et al., 2020b; Rehman et al., 2021 Ejaz et al., 2020, 2019), lung tumors (Khan et al., 2019e), and breast tumors (Mughal et al., 2017, 2018a; Sadad et al., 2018; Marie-Sainte et al., 2019a; Saba et al., 2019b) and other diseases such as Alzheimer's, skin, stomach, blood, retina, heart, and diabetes (Saba et al., 2019a; Marie-Sainte et al., 2019b Perveen et al., 2020; Ramzan et al., 2020a, 2020b). The use of image processing algorithms in diagnosing these diseases has been studied by Habibi and Harouni (2018), Ullah et al. (2019), Yousaf et al. (2019a), and Saba (2017). One of the main effects of coronavirus is on the lungs (Khan et al., 2021). Figure 12.1 shows a CT image of the lung of a person with coronavirus and illustrates a radiograph of the lung of a person infected with this challenging virus. Image-processing techniques can help diagnose coronavirus infection and even its progression. It can be argued that image-based methods can be a helpful clinical tool alongside a specialist in helping to save a person's life (Amin et al., 2019b, 2020; Hussain et al., 2016, 2020). The lungs of a person infected with the coronavirus are changed (Khan et al., 2021). This change is seen as a change in the lung tissue (Saba et al., 2019c).

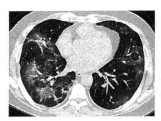 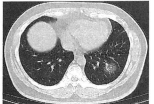

FIGURE 12.1   Lung affected by COVID-19.

Segmentation-based methods can be used to determine and identify these changes (Shi et al., 2020; Saba et al., 2018a, 2018b; Ejaz et al., 2018a, 2018b; Hussain et al., 2020; Iftikhar et al., 2017). Labeling different areas in each image is called image segmentation. The labeled areas in the segmentation are similar in meaning and concept. In another definition, labeling each pixel to specify different image areas is called segmentation (Yousaf et al., 2019b; Saba et al., 2020a). The sections and areas created after segmentation have common features such as illumination and statistics. There are several methods for slicing images. These methods are divided into two categories: supervised and unsupervised (Afza et al., 2020). In supervised methods, unique features are extracted for each pixel, through which the type of pixel is determined (Afza et al., 2019). In the case of determining the areas affected by the coronavirus, each pixel belongs to one of the two groups of coronavirus and health (Al-Ameen et al., 2015). In unsupervised methods, areas that are different from other sectors will be identified using methods such as superpixels, clustering (Amin et al., 2019c), and watershed and so on (Harouni et al., 2012a, 2012b). In unsupervised segmentation algorithms in medical images such as CT, there are four steps to segment the affected or desired region: (1) preprocessing; (2) region of interest (ROI) extraction; (3) determining the affected area using an algorithm; (4) representation using algorithms such as active contour (Chen et al., 2019; Saba, 2020). Figure 12.2 shows the steps of unsupervised image segmentation (Johnson & Xie, 2011).

There are other steps as follows: (1) preprocessing in supervised image segmentation methods; (2) determining a method for extracting a feature from each pixel; (3) pixel classification based on features; (4) labeling each pixel; (5) representation using an algorithm such as active contour. Figure 12.3 shows the steps of supervised methods (Liaqat et al., 2020).

In both supervised and unsupervised methods, evaluation is one of the main and determining steps (Harouni & Baghmaleki, 2020). The purpose of the evaluation is to determine the efficiency of the proposed segmentation method. A ground truth (GT) image is usually used for evaluation. Figure 12.4 shows a sample image of a lung infected with the coronavirus alongside a GT image. Researchers have used criteria such as Dice coefficient, accuracy, precision, sensitivity, F-measure, and call rate for this purpose (Lung et al., 2014; Ejaz et al., 2018, 2019, 2020; Harouni & Baghmaleki, 2020; Majid et al., 2020).

The coronavirus has been affecting human society for more than a year. Researchers have conducted numerous studies to combat this virus, and have

FIGURE 12.2 Steps of unsupervised image segmentation algorithms.

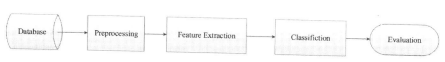

FIGURE 12.3 Steps of supervised image segmentation algorithms.

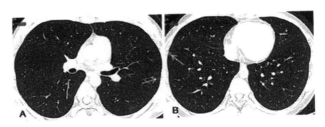

FIGURE 12.4   (a) a lung affected by COVID-19, (b) its GT (Chen et al., 2020b).

proposed various methods and algorithms to identify the lungs infected with the coronavirus in medical images. Although much research has been done, few people have presented an article on the diagnosis of coronavirus. This chapter provides a comprehensive overview of research into the diagnosis of coronavirus in a patient's lung. Therefore, the innovations of this chapter can be described as follows:

(1) A comprehensive review of methods for identifying and segmenting lung images to identify the lung involved with corona;
(2) Introducing databases provided by various organizations to identify the patient's lungs;
(3) Introducing all evaluation criteria in segmentation in medical images;
(4) Comparison of performed methods.

The rest of this article is divided as follows. Section 12.2 introduces the anatomy of the lungs in its entirety and the effect of the coronavirus on the lungs. Most of the methods presented in the segmentation of affected areas are based on deep learning (DL). Therefore, in Section 12.3 DL and its methods will be presented. In Section 12.4, the presented methods will be compared with each other, and finally, in Section 12.5, the chapter's conclusions will be presented.

12.2   LUNG ANATOMY

The lungs are a pair of spongy, air-filled organs located on either side of the chest (thorax), divided into three parts called lobes. The left lung has two lobes and is smaller since it shares space in the chest with the heart. Figure 12.5 shows the lung anatomy.

Air passes through the trachea into the lungs. The trachea is divided into bronchi, which are further divided into alveolar ducts that give rise to the alveolar sacs containing the alveoli. The alveoli absorb oxygen from the air and eliminate carbon dioxide; this is the main function of the lungs. A thin layer called pleura wraps around the lungs and facilitates the lungs' proper function during expansion and contraction. The lungs work in conjunction with the heart to circulate oxygen all through the body. When the heart circulates blood through the cardiovascular circle, the oxygen-poor blood is pumped into the lungs while returning to the heart. The pulmonary artery carries blood from the right side of the heart to the lungs to pick up a fresh oxygen supply. The aorta is the main artery that carries oxygen-rich blood from the

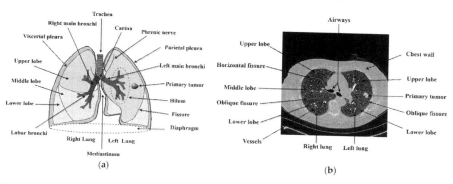

FIGURE 12.5 The lung anatomy details (Paing et al., 2019).

heart's left side to the body. Coronary arteries are the other important arteries attached to the heart. They carry oxygen-rich blood from the aorta to the heart muscle, which must have its own blood supply to function. The alveolar membrane makes the air inside the respiratory system liquid. The oxygen released all over the epithelium of the alveoli sacs is released into the capillaries. Carbon dioxide is also released into the alveolar sacs from the blood inside the capillaries. Now oxygen-rich blood returns to the heart through the pulmonary capillaries. Carbon dioxide is evacuated through the lungs by aspiration. Air reaches the lungs through the breathing process. The diaphragm plays a key role in breathing. The diaphragm is a muscle, which separates the chest and the abdominal cavity. It has a domed shape at rest, which limits the space in the chest. When the diaphragm is contracted, it moves down to the abdominal cavity, which causes the chest to expand. Accordingly, air pressure in the lungs is reduced and so the air is absorbed through the aspiratory airways; this is called aspiration. During exhalation, the diaphragm is relaxed, which decreases the volume of the lung cavity pushing air out. Inspiration is an automatic function of the neural system. It is controlled by a brain area called the medulla oblongata. The neurons of this area send signals to the diaphragm and chest muscles to regulate the contractions for aspiration (Qureshi et al., 2020; Javed et al., 2019a, 2019b, 2020a, 2020b).

COVID-19 is a new coronavirus disease that can cause pulmonary complications such as pneumonia and, in severe cases, acute respiratory distress syndrome (ARDS). Infectious poisoning due to the absorption of bacteria and perishable substances into the blood (sepsis) is another complication of COVID-19 that can cause permanent damage to the lungs and other organs (Xu et al., 2020c).

- COVID-19 pneumonia
  In pneumonia, the lungs became full of fluid and inflamed, making it very difficult to breathe. These breathing problems can be so severe for some people that they need to be treated in a hospital with oxygen or even ventilators. Pneumonia tends to persist in both lungs due to COVID-19. The lungs' air sacs are filled with fluid, resulting in limited ability to take in oxygen, which causes shortness of breath, coughing, and other symptoms. While most people recover from pneumonia without permanent lung damage, pneumonia with COVID-19

can be much more difficult. Even after the disease has passed, lung injuries can cause severe respiratory problems, which may take months to heal (Gattinoni et al., 2020).

- ARDS
  As COVID-19 pneumonia progresses, the air sacs become more filled with fluid leaking from the small arteries inside the lungs, which can eventually lead to shortness of breath, severe respiratory pain syndrome, and lung loss. Patients with ARDS will often be unable to breathe alone and will need respiratory support devices, including ventilators, to circulate oxygen throughout the body. Whether it happens at home or in hospital, ARDS can be fatal. Those who survive ARDS and recover from COVID-19 may have permanent lung ulcers.

- Infectious Poisoning
  Another possible serious complication of COVID-19 is sepsis, or infectious poisoning due to the absorption of bacteria and rotten substances into the blood. Sepsis occurs when an infection enters the bloodstream and spreads through it, causing tissue damage anywhere in the body. The lungs, heart, and other parts of the body system are like instruments in an orchestra. In sepsis, the coordination between the organs of the body is separated. The body's organs, including the lungs and heart, begin to fail one after another. Even if you survive sepsis, it can cause lasting damage to the lungs and other organs (Khan et al., 2020).

- Infection
  When a person becomes infected with COVID-19, the immune system works hard against the attackers. This can protect the body against infection with a bacterium or other virus stronger than COVID-19, and/or make it more vulnerable to further infection which can lead to additional damage to the lungs (Rad et al., 2013; Saba, 2019).

12.3 DL IN COVID-19

Segmentation in medical images assigns a label to pixels or regions so that the resulting areas make a meaning. The image in the segmented area is divided into separate sections, each of which has common features such as the same intensity distribution (Khan et al., 2017, 2019, 2020d; Lung et al., 2014a, 2014b). In medical imaging, accurate observation of tissue boundaries is crucial for tissue observation, especially in differentiating diseased tissue from healthy tissue. Although several methods have been proposed over the past year for the segmentation of lung images to identify affected areas on CT imaging, segmentation accuracy still needs to be improved, especially in the presence of common medical imaging challenges (Mashood Nasir et al., 2020; Mittal et al., 2020).

CT imaging of the lung can detect infection with the coronavirus and the spread of the disease. The effect of COVID-19 on the lung is more infectious than the original one. The lungs become infected with the coronavirus, causing regional changes in the lungs. Infected areas of the lung may be uniform in one place or spread locally throughout the lung. CT images are usually heterogeneous and of poor quality. As a result, the detection algorithm's performance is affected (Husham et al., 2016; Iqbal et al., 2017, 2018, 2019).

Although data mining and ML has been used in many fields and has been a great success, this disadvantage—the similar feature space in real-world training data and real-time test and validation data—is a major challenge when distributing feature changes in feature space. Statistical models able to estimate the new feature space have been proposed to overcome this learning challenge, but this is is, of course, very expensive. The challenge also demonstrates the need to learn to transfer time when educational data is insufficient. The labeled data is too small, or the data is not the same as the new data being recorded, especially in COVID-19 datasets. For example, images were recorded for the affected area in CT devices. In this case, two devices from two different manufacturers for recording MRI scans have different feature spaces or even have differences in the size and number of images. As a result, traditional methods of segmentation of COVID-19-affected lungs will be completely different in each device. In other words, COVID-19-affected lung segmentation of CT images using an algorithm may not be as efficient in one set of compatible images as in another. Or there may not be enough training images in one device, and as a result, the segmentation efficiency will be challenged. In this case, learning will definitely be beneficial (Jamal et al., 2017; Javed et al., 2019, 2020).

12.4 HISTORY OF TRANSFER LEARNING (TL)

There are many sets of TL in real-world learning. For example, when humans recognize an apple as a fruit, they also identify a pear with this identification. Similarly, learning to drive one car can help you drive other cars. The main motivation for research in TL began with the argument that knowledge that individuals learn in one area can solve other problems in similar areas more quickly and easily (Khan et al., 2019a, 2019b, 2019c, 2020a, 2020b, 2020c). But the main motivation for using TL was in the field of VIPS (Vision, Image Processing & Sound) 95 ML. Entitled "Learning to learn," the ML methods can be used for later applications. Figure 12.6 shows the difference between the traditional learning process and TL. Traditional learning methods can transfer each task in each set and use it in the same area. This is where TL methods use the learning of previous tasks in target learning. This is in case they have little training data.

TL is referred to as knowledge transfer, life-long learning, and meta-learning. There is even a discussion of a multitasking learning format that is similar to TL. The main task of TL is to identify and apply knowledge and expertise from previous work and apply it to new work. The purpose of TL is to learn and extract knowledge from one or more tasks and the source's work and use it in a goal task, though the task and the role of tasks and work in the source and purpose are not very symmetrical.

12.5 AIM OF TL

In TL, three issues are considered: (a) what to transfer, (b) how to transfer, (c) when to transfer. The question of "what is to be transferred" seeks to determine which part of the knowledge can be transferred in the field or task. Some knowledge is specific or used for a specific field. But some knowledge can be shared between different areas, and this sharing can also increase efficiency. Therefore, the required training algorithm

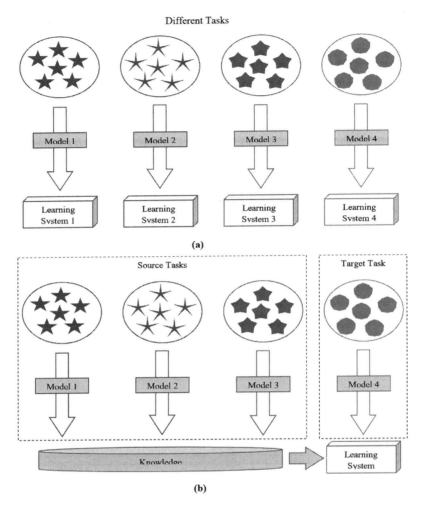

FIGURE 12.6   The difference between transfer and traditional learning: (a) traditional learning model, (b) TL model.

must be designed to identify the reaction to be transmitted. That is when the question of "how to transfer" arises. The question of how to transfer the training algorithm must be properly addressed. Finally, in the question of "when to transfer," the transfer position and specialization are considered. In fact, in this case, there is an interest in the issue of what knowledge is transferred and what knowledge is not transferred.

12.6 COMMON BLOCK DIAGRAM OF CONVOLUTIONAL NEURAL NETWORK (CNN)

In a CNN, there are usually a convolutional layer and ReLU activation function, fusion steps, and sampling steps. In this part, the image enters the CNN and the

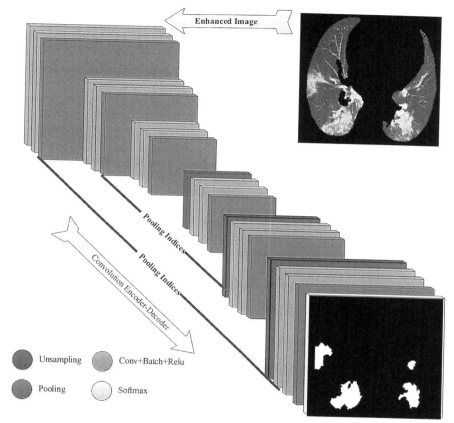

FIGURE 12.7   Segmentation of affected lung area in COVID-19 based on common CNN.

output is the specified area of the affected area. Figure 12.7 shows a designed example of CNN steps.

12.7   DISCUSSION

From one point of view, segmentation methods can be divided into traditional and DL methods (Khan et al., 2020). From another perspective, these methods can be divided into manual methods, e.g. semi-manual and automatic division. Detection and segmentation of CT images to identify lungs affected by COVID-19 are associated with challenges. These challenges are:

(1) The affected area, caused by the coronavirus, has a complex structure. This complex structure occurs due to reticulation, consolidation, and ground-glass opacity (GGO) (Ng et al., 2020).
(2) Changes in the appearance of COVID-19 in the lung and its effects have a completely random shape and size. The dependent has completely variable

dimensions and shapes at the time of its development. The boundaries of these tissues are completely random.

(3) CT images usually have low contrast. Also, noise is very effective in these images (Wang et al., 2020).

DL-based segmentation methods are proposed in the diagnosis of COVID-19. These methods typically use deep off-the-shelf camouflage models (Cao et al., 2020). Wang et al. (2020) used a noise-robust framework for automatic segmentation of the effects of COVID-19. The proposed method is called COVID-19 pneumonia lesion segmentation network (COPLE-Net). In the preprocessing phase, a two-dimensional histogram is applied to the image's brightness level and then the image quality is improved using a Gaussian filter. One of the non-supervisory methods uses cross-entropy to calculate the amount of Kapur's entropy and separates different lung areas using a threshold. Cao et al. (2020) used the U-Net structure for segmentation and was also able to estimate the percentage of virus progression and lung tissue destruction. In Khan et al. (2021), U-Net is used for segmentation. It has achieved good accuracy and has been able to use the base images well, but it is not generalizable. Jin et al. (2020) used U-Net and CNNs to determine areas destroyed by COVID-19. In Chakraborty and Mali (2020), an unsupervised method based on superpixels and hybrid segmentation was used. The evolutionary flower pollination algorithm with a reader in segmentation is also used. This proposed method is called superpixel-based fuzzy modified flower pollination algorithm (SuFMoFPA) and is fully automated. Chen et al. (2020a) provide a supervised segmentation method. This method, which uses DL techniques, uses a U-Net ++ model for segmentation. Wang et al. (2020), using a deep neural network and TL, identify the features needed to determine areas destroyed by COVID-19, which is in the category of supervised methods. Xu et al. (2020a) also used 3D CNNs with Bayesian functions to determine degraded areas. This uses two 3D CNN-based classification models. In this method, the ResNet-18 network and the location-attention-oriented model were combined, which combination was able to identify the affected areas with high accuracy.

12.8   EVALUATION

Evaluation criteria are used to evaluate segmentation methods (Mughal et al., 2018b; Rahim et al., 2017a, 2017b). These criteria are shown in Table 12.1 and detailed in Table 12.2 (Harouni & Baghmaleki, 2020).

12.9   COVID-19 DATABASE

Databases play a significant role in state-of-the-art experiments and comparisons of results (Rad et al., 2016). Despite researchers' efforts to create an image database of the lungs involved in the coronavirus, a comprehensive database has not yet been provided. Although different databases are provided, these images have been recorded in different conditions and each of them has been created for a specific purpose. Tables 12.3 and 12.4 summarize these image databases.

TABLE 12.1
Criteria for Evaluating Segmentation Methods (Javed et al., 2020a, 2020b)

Criteria	Equation
Dice	$\dfrac{2N_{TP}}{2N_{TP} + N_{FP} + N_{FN}}$
Precision	$\dfrac{N_{TP}}{N_{TP} + N_{FP}}$
Recall	$\dfrac{N_{TP}}{N_{TP} + N_{FN}}$
Accuracy	$\dfrac{TP + FP}{TP + TN + FP + FN}$
Sensitivity	(Harouni & Baghmaleki, 2020) $\dfrac{TP}{TP + FP}$
Specificity	$\dfrac{TN}{TN + FP}$

TABLE 12.2
Details of Equations

Abbreviation	Describe
TP	The pixels in the main image and the *GT* image are identified as an area affected by COVID-19.
TN	The pixels in the main image and the *GT* image are not recognized as an area affected by COVID-19.
FN	In this case, the corresponding pixel in the *GT* image is a vessel, but it is not detected in the image for lung segmentation.
FP	In this case, the pixel is detected in the segmented image, but it is not affected in the *GT* image.

TABLE 12.3
Summary of COVID-19 Image Databases (Fan et al., 2020)

Dataset	Modality	Number of Covid-19	Number of Non-Covid-19	Purpose
COVID-19 X-ray Collection	CT	229	0	Diagnosis
COVID-19 CT Collection	X-rays	20	0	Diagnosis
COVID-CT-Dataset	CT	288	1,000	Diagnosis
COVID-19 Patients' Lungs	CT	70	28	Diagnosis
COVID-19 Radiography	X-rays	219	286	Diagnosis
COVID-19 CT Segmentation	CT	110	0	Segmentation

TABLE 12.4
Statistics of Open-Source COVID-19 Datasets (Laradji et al., 2020)

Name	Number of COVID-19	Number of Slides	Number of Slides with Infections (Paing et al.)	Number of Infected Regions
COVID-19-A	60	98	100	77
COVID-19-B	9	829	44.9	1,488
COVID-19-C	20	3,520	52.3	5,608

12.10 CONCLUSION

COVID-19 has spread around the world within a year. It will cause a lot of psychological, social, economic, and cultural damage. The virus affects different parts of the body, including the lungs. The affected area can be seen on CT images. Segmentation-based methods are very efficient in image processing for detecting the affected area. Researchers have proposed several methods for slicing lung images. These methods are divided into two categories: supervised and unsupervised. From another point of view, these methods are classified as traditional-based methods and DL. This article has provided an overview of lung segmentation methods in determining the area affected by COVID-19. Image databases and evaluation criteria have also been introduced.

REFERENCES

Abbas, A., Saba, T., Rehman, A., Mehmood, Z., Javaid, N., Tahir, M., Khan, N.U., Ahmed, K.T., & Shah, R. (2019b) Plasmodium species aware based quantification of malaria, parasitemia in light microscopy thin blood smear, *Microscopy Research and Technique*, 82(7), 1198–1214. doi: 10.1002/jemt.23269.

Abbas, N., Saba, T., Mehmood, Z., Rehman, A., Islam, N., & Ahmed, K. T. (2019a). An automated nuclei segmentation of leukocytes from microscopic digital images. *Pakistan Journal of Pharmaceutical Sciences*, 32(5), 2123–2138.

Abbas, N. Saba, T. Mohamad, D. Rehman, A. Almazyad, A.S., & Al-Ghamdi, J.S. (2018) Machine aided malaria parasitemia detection in Giemsa-stained thin blood smears, *Neural Computing and Applications*, 29(3), 803818, doi:10.1007/s00521-016-2474-6.

Abbas, N., Saba, T., Rehman, A., Mehmood, Z., Kolivand, H., Uddin, M., & Anjum, A. (2019c). Plasmodium life cycle stage classification-based quantification of malaria parasitaemia in thin blood smears. *Microscopy Research and Technique*, 82(3), 283–295. doi:10.1002/jemt.23170.

Adeel, A., Khan, M. A., Akram, T., Sharif, A., Yasmin, M., Saba, T., & Javed, K. (2020). Entropy-controlled deep features selection framework for grape leaf diseases recognition. *Expert Systems*, 1(2), 1–25.

Afza, F., Khan, M. A., Sharif, M., & Rehman, A. (2019). Microscopic skin laceration segmentation and classification: A framework of statistical normal distribution and optimal feature selection. *Microscopy Research and Technique*, 82(9), 1471–1488.

Afza, F., Khan, M. A., Sharif, M., Saba, T., Rehman, A., & Javed, M. Y. (2020, October). *Skin Lesion Classification: An Optimized Framework of Optimal Color Features Selection*. In *2020 2nd International Conference on Computer and Information Sciences (ICCIS)* (pp. 1–6). IEEE.

Al-Ameen, Z. Sulong, G. Rehman, A., Al-Dhelaan, A. Saba, T., Al-Rodhaan, M. (2015) An innovative technique for contrast enhancement of computed tomography images using normalized gamma-corrected contrast-limited adaptive histogram equalization. *EURASIP Journal on Advances in Signal Processing:32*, 32, 1–12. doi:10.1186/s13634-015-0214-1

Amin, J., Sharif, M., Raza, M., Saba, T., & Anjum, M. A. (2019c). Brain tumor detection using statistical and machine learning method. *Computer Methods and Programs in Biomedicine*, 177, 69–79.

Amin, J., Sharif, M., Raza, M., Saba, T., & Rehman, A. (2019a). *Brain Tumor Classification: Feature Fusion*. In *2019 International Conference on Computer and Information Sciences (ICCIS)* (pp. 1–6). IEEE.

Amin, J., Sharif, M., Raza, M., Saba, T., Sial, R., & Shad, S. A. (2020). Brain tumor detection: A long short-term memory (LSTM)-based learning model. *Neural Computing and Applications*, 32, 15965–15973.

Amin, J., Sharif, M., Rehman, A., Raza, M., & Mufti, M. R. (2018). Diabetic retinopathy detection and classification using hybrid feature set. *Microscopy Research and Technique*, 81(9), 990–996.

Amin, J., Sharif, M., Yasmin, M. T Saba, M Raza (2019b) "Use of machine intelligence to conduct analysis of human brain data for detection of abnormalities in its cognitive functions" *Multimed Tools and Application*, 79(15), 10955–10973. doi:10.1007/s11042-019-7324-y

Benvenuto, D., Giovanetti, M., Salemi, M., Prosperi, M., De Flora, C., Junior Alcantara, L. C., Angeletti, S., & Ciccozzi, M. 2020. The global spread of 2019-nCoV: A molecular evolutionary analysis. *Pathogens and Global Health*, 114, 64–67.

Cao, Y., Xu, Z., Feng, J., Jin, C., Han, X., Wu, H. & Shi, H. 2020. Longitudinal assessment of covid-19 using a deep learning–based quantitative ct pipeline: Illustration of two cases. *Radiology: Cardiothoracic Imaging*, 2, e200082.

Chakraborty, S., & Mall, K. 2020. SuFMoFPA: A superpixel and meta-heuristic based fuzzy image segmentation approach to explicate COVID-19 radiological images. *Expert Systems with Applications*, 167(11), 114142.

Chen, J., Wu, L., Zhang, J., Zhang, L., Gong, D., Zhao, Y., Hu, S., Wang, Y., Hu, X. & Zheng, B. 2020b. Deep learning-based model for detecting 2019 novel coronavirus pneumonia on high-resolution computed tomography: a prospective study. *MedRxiv*.

Chen, X., Williams, B. M., Vallabhaneni, S. R., Czanner, G., Williams, R. & Zheng, Y. Learning active contour models for medical image segmentation. *Proceedings of the IEEE Conference on Computer Vision and Pattern Recognition*, 2019. 11632–11640.

Chen, Z., Fan, H., Cai, J., Li, Y., Wu, B., Hou, Y., Xu, S., Zhou, F., Liu, Y. & Xuan, W. 2020a. High-resolution computed tomography manifestations of COVID-19 infections in patients of different ages. *European Journal of Radiology*, 126, 108972.

Ejaz, K., Rahim, D. M. S. M., Rehman, D. A., Ejaz, E. F. (2018b). An image-based multimedia database and efficient detection though features. *VFAST Transactions on Software Engineering*, 14(1), 6–15.

Ejaz, K., Rahim, M. S. M., Bajwa, U. I., Chaudhry, H., Rehman, A., Ejaz, F. (2020) Hybrid segmentation method with confidence region detection for tumor identification. *IEEE Access*, doi:10.1109/ACCESS.2020.3016627

Ejaz, K., Rahim, M. S. M., Bajwa, U. I., Rana, N., & Rehman, A. (2019). An *Unsupervised Learning with Feature Approach for Brain Tumor Segmentation Using Magnetic Resonance Imaging*. In *Proceedings of the 2019 9th International Conference on Bioscience, Biochemistry and Bioinformatics* (pp. 1–7).

Ejaz, K., Rahim, M. S. M., Rehman, A., Chaudhry, H., Saba, T., & Ejaz, A. (2018a). Segmentation method for pathological brain tumor and accurate detection using MRI. *International Journal of Advanced Computer Science and Applications*, 9(8), 394–401.

Fahad H. M., Khan M. U. G., Saba T., Rehman A., & Iqbal S. (2018) Microscopic abnormality classification of cardiac murmurs using ANFIS and HMM.. *Microscopy Research and Technique*, 81(5), 449–457. doi: 10.1002/jemt.22998.

Fan, D. P., Zhou, T., Ji, G.-P., Zhou, Y., Chen, G., Fu, H., Shen, J., & Shao, L. 2020. Inf-Net: Automatic COVID-19 lung infection segmentation from CT images. *IEEE Transactions on Medical Imaging*, 39(8), 2626–2637.

Gattinoni, L., Chiumello, D. & Rossi, S. 2020. COVID-19 pneumonia: ARDS or not? : *BioMed Central*, 24, 154.

Habibi, N., & Harouni, M. 2018. Estimation of Re-hospitalization Risk of Diabetic Patients based on Radial Base Function (RBF) *Neural Network Method Combined with Colonial Competition Optimization Algorithm*. 12, 109–116.

Harouni, M., & Baghmaleki, H. 2020. Color image segmentation metrics. arXiv:2010.09907.

Harouni, M., Mohamad, D., Rahim, M. S. M., & Halawani, S. M. 2012b. Finding critical points of handwritten Persian/Arabic character. *International Journal of Machine Learning and Computing*, 2, 573.Harouni, M., Rahim, M., Al-Rodhaan, M., Saba, T., Rehman, A., & Al-Dhelaan, A. 2014. Online Persian/Arabic script classification without contextual information. *The Imaging Science Journal*, 62, 437–448.

Harouni, M., Mohamad, D., Rahim, M. S. M., Halawani, S. M., & Afzali, M. L. 2012a. Handwritten Arabic character recognition based on minimal geometric features. *International Journal of Machine Learning and Computing*, 2, 578.

Harouni, M., Mohamad, D., & Rasouli, A. *Deductive method for recognition of on-line hand-written Persian/Arabic characters. The 2nd International Conference on Computer and Automation Engineering (ICCAE)*, 2010. IEEE, 791–795.

Harouni, M., Rahim, M.S.M., Al-Rodhaan, M., Saba, T., Rehman, A., Al-Dhelaan, A. (2014) Online Persian/Arabic script classification without contextual information, *The Imaging Science Journal*, 62(8), 437–448, doi. 10.1179/1743131X14Y.0000000083

Husham, A., Alkawaz, M. H., Saba, T., Rehman, A., & Alghamdi, J.S. (2016) Automated nuclei segmentation of malignant using level sets, *Microscopy Research and Technique*, 79(10), 993–997, doi. 10.1002/jemt.22733.

Hussain, N., Khan, M. A., Sharif, M., Khan, S. A., Albesher, A. A., Saba, T., & Armaghan, A. (2020). A deep neural network and classical features based scheme for objects recognition: An application for machine inspection. *Multimed Tools and Application*. doi:10.1007/s11042-020-08852-3.

Iftikhar, S. Fatima, K. Rehman, A. Almazyad, A.S., & Saba, T. (2017) An evolution based hybrid approach for heart diseases classification and associated risk factors identification. *Biomedical Research* 28 (8), 3451–3455.

Iqbal, S. Ghani, M.U. Saba, T., & Rehman, A. (2018). Brain tumor segmentation in multi-spectral MRI using convolutional neural networks (CNN). *Microscopy Research and Technique* 81(4), 419–427. doi: 10.1002/jemt.22994.

Iqbal, S., Khan, M.U.G., Saba, T. Mehmood, Z. Javaid, N., Rehman, A., & Abbasi, R. (2019) Deep learning model integrating features and novel classifiers fusion for brain tumor segmentation. *Microscopy Research and Technique*, 82(8), 1302–1315, doi:10.1002/jemt.23281

Iqbal, S., Khan, M. U. G., Saba, T., & Rehman, A. (2017). Computer assisted brain tumor type discrimination using magnetic resonance imaging features. *Biomedical Engineering Letters*, 8(1), 5–28. doi: 10.1007/s13534-017-0050-3.

Jamal A, Hazim Alkawaz M, Rehman A, & Saba T. (2017). Retinal imaging analysis based on vessel detection. *Microscopy Research and Technique*, 80 (17), 799–811. doi:10.1002/jemt.

Javed, R., Rahim, M.S.M., & Saba, T. (2019b) An improved framework by mapping salient features for skin lesion detection and classification using the optimized hybrid features. *International Journal of Advanced Trends in Computer Science and Engineering*, 8(1), 95–101.

Javed, R., Rahim, M. S. M., Saba, T., & Rashid, M. (2019a). Region-based active contour JSEG fusion technique for skin lesion segmentation from dermoscopic images. *Biomedical Research*, 30(6), 1–10.

Javed, R., Rahim, MSM, Saba, T., & Rehman, A. (2020a) A comparative study of features selection for skin lesion detection from dermoscopic images. *Network Modeling Analysis in Health Informatics and Bioinformatics*, 9 (1), 4.

Javed, R., Saba, T., Shafry, M., & Rahim, M. (2020b). *An Intelligent Saliency Segmentation Technique and Classification of Low Contrast Skin Lesion Dermoscopic Images Based on Histogram Decision*. In *2019 12th International Conference on Developments in eSystems Engineering (DeSE)* (pp. 164–169).

Jin, S., Wang, B., Xu, H., Luo, C., Wei, L., Zhao, W., Hou, X., Ma, W., Xu, Z., & Zheng, Z. 2020. AI-assisted CT imaging analysis for COVID-19 screening: Building and deploying a medical AI system in four weeks. *medRxiv*.

Johnson, B., & Xie, Z. 2011. Unsupervised image segmentation evaluation and refinement using a multi-scale approach. *ISPRS Journal of Photogrammetry and Remote Sensing*, 66, 473–483.

Khan, M. A., Akram, T., Sharif, M., Javed, K., Raza, M., & Saba, T. (2020d). An automated system for cucumber leaf diseased spot detection and classification using improved saliency method and deep features selection. *Multimedia Tools and Applications*, 1(1), 1–30.

Khan, M. A.; Akram, T. Sharif, M., Saba, T., Javed, K., Lali, I.U., Tanik, U.J., & Rehman, A. (2019b). Construction of saliency map and hybrid set of features for efficient segmentation and classification of skin lesion, *Microscopy Research and Technique*, 82(5), 741–763, doi:10.1002/jemt.23220

Khan, M. A., Ashraf, I., Alhaisoni, M., Damaševičius, R., Scherer, R., Rehman, A., & Bukhari, S.A.C. (2020a) Multimodal brain tumor classification using deep learning and Robust feature selection: A machine learning application for radiologists. *Diagnostics*, 10, 565.

Khan, M. A., Javed, M. Y., Sharif, M., Saba, T., & Rehman, A. (2019c). *Multi-Model Deep Neural Network Based Features Extraction And Optimal Selection Approach For Skin Lesion Classification*. In *2019 International Conference On Computer And Information Sciences (ICCIS)* (pp. 1–7). IEEE.

Khan, M. A. Kadry, S., Zhang, Y. D., Akram, T.,Sharif, M., Rehman, A., Saba, T. (2021) Prediction of COVID-19 - pneumonia based on selected deep features and one class kernel extreme learning machine. *Computers & Electrical Engineering*, 1(90), 106960–106970.

Khan, M.A., Lali, I.U. Rehman, A. Ishaq, M. Sharif, M. Saba, T., Zahoor, S., & Akram, T. (2019e) Brain tumor detection and classification: A framework of marker-based watershed algorithm and multilevel priority features selection, *Microscopy Research and Technique*, 82(6), 909–922, doi:10.1002/jemt.23238

Khan, M. A., Sharif, M. Akram, T., Raza, M., Saba, T., & Rehman, A. (2020c) Hand-crafted and deep convolutional neural network features fusion and selection strategy: An application to intelligent human action recognition. *Applied Soft Computing* 87, 105986

Khan, M. A., Sharif, M. I., Raza, M., Anjum, A., Saba, T., & Shad, S. A. (2019a). Skin lesion segmentation and classification: A unified framework of deep neural network features fusion and selection. *Expert Systems*, e12497.

Khan, M. W., Sharif, M., Yasmin, M., & Saba, T. (2017). CDR based glaucoma detection using fundus images: A review. *International Journal of Applied Pattern Recognition*, 4(3), 261–306.

Khan, M. Z., Jabeen, S., Khan, M. U. G., Saba, T., Rehmat, A., Rehman, A., & Tariq, U. (2020b). A realistic image generation of face from text description using the fully trained generative adversarial networks. *IEEE Access*, 81, 106524.

Khan, S. A., Nazir, M., Khan, M. A., Saba, T., Javed, K., Rehman, A., ... & Awais, M. (2019d). Lungs nodule detection framework from computed tomography images using support vector machine. *Microscopy Research and Technique*, 82(8), 1256–1266.

Laradji, I., Rodriguez, P., Manas, O., Lensink, K., Law, M., Kurzman, L., Parker, W., Vazquez, D. & Nowrouzezahrai, D. 2020. A weakly supervised consistency-based learning method for covid-19 segmentation in ct images. arXiv preprint arXiv:2007.02180.

Liaqat, A., Khan, M. A., Sharif, M., Mittal, M., Saba, T., Manic, K. S., & Al Attar, F. N. H. (2020). Gastric tract infections detection and classification from wireless capsule endoscopy using computer vision techniques: A review. *Current Medical Imaging*, 16(10), 1229–1242.

Lung, J. W. J., Salam, M. S. H., Rehman, A., Rahim, M. S. M., & Saba, T. (2014a) Fuzzy phoneme classification using multi-speaker vocal tract length normalization, *IETE Technical Review*, 31(2), 128–136, doi: 10.1080/02564602.2014.892669

Lung, J.W.J., Salam, M.S.H., Rehman, A., Rahim, M.S.M., & Saba, T. (2014b) Fuzzy phoneme classification using multi-speaker vocal tract length normalization. *IETE Technical Review*, 31(2), 128–136, doi: 10.1080/02564602.2014.892669.

Majid, A., Khan, M. A., Yasmin, M., Rehman, A., Yousafzai, A., & Tariq, U. (2020). Classification of stomach infections: A paradigm of convolutional neural network along with classical features fusion and selection. *Microscopy Research and Technique*, 83(5), 562–576.

Marie-Sainte, S. L. Aburahmah, L., Almohaini, R., & Saba, T. (2019a). Current techniques for diabetes prediction: Review and case study. *Applied Sciences*, 9(21), 4604.

Marie-Sainte, S. L., Saba, T., Alsaleh, D., Alotaibi, A., & Bin, M. (2019b). An improved strategy for predicting diagnosis, survivability, and Recurrence of Breast Cancer. *Journal of Computational and Theoretical Nanoscience*, 16(9), 3705–3711.

Mashood Nasir, I., Attique Khan, M., Alhaisoni, M., Saba, T., Rehman, A., & Iqbal, T. (2020). A hybrid deep learning architecture for the classification of superhero fashion products: An application for medical-tech classification. *Computer Modeling in Engineering & Sciences*, 124(3), 1017–1033.

Meethongjan, K. Dzulkifli, M. Rehman, A. Altameem, A., & Saba, T. (2013) An intelligent fused approach for face recognition, *Journal of Intelligent Systems*, 22(2), 197–212. doi: 10.1515/jisys-2013-0010

Mittal, A., Kumar, D., Mittal, M., Saba, T., Abunadi, I., Rehman, A., & Roy, S. (2020). Detecting pneumonia using convolutions and dynamic capsule routing for chest X-ray images. *Sensors*, 20(4), 1068.

Mughal, B., Muhammad, N., Sharif, M., Rehman, A., & Saba, T. (2018a). Removal of pectoral muscle based on topographic map and shape-shifting silhouette. *BMC Cancer*, 18(1), 1–14.

Mughal, B. Muhammad, N. Sharif, M. Saba, T., & Rehman, A. (2017) Extraction of breast border and removal of pectoral muscle in wavelet domain. *Biomedical Research*, 28 (11), 5041–5043.

Mughal, B., Sharif, M., Muhammad, N., & Saba, T. (2018b). A novel classification scheme to decline the mortality rate among women due to breast tumor. *Microscopy Research and Technique*, 81(2), 171–180.

Nazir, M., Khan, M. A., Saba, T., & Rehman, A. (2019). *Brain Tumor Detection from MRI images using Multi-level Wavelets*. In *2019, IEEE International Conference on Computer and Information Sciences (ICCIS)* (pp. 1–5).

Ng, M.-Y., Lee, E. Y., Yang, J., Yang, F., Li, X., Wang, H., Lui, M. M.-S., Lo, C. S.-Y., Leung, B. & Khong, P.-L. 2020. Imaging profile of the COVID-19 infection: radiologic findings and literature review. *Radiology: Cardiothoracic Imaging*, 2, e200034.

Norbash, A. M., Van Moore Jr, A., Recht, M. P., Brink, J. A., Hess, C. P., Won, J. J., Jain, S., Sun, X., Brown, M., & Enzmann, D. 2020. Early-stage radiology volume effects and considerations with the coronavirus disease 2019 (COVID-19) pandemic: adaptations, risks and lessons learned. *Journal of the American College of Radiology*, 17, 1086–1095.

Paing, M. P., Hamamoto, K., Tungjitkusolmun, S., & Pintavirooj, C. J. A. S. 2019. Automatic detection and staging of lung tumors using locational features and double-staged classifications. Applied Sciences, 9, 2329.

Perveen, S., Shahbaz, M., Saba, T., Keshavjee, K., Rehman, A., & Guergachi, A. (2020). Handling irregularly sampled longitudinal data and prognostic modeling of diabetes using machine learning technique. IEEE Access, 8, 21875–21885.

Qureshi, I. Khan, M. A. A., Sharif, M., Saba, T., & Ma, J. (2020) Detection of glaucoma based on cup-to-disc ratio using fundus images. International Journal of Intelligent Systems Technologies and Applications, 19(1), pp. 1–16, doi:10.1504/IJISTA.2020.105172

Rad, A. E., Rahim, M. S. M., Rehman, A. Altameem, A., & Saba, T. (2013) Evaluation of current dental radiographs segmentation approaches in computer-aided applications IETE Technical Review, 30(3), 210–222.

Rad, A. E., Rahim, M. S. M., Rehman, A., & Saba, T. (2016) Digital dental X-ray database for caries screening. 3D Research, 7(2), 1–5, doi:10.1007/s13319-016-0096-5

Rahim, M. S. M., Norouzi, A. Rehman, A., & Saba, T. (2017a) 3D bones segmentation based on CT images visualization. Biomedical Research, 28(8), 3641–3644.

Rahim, M. S. M., Rehman, A. Kurniawan, F., & Saba, T. (2017b) Ear biometrics for human classification based on region features mining, Biomedical Research, 28 (10), 4660–4664.

Ramzan, F., Khan, M. U. G., Iqbal, S., Saba, T., & Rehman, A. (2020b). Volumetric Segmentation of Brain Regions From MRI Scans Using 3D Convolutional Neural Networks. IEEE Access, 8, 103697–103709.

Ramzan, F., Khan, M. U. G., Rehmat, A., Iqbal, S., Saba, T., Rehman, A., & Mehmood, Z. (2020a). A deep learning approach for automated diagnosis and multi-class classification of Alzheimer's disease stages using resting-state fMRI and residual neural networks. Journal of Medical Systems, 44(2), 37.

Rehman, A. Abbas N. Saba, T. Mahmood, T., & Kolivand, H. (2018a). Rouleaux red blood cells splitting in microscopic thin blood smear images via local maxima, circles drawing, and mapping with original RBCs. Microscopic Research and Technique, 81(7), 737–744. doi: 10.1002/jemt.23030

Rehman, A. Abbas, N. Saba, T. Mehmood, Z. Mahmood, T., & Ahmed, K. T. (2018b) Microscopic malaria parasitemia diagnosis and grading on benchmark datasets, Microscopic Research and Technique, 81(9), 1042–1058. doi: 10.1002/jemt.23071

Rehman, A., Abbas, N., Saba, T., Rahman, S.I.U., Mehmood, Z., & Kolivand, K. (2018c) Classification of acute lymphoblastic leukemia using deep learning. Microscopy Research & Technique, 81(11), 1310–1317. doi: 10.1002/jemt.23139

Rehman, A., & Harouni, M., Saba (2020a). Cursive multilingual characters recognition based on hard geometric features. International Journal of Computational Vision and Robotics, 10, 213–222.

Rehman, A. Khan, M. A. Mehmood, Z. Saba, T. Sardaraz, M., & Rashid, M. (2020b) Microscopic melanoma detection and classification: A framework of pixel-based fusion and multilevel features reduction, Microscopy Research and Technique, 83(4), 410–423, doi: 10.1002/jemt.23429

Rehman, A., Khan, M. A., Saba, T., Mehmood, Z., Tariq, U., & Ayesha, N. (2021). Microscopic brain tumor detection and classification using 3D CNN and feature selection architecture. Microscopy Research and Technique, 84(1), 133–149. doi:10.1002/jemt.23597.

Reusken, C. B., Broberg, E. K., Haagmans, B., Meijer, A., Corman, V. M., Papa, A., Charrel, R., Drosten, C., Koopmans, M. & Leitmeyer, K. (2020). Laboratory readiness and response for novel coronavirus (2019-nCoV) in expert laboratories in 30 EU/EEA countries. Eurosurveillance, 25, 2000082.

Saba, T. (2017) Halal food identification with neural assisted enhanced RFID antenna. Biomedical Research, 28(18), 7760–7762.

Saba, T. (2019). Automated lung nodule detection and classification based on multiple classi-fiers voting. *Microscopy Research and Technique*, 82(9), 1601–1609.

Saba, T. (2020). Recent advancement in cancer detection using machine learning: Systematic survey of decades, comparisons and challenges. *Journal of Infection and Public Health*, 13(9), 1274–1289.

Saba, T., Bokhari, S. T. F., Sharif, M., Yasmin, M., & Raza, M. (2018a). Fundus image clas-sification methods for the detection of glaucoma: A review. *Microscopy Research and Technique*, 81(10), 1105–1121.

Saba, T., Haseeb, K., Ahmed, I., & Rehman, A. (2020b). Secure and energy-efficient frame-work using Internet of Medical Things for e-healthcare. *Journal of Infection and Public Health*, 13(10), 1567–1575.

Saba, T., Khan, M. A., Rehman, A., & Marie-Sainte, S. L. (2019a). Region extraction and clas-sification of skin cancer: A heterogeneous framework of deep CNN features fusion and reduction. *Journal of Medical Systems*, 43(9), 289.

Saba, T., Khan, S.U., Islam, N., Abbas, N., Rehman, A., Javaid, N., & Anjum, A., (2019b). Cloud-based decision support system for the detection and classification of malignant cells in breast cancer using breast cytology images. *Microscopy Research and Technique*, 82(6), pp. 775–785.

Saba, T., Mohamed, A. S., El-Affendi, M. Amin, J., & Sharif, M. (2020a). Brain tumor detec-tion using fusion of hand crafted and deep learning features. *Cognitive Systems Research*, 59, 221–230.

Saba, T., Rehman, A. Mehmood, Z., Kolivand, H., & Sharif, M. (2018b). Image enhancement and segmentation techniques for detection of knee joint diseases: A survey. *Current Medical Imaging Reviews*, 14(5), 704–715, doi. 10.2174/1573405613666170912164546

Saba, T., Sameh, A., Khan, F., Shad, S. A., & Sharif, M. (2019c). Lung nodule detection based on ensemble of hand crafted and deep features. *Journal of Medical Systems*, 43 (12), 332.

Sadad, T. Munir, A. Saba, T., & Hussain, A. (2018) Fuzzy C-means and region growing based classification of tumor from mammograms using hybrid texture feature. *Journal of Computational Science*, 29, 34–45.

Sharif, M. Khan, M. A., Akram, T. Javed, M.Y. Saba, T., & Rehman, A (2017) A framework of human detection and action recognition based on uniform segmentation and combination of Euclidean distance and joint entropy-based features selection. *EURASIP Journal on Image and Video Processing*, 89(1), 1–18.

Shi, F., Wang, J., Shi, J., Wu, Z., Wang, Q., Tang, Z., He, K., Shi, Y., & Shen, D. 2020. Review of artificial intelligence techniques in imaging data acquisition, segmentation and diag-nosis for covid-19. *IEEE Reviews in Biomedical Engineering*, 14, 4–15.

Ullah, H., Saba, T., Islam, N., Abbas, N., Rehman, A., Mehmood, Z., & Anjum, A. (2019). An ensemble classification of exudates in color fundus images using an evolutionary algo-rithm based optimal features selection. *Microscopy Research and Technique*, 82(4), 361–372.

Wang, G., Liu, X., Li, C., Xu, Z., Ruan, J., Zhu, H., Meng, T., Li, K., Huang, N.. & Zhang, S. 2020. A noise-robust framework for automatic segmentation of COVID-19 pneumonia lesions from CT images. *IEEE Transactions on Medical Imaging*, 39, 2653–2663.

Wu, W., Zhang, Y., Wang, P., Zhang, L., Wang, G., Lei, G., Xiao, Q., Cao, X., Bian, Y., & Xie, S. 2020. Psychological stress of medical staffs during outbreak of COVID-19 and adjustment strategy. *Journal of Medical Virology*, 92(10), 1962–1970.

Xu, X., Jiang, X., Ma, C., Du, P., Li, X., Lv, S., Yu, L., Ni, Q., Chen, Y., & Su, J. 2020a. A deep learning system to screen novel coronavirus disease 2019 pneumonia. *Engineering*, 6(10), 1122–1129.

Xu, Y., Xiao, M., Liu, X., Xu, S., Du, T., Xu, J., Yang, Q., Xu, Y., Han, Y., & Li, T. 2020b. Significance of serology testing to assist timely diagnosis of SARS-CoV-2 infections: Implication from a family cluster. *Emerging Microbes & Infections*, 9, 924–927.

Xu, Z., Shi, L., Wang, Y., Zhang, J., Huang, L., Zhang, C., Liu, S., Zhao, P., Liu, H., & Zhu, L. 2020c. Pathological findings of COVID-19 associated with acute respiratory distress syndrome. *The Lancet Respiratory Medicine*, 8, 420–422.

Yaseen, S., Abbas, S. M. A., Anjum, A., Saba, T., Khan, A., Malik, S. U. R., … & Bashir, A. K. (2018). Improved generalization for secure data publishing. *IEEE Access*, 6, 27156–27165.

Yousaf, K., Mehmood, Z., Awan, I. A., Saba, T., Alharbey, R., Qadah, T., & Alrige, M. A. (2019a). A comprehensive study of mobile-health based assistive technology for the healthcare of dementia and Alzheimer's disease (AD). *Health Care Management Science*, 1–23.

Yousaf, K. Mehmood, Z. Saba, T. Rehman, A. Munshi, A.M. Alharbey, R. Rashid, M. (2019b). Mobile-health applications for the efficient delivery of health care facility to people with dementia (PwD) and support to their carers: A survey. *BioMed Research International*, 2019, 1–26.

13 A Review of Feature Selection Algorithms in Determining the Factors Affecting COVID-19

Sogand B. Jaferi, Ziafat Rahmati, Shadi Rafieipour, Nakisa Tavakoli, and Shima Zarrabi Baboldasht

CONTENTS

13.1 INTRODUCTION

The unknown COVID-19 virus is spreading around the world. The outbreak of this disease and this virus is spreading rapidly and many countries are affected by this virus. Numerous studies have been performed to determine the causes of COVID-19 outbreaks. Some of the most up-to-date research in diagnosing this virus concerns the use of data-mining methods. Any technique that gives us new insights into data can be considered data mining. In short, data mining has acted as a bridge between computer science, statistics, artificial intelligence (AI), modeling, machine learning (ML), and visual data representation (Pang et al., 2021). It can be said that data mining, by combining database theories, ML, and AI, as well as statistical science, provides a new field of application (Habibi & Harouni, 2018(. Data mining is one of the recent advances in the industry of utilizing hidden knowledge of data (Harouni et al., 2010(. In essence, data mining is a set of different techniques that enable a person to move beyond ordinary data processing and to explore information in the mass of data and to gain its hidden knowledge (Kenari et al., 2010(. Much data has been generated in the last year with the outbreak of COVID-19. The available data, both in number and size, has been significantly increased in many COVID-19 ML applications. Based on knowledge acquisition, studying how to use this large-scale data is very important and necessary. The sheer volume of large-scale data has posed a significant challenge to ML methods (Xia & Chen, 2021). Due to the presence of noisy, irrelevant, and extra data, not only do learning algorithms become very slow and reduce the efficiency of learning tasks, but they also make the model difficult to interpret. Feature selection, by removing noisy, irrelevant, and extra features, is able to select a small subset of related features from among the main features (Kenthapadi et al., 2019(. The exact definition of feature selection depends on the context in which it is used. But the most widely used definition is that feature selection is the selection of a subset of features with the best result of the classification function (Wu et al., 2019(. The reason for this definition is that additional or irrelevant features often have similar noise in the data, which causes the classifier to be mistaken and the classifier function to be degraded. Eliminating such attributes causes the resulting attributes to have the same or higher classification function than the total attributes. As a direct result, fewer features are required for storage, and thus the classification operation is accelerated. In addition, reducing the number of features helps the expert to focus on a subset of the relevant features, which gives them a better view of the process described by the data (Harouni et al., 2012c; Sun et al., 2020; Rehman et al., 2018c; Harouni et al., 2012a). In general, feature selection helps to better understand the data, reduce computational requirements, reduce the detrimental effect of dimensions, and improve predicted performance. Feature selection focuses on selecting a subset of input variables that can effectively describe input data, reduce noise effects and irrelevant variables, while providing well-predicted results (Sun et al., 2020).

- Classification
  In supervised learning, which is a general method of ML, a system is given a set of output/input pairs and the system tries to learn a function from input to output (Harouni et al., 2012a). Supervised learning requires a number of input data in order to train the system (Harouni et al., 2012b). So in fact when we use the learning algorithm we have a set of learning examples that, for each input, output value or the corresponding function, is also specified. The purpose of the learning system is to obtain a hypothesis that guesses the function or relationship between input and output, which is called supervised learning. This classification has been considered in many studies of ML and pattern recognition (Harouni et al., 2012a; Harouni et al., 2014; Kenari et al., 2010; Mohammadi Dashti & Harouni, 2018; Rehman et al., 2020), including identity and gender identification (Kenari et al., 2010 (and disease diagnosis (Mohammadi Dashti & Harouni, 2018).
- *Artificial Neural Networks (ANN)*
  Neural networks are a data-processing system that takes ideas from the human brain and entrusts data processing to many small processors that act as interconnected and parallel networks to solve a problem. In these networks, with the help of programming knowledge, a data structure is designed that can act like a neuron; this structure is called neurons. They then train the network by creating a network between these neurons and applying a training algorithm to it. There are different types of neural networks that can be used, based on their structure (Harouni et al., 2012b; Rehman et al., 2020; Harouni et al., 2010). One common model is the multilayer perceptron neural network (MLP) (Harouni et al., 2012b).

- *K-Nearest Neighbors (KNN)*

  One of the best classifications in the field of pattern recognition, especially iden-tification, is the K class of the nearest neighbors (Rehman et al., 2020). This classification considers the test sample to belong to the class that has the highest number of votes among its nearest neighbors. To obtain the nearest neighbors of a sample, different methods of neighborhood calculation are usually used. In KNN classifiers, the distance between samples in data is calculated, and then, based on the value of K parameter, the nearest data are categorized in a class (Harouni et al., 2010). In Figure 13.1 classification using KNN is represented.

- *Support Vector Machine (SVM)*

  The mechanism of action of SVM is based on the rule that structural risk mini-mization is based on reducing the decision error (for example, the error created by the learning algorithm for data forgotten during the learning process) as well as reducing the mean square error. It is basically the SVM method of associative learning algorithm to assign each instance in the data space to the desired class. In SVM there are support vectors, and also the discriminative hyperplain. The aim of SVM classifiers is to maximize the distance between support vectors and the discriminative hyperplain. Figure 13.2 shows the oper-ation of SVM (Niazkar & Niazkar, 2020; Zhang & Guan, 2012).

- *Random forest (RF)*

  RF is a classification and regression technique that is very popular due to its efficiency and simplicity. RF consists of decision trees. Each decision tree is normally a binary tree, and data is propagated between each tree starting at the root of the tree (da Silva et al., 2020; Jakkula, 2006).

- *Deep learning (DL)*

  In a general definition, DL is ML, in that it performs representations or abstrac-tions of learning for the machine at different levels. By doing this, the machine will have a better understanding of the reality of data existence and can identify

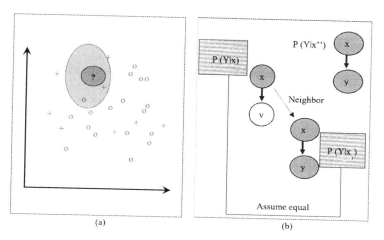

(a)   (b)

FIGURE 13.1 The operation of KNN classifiers: (a) Convergence of 1-NN; (b) KNN and irrelevant features.

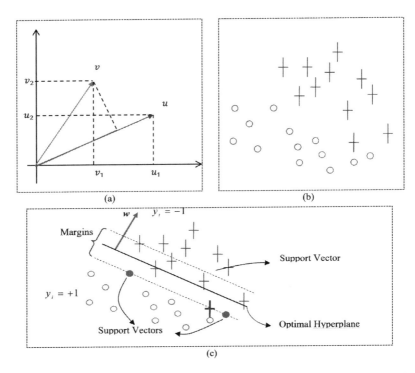

FIGURE 13.2 SVM operation: (a) Vector inner product, (b) Feature space, (c) SVM decision boundary.

different patterns. Identifying DL is inextricably linked to neural networks. According to a popular definition, DL is actually learning by neural networks, which have many hidden layers. The deeper we go in the layers of a neural network, the more complex and complete the model becomes (Singh et al., 2020; Giannakaki et al., 2017).

This study seeks to review the feature selection methods in determining the causes of the spread and epidemic of COVID-19. In this regard, the innovations of this research are expressed as follows: (1) defining feature selection in COVID-19; (2) introducing feature selection methods in COVID-19; (3) comparison of feature selection methods in COVID-19. The main purpose of this research is to present a new method in feature selection with the help of reinforcement learning. In the continuation of this chapter, in Section 13.2, the concepts of feature selection will be stated. In Section 13.3, various feature selection methods will be introduced and reviewed. In Section 13.4, these methods will be compared on two databases. Section 13.5 concludes the chapter.

13.2 FEATURE SELECTION

Features display the properties of objects, and the key solution to identifying objects and classifying them is to efficiently select a combination of features (Wang et al.,

2020). These attributes are given to the categorizer, and the performance of the categorizer is highly dependent on the efficiency of these features (Shahin et al., 2019). Extracting suitable features is difficult due to many factors such as noise. In detection and detection systems, the purpose of selecting a feature is to find a subset of features that result in the best detection and detection performance and require the least computational effort (Jamshidi et al., 2020; Abbas et al., 2019a; Afza et al., 2020; Al-Ameen et al., 2015; Amin et al., 2019; Ejaz et al., 2018; Fahad et al., 2018; Husham et al., 2016; Iftikhar et al., 2017; Iqbal et al., 2019; Khan et al., 2020a). Feature selection is generally important for detection and detection systems for the following three reasons:

- Sometimes, many features are available for detection or a detection system. However, these features are not independent and may be interdependent. A bad feature can drastically reduce system performance. Using more features can also increase the complexity of the system, but may not lead to high detection or detection accuracy. Therefore, it is important to select a subset of good features (Mohammadi Dashti & Harouni, 2018).
- Features are selected by a learning algorithm during the training phase. The selected features are used as a model to describe the training data. Selecting multiple features means that a complex model has been used to approximate educational data. According to the principle of minimum description length, a simple model is better than a complex model. Educational data may be corrupted by various noises, and a complex model may be sensitive to noise in educational data and perform poorly on experimental data (Too & Mirjalili, 2020).
- Using fewer features can reduce computational costs, which is important for real-time applications. Reducing features may also improve classification accuracy (Hancer et al., 2020; Saranya & Pravin, 2021).

Ideally, the goal is to use features that have better separation power (Wang et al., 2016).

As data volumes increase, so does the ability of learning algorithms. Increasing the volume of data requires a high amount of memory as well as high processing power. Feature selection is one way to overcome the challenges posed by high-volume data (Lung et al., 2014; Mashood Nasir et al., 2020; Mittal et al., 2020; Mughal et al., 2018; Nazir et al., 2019; Perveen et al., 2020; Rahim et al., 2017a; Ramzan et al., 2020a; Rehman et al., 2018a; Saba et al., 2020; Ullah et al., 2019). In feature selection, irrelevant and duplicate features are removed, and as a result of this deletion, the dimensions of the data are reduced. Feature selection is used in many fields such as intelligent expert systems, data mining, and ML. Although several methods for feature selection have been proposed so far, new methods for feature selection are still needed (Hu et al., 2018).

In areas such as signal processing (Harouni et al., 2014), image processing (Pereira et al., 2018), data processing, and working data (Alashwal et al., 2020), new information and data are generated every day. Rising computational costs, over-training, and low efficiency are some of the disadvantages of high-volume data processing

(Lin et al., 2019). Choosing a subset of the most knowledgeable features in high-volume data is very challenging in pattern recognition and ML, due to the presence of thousands of features. Filtering duplicate and unrelated features can reduce over-training, minimize computational costs, and ultimately improve the efficiency of the classification process. Therefore, it seems that efficient and appropriate feature selection algorithms that reduce the dimensions of the data and at the same time increase efficiency are necessary.

Feature selection is a hybrid optimization problem that is used in many applications and research fields. Feature selection is now considered successfully in many expert systems such as text processing or text mining (Emadi & Emadi, 2020), image processing (Alharan et al., 2019; Srividya & Arulmozhi, 2018; Deng et al., 2019), medical information (Alharan et al., 2019), industrial applications (Harouni et al., 2012b), medical signal processing (Harouni et al., 2012c), surveillance systems, and gender recognition (Remeseiro & Bolon-Canedo, 2019).

In signal processing, due to the large number of features, it is very difficult to choose a suitable feature selection method. In practice, the type of feature selection depends on the application. Feature selection can improve the accuracy of detecting an error in an industrial or medical process. There are several methods for selecting a feature, as one method for selecting a feature may not be suitable for all applications. Each process, depending on the application, requires an appropriate feature selection method.

13.3   REVIEW OF FEATURE SELECTION ALGORITHMS

Several categories are provided for the various methods offered in feature selection. These categories are:

- Classification based on data labels
- Classification based on the relationship between learning models
- Relationship and relevance and redundancy between filter models
- Relevance and diversity in the filter model
- Categorization based on search strategies (Andreasson, 2019).

These categories are presented in Figure 13.3.

Based on the availability of data labels, feature selection methods are divided into three groups, which are supervised, semi-supervised, and unsupervised (Aryanmehr et al., 2018; Abbas et al., 2019b; Iqbal et al., 2018a; Khan et al., 2020a; Khan et al., 2019a; Mughal et al., 2017; Rahim et al., 2017b; Rehman et al., 2018b; Rehman et al., 2018c; Saba et al., 2019a; Saba et al., 2018). In supervised feature selection methods, all data are labeled. However, in selecting the unsupervised feature, none of the primary data is labeled. But in semi-supervised methods, some data are labeled, while others are not labeled. It should be noted that real-world data is more than the third type. It seems that semi-regulatory methods can be of great importance (Rakhmatulin, 2020).

- Classification based on the relationship between learning models
  Feature selection algorithms are divided into four groups, based on different relationships and evaluations between learning models: filter, wrapper, embedded and hybrid methods.

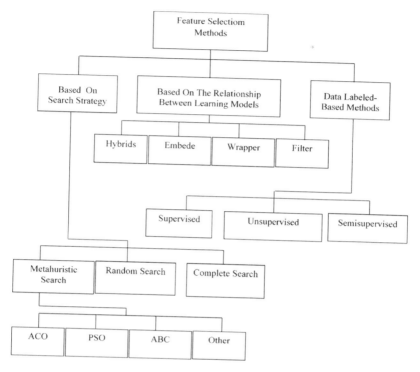

FIGURE 13.3 Classification of feature selection methods (Andreasson, 2019).

- *Filter-based methods*
  Filter-based methods use feature statistical information. As a result, these methods do not require any learning model. These methods are fast and efficient, and at the same time they are computationally light (Rakhmatulin, 2020).
- *Wrapper-based methods*
  Wrapper-based methods select a subset of features based on a learning algorithm. The type of features selected depends a lot on the type of learning algorithm. These methods are more effective and better than filter-based methods, but they are computationally complex (Khairi et al., 2020).
- *Embedded methods*
  The placement methods select the selected set based on the placement of the features in a learning algorithm. The selected subset of features is selected based on the category created. In this model, there is a deeper connection between the learning model and the selected features. These methods are used for high-dimensional data and, although they have good efficiency, they are very complex (Sheikhpour et al., 2017).
- *Hybrid methods*
  These algorithems use a balanced combination of filter- and wrapper-based methods to select the optimal subset of features. Hybrid methods take advantage of both types of feature selection methods, while trying not to include the

weaknesses of either methods. Hybrid-based methods reduce the correlation between attributes and, in turn, improve the correlation between attributes and class labels. Filter-based methods are divided into two categories (Guan et al., 2014).

- *Methods based on Relevance and Redundancy*
  Redundancy and relevance analysis is the basis of this category of methods. One of the most stable and efficient methods is the method of Maximum Relevance and Minimum Redundancy (MRMR) (Wang et al., 2017). Numerous methods have been proposed to reduce redundancy and correlation or increase correlation. When a duplicate attribute is considered, it means that the information referenced to that attribute is similar to one or more other attributes that have a negative effect on algorithm learning, so a small number of attributes can be considered as part of the final subset. On the other hand, features that are unrelated or have the potential to be unrelated are almost insignificant in learning algorithms and have no information in clustering or classification. In high-dimensional data, these types of features are common, so the correlation between features becomes very complex. It seems that the repetition between them should be examined more carefully. Correlation between features can be used to remove duplicate features. There are not many study methods for investigating the duplication of features; also, the methods presented in this field have a high computational complexity (Lu and Lysecky, 2019; Khan et al., 2020c; Khan et al., 2020d; Khan et al., 2019b; Rehman et al., 2018c; Rehman et al., 2021; Saba et al., 2019a).
- *Methods based on diversity and relation*
  Feature selection methods based on feature diversity have also been examined. In the problem of optimal subset feature selection, including the most relevance and maximum diversity in this category is examined. As shown in Figure 13.4, these methods consist of three steps:
  - Selecting a suitable distance criterion in forming the feature space
  - Classification of features by clustering methods
  - Selecting related properties in each cluster to obtain the optimal subset (Berbar, 2018).
- *Categorization based on search strategies*
  Many feature selection algorithms based on search strategy for feature selection are presented (Abbas et al., 2019c; Iqbal et al., 2018a; Iqbal et al., 2018b;

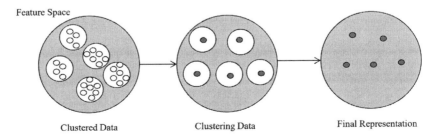

FIGURE 13.4   Cluster-based feature selection method.

Khan et al., 2020b; Khan et al., 2019a; Ramzan et al., 2020b; Rehman et al., 2020; Rehman et al., 2021; Saba et al., 2019b). Based on search strategies, feature selection and its methods are divided into three categories: complete, randomized, and evolutionary methods (Bugata and Drotar, 2020).

Complete search methods include searching for the best subset of features in the entire 2^n search space, n being the number of features. As a result, it is almost impossible to identify the best subset; especially if the data volume is high or the data is large, it cannot select the best subset in a reasonable time. In random search, in a limited space of the entire feature space, the best subset is searched, the size of the selected subset depending on the stop criterion such as maximum repetition, or cost function. Although the interaction between the convergence speed and the optimization of the search algorithm is done using parameter setting, the algorithms used have the ability to get stuck in the global minimum. In meta-heuristic feature selection algorithms, a set is added to or subtracted from a subset of selected features (Agarwal & Mittal, 2016). These methods are very computationally suitable and many such studies have been presented so far. Methods have been proposed based on the particle swarm optimization algorithm (PSO) (Ismi et al., 2016), the artificial bee colony algorithm (ABC) (Liu and Wang, 2019), and ant colony optimization (ACO) (Harouni et al., 2014). The ability to search these algorithms globally has caused them to be highly regarded, although there is still the possibility of learning at a local minimum in these types of algorithm.

- *Reinforcement learning*

Use of reinforcement learning is one of the ML trends that is inspired by behavioral psychology. This method focuses on the behaviors that the machine must perform to maximize its reward. This issue is investigated due to the scope of this method in various fields, such as game theory, control theory, operations research, information theory, multitasking systems, congestion intelligence, statistics, genetic algorithms, and simulation-based optimization. In the field of operations research and in the control literature, the field in which the reinforcement learning method is studied is called approximate dynamic programming (Rostami et al., 2020).

- *Challenges in feature selection methods*

There are challenges to classifying data using extracted or selected features, some of which are:

- In the worst case, class labels do not exist in real-world data, so non-regulatory methods seem to be useful.
- Filter-based methods are effective and fast, especially if the data volume is high.
- As the number of features in high-dimensional data increases, the correlation between features becomes very complex, so the repetition between these features also increases and should be examined more carefully.
- Cluster-based feature selection methods are sensitive to clustering method, and the selection of the appropriate feature depends on the selection of the appropriate cluster (Andrushia & Patricia, 2020).

13.4   COMPARISON OF FEATURE SELECTION METHODS

The proposed method of selecting features based on reinforcement learning can be used most in real-world data (V.J Sara & Kalaiselvi, 2019). The proposed method of feature selection will be simulated on two databases, COVID-19[1] and WHO COVID-2019.[2] It should be noted that 70% of the total data is considered as educational data and 30% as experimental data. The final step in a database classification system is to calculate the precision, sensitivity, positive predictive value (PPV0, and negative predictive value (NPV) in the classification. Positive and negative predictive values are the true positive and negative probabilities reported by an experiment, respectively. In classification issues, the most common tools for evaluation are the proposed criteria. These criteria are calculated as Figure 13.5.

In Figure 13.5 are true negative (TN) (Smadi et al. 2018), false negative (FN) (Mostafa et al. 2017), true positive (TP), and false positive (FP); the following equations show these criteria.

$$\text{Accuracy} = \frac{\text{TP} + \text{TN}}{\text{TP} + \text{TN} + \text{FP} + \text{FN}} \tag{13.1}$$

$$\text{Sensitivity} = \frac{\text{TP}}{\text{TP} + \text{FN}} \tag{13.2}$$

$$\text{Positive Predictive Value} = \frac{\text{TP}}{\text{TP} + \text{FP}} \tag{13.3}$$

$$\text{Negative Predictive Value} = \frac{\text{TP}}{\text{TP} + \text{FN}} \tag{13.4}$$

The geometric mean (gmean) criterion is one of the most important evaluation criteria for feature selection. Due to the large difference between the number of samples in the normal class and the number of samples in each of the attack classes, this dataset can be considered unbalanced, so the gmean criterion can be an evaluation criterion in the problem under study.

- *Precision*
  The results of classification precision evaluation with the help of three classifiers, RF, KNN, and SVM, on the two databases used in the chapter are shown in Table 13.1 in the three modes of selecting the Laplacian supervisory feature, selecting the proposed reinforcement learning feature, and without selecting the feature.

		Real Value	
		Positive	Negative
Predicted	Positive	TP	FP
Value	Negative	FN	TN

FIGURE 13.5   Evaluation of classification data in COVID-19.

TABLE 13.1

Results of Classification of Precision Evaluation Criteria

Database	COVID-19			WHO COVID-2019		
Classifier	RF	KNN	SVM	RF	KNN	SVM
Reinforcement Learning Feature Selection	87.25	82.11	90.9	91.73	85.21	100
Laplacian Feature Selection	83.65	76.68	89.47	73.12	86.12	83.11
Without Feature Selection	73.85	74.21	87.21	78.26	70.28	86.7

TABLE 13.2

Results of Sensitivity Classification Evaluation Criteria

Database	COVID-19			WHO COVID-2019		
Classifier	RF	KNN	SVM	RF	KNN	SVM
Reinforcement Learning Feature Selection	86.54	81.11	83.25	84.25	95.19	100
Laplacian Feature Selection	73.68	82.57	87.5	63.25	65.6	91.2
Without Feature Selection	66.66	90.42	80	75.61	83.54	88.68

With the implementation of the RF classifier on the COVID-19 database and the WHO COVID-2019 database, the classification precision in the reinforcement learning feature selection mode has been reduced compared to the two modes of Laplacian supervised feature selection and without feature selection (Harouni & Baghmaleki, 2020). Classification precision on the COVID-19 database has improved in the enhanced learning feature selection mode compared to the no feature selection mode.

- *Sensitivity*

Classification sensitivity was evaluated with the help of three classifiers, RF, KNN, and SVM, on the two databases used in the chapter in three modes: selection of Laplacian supervised feature, reinforcement learning feature selection, and no feature selection. Table 13.2 shows the results obtained.

The sensitivity of classification using RF classifier on the two databases had the highest values in the mode of selective supervised feature. This value reached 100 on the COVID-19 database.

- *PPV*

The PPV of the classification was evaluated with the help of three classifiers, RF, KNN, and SVM, on the two databases used in the chapter in the three modes of selecting the Laplacian supervised feature, selecting the

TABLE 13.3
Results of PPV Classification Evaluation Criteria

Database	COVID-19			WHO COVID-2019		
Classifier	RF	KNN	SVM	RF	KNN	SVM
Reinforcement Learning Feature Selection	89.36	98.2	90.21	68.53	52.43	100
Laplacian Feature Selection	87.5	100	87.5	63.24	72.59	92.1
Without Feature Selection	75	100	83.3	73.33	88.66	83.61

TABLE 13.4
Results of NPV Classification Evaluation Criteria

Database	COVID-19			WHO COVID-2019		
Classifier	RF	KNN	SVM	RF	KNN	SVM
Reinforcement Learning Feature Selection	69.38	95.1	89.12	78.01	72.29	100
Laplacian Feature Selection	69.89	100	91.25	78.63	78.58	90.01
Without Feature Selection	72.25	68.23	86.64	80	75.84	87.29

reinforcement learning feature, and without selecting the feature. Table 13.3 shows the PPV results.

The number 100 obtained indicates a PPV in the COVID-19 database and in the case of selecting the reinforcement learning feature. On the COVID-19 database, this classifier has improved in the enhanced learning feature selection mode compared to the no feature selection mode. There are also acceptable results on the COVID-19 database when selecting the reinforcment learning feature.

- *NPV*

The NPV of the classification was evaluated with the help of three classifiers, RF, KNN, and SVM, on the two databases used in the chapter in the three modes of selecting the Laplacian supervised feature, selecting the reinforcement learning feature, and without selecting the feature. Table 13.4 shows the results for the NPV.

The NPV in the COVID-19 database has shown its superiority in the reinforcement learning feature selection mode compared to the Laplacian supervised feature selection mode and also in the no feature selected mode.

TABLE 13.5
Results of gmean of Classification Evaluation Criteria

Database	COVID-19			WHO COVID-2019		
Classifier	RF	KNN	SVM	RF	KNN	SVM
Reinforcement Learning Feature Selection	79.75	73.15	81.8	99	84.73	84.82
Laplacian Feature Selection	94.23	68.14	79.94	98.05	71.67	81.25
Without Feature Selection	84.21	73.68	68.25	87.41	75.77	86.87

- *Gmean*

The gmean of the classification was evaluated with the help of three classifiers, RF, KNN, and SVM, on the two databases used in the chapter in the three modes of selecting the Laplacian supervised feature, selecting the reinforcement learning feature, and without selecting the feature. Table 13.5 shows the gmean results obtained.

13.5 CONCLUSION

This chapter has presented the value of correct feature selection in the classification process. Precise and discriminative feature selection can improve the pattern recognition accuracy and COVID-19 data classification rate. Using an appropriate feature selection method to classify and determine the causes of the spread of COVID-19 could improve the efficiency and evaluation criteria of state-of-the-art detection approaches to COVID-19. In this chapter, various feature selection methods have been examined and compared, using the COVID-19-related databases. To evaluate the criteria, precision, sensitivity, PPV, NPV, and gmean were calculated. At the end, the results were compared with the baseline methodology.

NOTES

1 https://github.com/CSSEGISandData/ COVID-19
2 https://www.who.int/emergencies/diseases/novel-coronavirus-2019/global-research-on-novel-coronavirus-2019-ncov

REFERENCES

Abbas, N., Saba, T., Mehmood, Z., Rehman, A., Islam, N. & Ahmed, K. T. 2019a. An automated nuclei segmentation of leukocytes from microscopic digital images. *Pakistan Journal of Pharmaceutical Sciences*, 32e, no 5.
Abbas, N., Saba, T., Rehman, A., Mehmood, Z., Javaid, N., Tahir, M., Khan, N. U., Ahmed, K. T. & Shah, R. 2019b. Plasmodium species aware based quantification of malaria parasitemia in light microscopy thin blood smear. *Microscopy Research and Technique*, 82, 1198–1214.

Abbas, N., Saba, T., Rehman, A., Mehmood, Z., Kolivand, H., Uddin, M. & Anjum, A. 2019c. Plasmodium life cycle stage classification based quantification of malaria parasitaemia in thin blood smears. *Microscopy Research and Technique*, 82, 283–295.

Afza, F., Khan, M. A., Sharif, M., Saba, T., Rehman, A. & Javed, M. Y. *Skin Lesion Classification: An Optimized Framework of Optimal Color Features Selection. 2020 2nd International Conference on Computer and Information Sciences (ICCIS)*, 2020. IEEE, 1–6.

Agarwal, B. & Mittal, N. 2016. Prominent feature extraction for review analysis: an empirical study. *Journal of Experimental & Theoretical Artificial Intelligence*, 28, 485–498.

Al-Ameen, Z., Sulong, G., Rehman, A., Al-Dhelaan, A., Saba, T. & Al-Rodhaan, M. 2015. An innovative technique for contrast enhancement of computed tomography images using normalized gamma-corrected contrast-limited adaptive histogram equalization. *EURASIP Journal on Advances in Signal Processing*, 2015, 1–12.

Alashwal, H., Abdalla, A., Halaby, M. E. & Moustafa, A. A. *Feature Selection for the Classification of Alzheimer's Disease Data. Proceedings of the 3rd International Conference on Software Engineering and Information Management*, 2020. 41–45.

Alharan, A. F., Fatlawi, H. K. & Ali, N. S. 2019. A cluster-based feature selection method for image texture classification. *Indonesian Journal of Electrical Engineering and Computer Science*, 14, 1433–1442.

Amin, J., Sharif, M., Raza, M., Saba, T. & Rehman, A. *Brain Tumor Classification: Feature Fusion. 2019 International Conference on Computer and Information Sciences (ICCIS)*, 2019. IEEE, 1–6.

Andrushia, A. D. & Patricia, A. T. 2020. Artificial bee colony optimization (ABC) for grape leaves disease detection. *Evolving Systems*, 11, 105–117.

Aryanmehr, S., Karimi, M. & Boroujeni, F. Z. *CVBL IRIS Gender Classification Database Image Processing and Biometric Research, Computer Vision and Biometric Laboratory (CVBL). 2018 IEEE 3rd International Conference on Image, Vision and Computing (ICIVC)*, 2018. IEEE, 433–438.

Berbar, M. A. 2018. Hybrid methods for feature extraction for breast masses classification. *Egyptian informatics journal*, 19, 63–73.

Bugata, P. & Drotar, P. 2020. On some aspects of minimum redundancy maximum relevance feature selection. *Science China Information Sciences*, 63, 1–15.

Da Silva, R. G., Ribeiro, M. H. D. M., Mariani, V. C. & Dos Santos Coelho, L. 2020. Forecasting Brazilian and American COVID-19 cases based on artificial intelligence coupled with climatic exogenous variables. *Chaos, Solitons & Fractals*, 139, 110027.

Deng, X., Li, Y., Weng, J. & Zhang, J. 2019. Feature selection for text classification: A review. *Multimedia Tools and Applications*, 78, 3797–3816.

Ejaz, K., Rahim, M. S. M., Rehman, A., Chaudhry, H., Saba, T. & Ejaz, A. 2018. Segmentation method for pathological brain tumor and accurate detection using MRI. *International Journal of Advanced Computer Science and Applications*, 9, 394–401.

Emadi, M. & Emadi, M. 2020. Human face detection in color images using fusion of Ada Boost and LBP feature. *Majlesi Journal of Telecommunication Devices*, 9(1), 23–33.

Fahad, H., Ghani Khan, M. U., Saba, T., Rehman, A. & Iqbal, S. 2018. Microscopic abnormality classification of cardiac murmurs using ANFIS and HMM. *Microscopy Research and Technique*, 81, 449–457.

Giannakaki, K., Giannakakis, G., Farmaki, C. & Sakkalis, V. *Emotional state recognition using advanced machine learning techniques on EEG data. 2017 IEEE 30th International Symposium on Computer-Based Medical Systems (CBMS)*, 2017. IEEE, 337–342.

Guan, D., Yuan, W., Lee, Y.-K., Najeebullah, K. & Rasel, M. K. 2014. A review of ensemble learning based feature selection. *IETE Technical Review*, 31, 190–198.

Habibi, N. & Harouni, M. 2018. Estimation of Re-hospitalization Risk of Diabetic Patients based on Radial Base Function (RBF) Neural Network Method Combined with Colonial Competition Optimization Algorithm. *Majlesi Journal of Electrical Engineering*, 12, 109–116.

Hancer, E., Xue, B. & Zhang, M. 2020. A survey on feature selection approaches for clustering. *Artificial Intelligence Review*, 1–27.

Harouni, M. & Baghmaleki, H. Y. 2020. Color image segmentation metrics. arXiv preprint arXiv:.09907.

Harouni, M., Mohamad, D., Rahim, M. S. M. & Halawani, S. M. 2012a. Finding critical points of handwritten Persian/Arabic character. *International Journal of Machine Learning Computing*, 2, 573.

Harouni, M., Mohamad, D., Rahim, M. S. M., Halawani, S. M. & Afzali, M. 2012b. Handwritten Arabic character recognition based on minimal geometric features. *International Journal of Machine Learning Computing*, 2, 578.

Harouni, M., Mohamad, D. & Rasouli, A. *Deductive method for recognition of on-line handwritten Persian/Arabic characters.* 2010 The 2nd International Conference on Computer and Automation Engineering (ICCAE), 2010. IEEE, 791–795.

Harouni, M., Rahim, M., Al-Rodhaan, M., Saba, T., Rehman, A. & Al-Dhelaan, A. 2014. Online Persian/Arabic script classification without contextual information. *The Imaging Science Journal*, 62, 437–448.

Harouni, M., Rahim, M., Mohamad, D., Rehman, A. & Saba, T. 2012c. Online cursive Persian/Arabic character recognition by detecting critical points. *International Journal of Academic Research*, 4.

Hu, X., Zhou, P., Li, P., Wang, J. & Wu, X. 2018. A survey on online feature selection with streaming features. *Frontiers of Computer Science*, 12, 479–493.

Husham, A., Hazim Alkawaz, M., Saba, T., Rehman, A. & Saleh Alghamdi, J. 2016. Automated nuclei segmentation of malignant using level sets. *Microscopy Research and Technique*, 79, 993–997.

Iftikhar, S., Fatima, K., Rehman, A., Almazyad, A. S. & Saba, T. 2017. An evolution based hybrid approach for heart diseases classification and associated risk factors identification, *Biomedical Research*, 28(8), 3451–3455.

Iqbal, S., Ghani Khan, M. U., Saba, T., Mehmood, Z., Javaid, N., Rehman, A. & Abbasi, R. 2019. Deep learning model integrating features and novel classifiers fusion for brain tumor segmentation. *Microscopy Research and Technique*, 82, 1302–1315.

Iqbal, S., Ghani, M. U., Saba, T. & Rehman, A. 2018a. Brain tumor segmentation in multi-spectral MRI using convolutional neural networks (CNN). *Microscopy Research and Technique*, 81, 419–427.

Iqbal, S., Khan, M. U. G., Saba, T. & Rehman, A. 2018b. Computer-assisted brain tumor type discrimination using magnetic resonance imaging features. *Biomedical Engineering Letters*, 8, 5–28.

Ismi, D. P., Panchoo, S. & Murinto, M. 2016. K-means clustering based filter feature selection on high dimensional data. *International Journal of Advances in Intelligent Informatics*, 2, 38–45.

Jakkula, V. 2006. Tutorial on support vector machine (svm). *School of EECS, Washington State University*, 37.

Jamshidi, M., Lalbakhsh, A., Talla, J., Peroutka, Z., Hadjilooei, F., Lalbakhsh, P., Jamshidi, M., La Spada, L., Mirmozafari, M. & Dehghani, M. 2020. Artificial intelligence and COVID-19: Deep learning approaches for diagnosis and treatment. *IEEE Access*, 8, 109581–109595.

Kenari, A. R., Hosseinkhani, J., Shamsi, M. & Harouni, M. *A robust and high speed E-voting algorithm using elgammel cryptosystem.* 2010 The 2nd International Conference on Computer and Automation Engineering (ICCAE), 2010. IEEE, 812–816.

Kenthapadi, K., Mironov, I. & Thakurta, A. G. Privacy-preserving data mining in industry. *Proceedings of the Twelfth ACM International Conference on Web Search and Data Mining*, 2019. 840–841.

Khairi, N. I., Mohamed, A. & Yusof, N. N. *Feature Selection Methods in Sentiment Analysis: A Review. Proceedings of the 3rd International Conference on Networking, Information Systems & Security*, 2020. 1–7.

Khan, M. A., Ashraf, I., Alhaisoni, M., Damaševičius, R., Scherer, R., Rehman, A. & Bukhari, S. A. C. 2020a. Multimodal brain tumor classification using deep learning and robust feature selection: A machine learning application for radiologists. *Diagnostics*, 10, 565.

Khan, M. A., Kadry, S., Zhang, Y.-D., Akram, T., Sharif, M., Rehman, A. & Saba, T. 2020b. Prediction of COVID-19-pneumonia based on selected deep features and one class kernel extreme learning machine. *Computers & Electrical Engineering*, 90, 106960.

Khan, M. A., Lali, I. U., Rehman, A., Ishaq, M., Sharif, M., Saba, T., Zahoor, S. & Akram, T. 2019a. Brain tumor detection and classification: A framework of marker-based watershed algorithm and multilevel priority features selection. *Microscopy Research and Technique*, 82, 909–922.

Khan, M. A., Sharif, M., Akram, T., Raza, M., Saba, T. & Rehman, A. 2020c. Hand-crafted and deep convolutional neural network features fusion and selection strategy: An application to intelligent human action recognition. *Applied Soft Computing*, 87, 105986.

Khan, M. Z., Jabeen, S., Khan, M. U. G., Saba, T., Rehmat, A., Rehman, A. & Tariq, U. 2020d. A realistic image generation of face from text description using the fully trained generative adversarial networks. *IEEE Access*, 9, 1250–1260.

Khan, S. A., Nazir, M., Khan, M. A., Saba, T., Javed, K., Rehman, A., Akram, T. & Awais, M. 2019b. Lungs nodule detection framework from computed tomography images using support vector machine. *Microscopy Research and Technique*, 82, 1256–1266.

Lin, L., Wang, B., Qi, J., Chen, L. & Huang, N. 2019. A novel mechanical fault feature selection and diagnosis approach for high-voltage circuit breakers using features extracted without signal processing. *Sensors*, 19, 288.

Liu, W. & Wang, J. *A brief survey on nature-inspired metaheuristics for feature selection in classification in this decade. 2019 IEEE 16th International Conference on Networking, Sensing and Control (ICNSC)*, 2019. IEEE, 424–429.

Lu, S. & Lysecky, R. 2019. Data-driven anomaly detection with timing features for embedded systems. *ACM Transactions on Design Automation of Electronic Systems (TODAES)*, 24, 1–27.

Lung, J. W. J., Salam, M. S. H., Rehman, A., Rahim, M. S. M. & Saba, T. 2014. Fuzzy phoneme classification using multi-speaker vocal tract length normalization. *IETE Technical Review*, 31, 128–136.

Mashood Nasir, I., Attique Khan, M., Alhaisoni, M., Saba, T., Rehman, A. & Iqbal, T. 2020. A hybrid deep learning architecture for the classification of superhero fashion products: An application for medical-tech classification. *Computer Modeling in Engineering & Sciences*, 124, 1017–1033.

Mittal, A., Kumar, D., Mittal, M., Saba, T., Abunadi, I., Rehman, A. & Roy, S. 2020. Detecting pneumonia using convolutions and dynamic capsule routing for chest X-ray images. *Sensors*, 20, 1068.

Mohammadi Dashti, M. & Harouni, M. 2018. Smile and laugh expressions detection based on local minimum key points. *Signal Data Processing*, 15, 69–88.

Mostafa, A., Hassanien, A. E., Houseni, M. & Hefny, H. 2017. Liver segmentation in MRI images based on whale optimization algorithm. *Multimedia Tools and Applications*, 76, 24931–24954.

Mughal, B., Muhammad, N., Sharif, M., Rehman, A. & Saba, T. 2018. Removal of pectoral muscle based on topographic map and shape-shifting silhouette. *BMC Cancer*, 18, 778.

Mughal, B., Muhammad, N., Sharif, M., Saba, T. & Rehman, A. 2017. Extraction of breast border and removal of pectoral muscle in wavelet domain, *Biomedical Research*, 28(11)

Nazir, M., Khan, M. A., Saba, T. & Rehman, A. *Brain tumor detection from MRI images using multi-level wavelets. 2019 International Conference on Computer and Information Sciences (ICCIS)*, 2019. IEEE, 1–5.

Niazkar, H. R. & Niazkar, M. 2020. Application of artificial neural networks to predict the COVID-19 outbreak. *Global Health Research and Policy*, 5, 1–11.

Pang, Z., Zhou, G., Chong, J. & Xia, J. 2021. Comprehensive meta-analysis of COVID-19 global metabolomics datasets. *Metabolites*, 11, 44.

Pereira, R. B., Plastino, A., Zadrozny, B. & Merschmann, L. H. 2018. Categorizing feature selection methods for multi-label classification. *Artificial Intelligence Review*, 49, 57–78.

Perveen, S., Shahbaz, M., Saba, T., Keshavjee, K., Rehman, A. & Guergachi, A. 2020. Handling irregularly sampled longitudinal data and prognostic modeling of diabetes using machine learning technique. *IEEE Access*, 8, 21875–21885.

Rahim, M. S. M., Norouzi, A., Rehman, A. & Saba, T. 2017a. 3D bones segmentation based on CT images visualization. *Biomedical Research*, 28(8), 3641–3644.

Rahim, M. S. M., Rehman, A., Kurniawan, F. & Saba, T. 2017b. Ear biometrics for human classification based on region features mining. *Biomedical Research*, 28(10), 4660–4664.

Rakhmatulin, I. 2020. Review of EEG feature selection by neural networks. *International Journal of Science and Business*, 4, 101–112.

Ramzan, F., Khan, M. U. G., Iqbal, S., Saba, T. & Rehman, A. 2020a. Volumetric segmentation of brain regions from MRI scans using 3D convolutional neural networks. *IEEE Access*, 8, 103697–103709.

Ramzan, F., Khan, M. U. G., Rehmat, A., Iqbal, S., Saba, T., Rehman, A. & Mehmood, Z. 2020b. A deep learning approach for automated diagnosis and multi-class classification of Alzheimer's disease stages using resting-state fMRI and residual neural networks. *Journal of Medical Systems*, 44, 37.

Rehman, A., Abbas, N., Saba, T., Mahmood, T. & Kolivand, H. 2018a. Rouleaux red blood cells splitting in microscopic thin blood smear images via local maxima, circles drawing, and mapping with original RBCs. *Microscopy Research and Technique*, 81, 737–744.

Rehman, A., Abbas, N., Saba, T., Mehmood, Z., Mahmood, T. & Ahmed, K. T. 2018b. Microscopic malaria parasitemia diagnosis and grading on benchmark datasets. *Microscopy Research and Technique*, 81, 1042–1058.

Rehman, A., Abbas, N., Saba, T., Rahman, S. I. U., Mehmood, Z. & Kolivand, H. 2018c. Classification of acute lymphoblastic leukemia using deep learning. *Microscopy Research and Technique*, 81, 1310–1317.

Rehman, A., Harouni, M. & Saba, T. 2020. Cursive multilingual characters recognition based on hard geometric features. *International Journal of Computational Vision Robotics*, 10, 213–222.

Rehman, A., Khan, M. A., Saba, T., Mehmood, Z., Tariq, U. & Ayesha, N. 2021. Microscopic brain tumor detection and classification using 3D CNN and feature selection architecture. *Microscopy Research and Technique*, 84, 133–149.

Remeseiro, B. & Bolon-Canedo, V. 2019. A review of feature selection methods in medical applications. *Computers in Biology and Medicine*, 112, 103375.

Rostami, M., Forouzandeh, S., Berahmand, K. & Soltani, M. 2020. Integration of multi-objective PSO based feature selection and node centrality for medical datasets. *Genomics*, 112, 4370–4384.

Saba, T., Haseeb, K., Ahmed, I. & Rehman, A. 2020. Secure and energy-efficient framework using Internet of Medical Things for e-healthcare. *Journal of Infection and Public Health*, 13, 1567–1575.

Saba, T., Khan, M. A., Rehman, A. & Marie-Sainte, S. L. 2019a. Region extraction and classification of skin cancer: A heterogeneous framework of deep CNN features fusion and reduction. *Journal of Medical Systems*, 43, 289.

Saba, T., Khan, S. U., Islam, N., Abbas, N., Rehman, A., Javaid, N. & Anjum, A. 2019b. Cloud-based decision support system for the detection and classification of malignant cells in breast cancer using breast cytology images. *Microscopy Research and Technique*, 82, 775–785.

Saba, T., Rehman, A., Mehmood, Z., Kolivand, H. & Sharif, M. 2018. Image enhancement and segmentation techniques for detection of knee joint diseases: A survey. *Current Medical Imaging*, 14, 704–715.

Saranya, G. & Pravin, A. Feature selection techniques for disease diagnosis system: A survey. *Artificial Intelligence Techniques for Advanced Computing Applications*. Springe, 2021.

Shahin, I., Nassif, A. B. & Hamsa, S. 2019. Emotion recognition using hybrid Gaussian mixture model and deep neural network. *IEEE Access*, 7, 26777–26787.

Sheikhpour, R., Sarram, M. A., Gharaghani, S. & Chahooki, M. A. Z. 2017. A survey on semi-supervised feature selection methods. *Pattern Recognition*, 64, 141–158.

Singh, V., Poonia, R. C., Kumar, S., Dass, P., Agarwal, P., Bhatnagar, V. & Raja, L. 2020. Prediction of COVID-19 corona virus pandemic based on time series data using Support Vector Machine. *Journal of Discrete Mathematical Sciences and Cryptography*, 23, 1583–1597.

Smadi, S., Aslam, N. & Zhang, L. 2018. Detection of online phishing email using dynamic evolving neural network based on reinforcement learning. *Decision Support Systems*, 107, 88–102.

Srividya, T. & Arulmozhi, V. *Feature Selection Classification of Skin Cancer using Genetic Algorithm. 2018 3rd International Conference on Communication and Electronics Systems (ICCES)*, 2018. IEEE, 412–417.

Sun, L., Mo, Z., Yan, F., Xia, L., Shan, F., Ding, Z., Shao, W., Shi, F., Yuan, H. & Jiang, H. 2020. Adaptive feature selection guided deep forest for COVID-19 classification with chest CT. arXiv preprint arXiv:2005.03264.

Too, J. & Mirjalili, S. 2020. A hyper learning binary dragonfly algorithm for feature selection: A COVID-19 case study. *Knowledge-Based Systems*, 212, 106553.

Ullah, H., Saba, T., Islam, N., Abbas, N., Rehman, A., Mehmood, Z. & Anjum, A. 2019. An ensemble classification of exudates in color fundus images using an evolutionary algorithm based optimal features selection. *Microscopy Research and Technique*, 82, 361–372.

Vj Sara, S. B. & Kalaiselvi, K. 2019. Ant colony optimization (ACO) based feature selection and extreme learning machine (ELM) for chronic kidney disease detection. *International Journal of Advanced Studies of Scientific Research*, 4.

Wang, A., An, N., Yang, J., Chen, G., Li, L. & Alterovitz, G. 2017. Wrapper-based gene selection with Markov blanket. *Computers in Biology and Medicine*, 81, 11–23.

Wang, J., Yu, H., Hua, Q., Jing, S., Liu, Z., Peng, X. & Luo, Y. 2020. A descriptive study of random forest algorithm for predicting COVID-19 patients outcome. *PeerJ*, 8, e9945.

Wang, L., Wang, Y. & Chang, Q. 2016. Feature selection methods for big data bioinformatics: A survey from the search perspective. *Methods*, 111, 21–31.

Wu, T.-Y., Lin, J. C.-W., Zhang, Y. & Chen, C.-H. 2019. A grid-based swarm intelligence algorithm for privacy-preserving data mining. *Applied Sciences*, 9, 774.

Xia, Z. & Chen, J. 2021. Mining the relationship between COVID-19 sentiment and market performance. arXivpreprint arXiv:2101.02587.

Zhang, H. & Guan, X. *Iris recognition based on grouping KNN and rectangle conversion. Software Engineering and Service Science (ICSESS), 2012 IEEE 3rd International Conference on*, 2012. IEEE, 131–134.

14 Industry 4.0 Technology-based Diagnosis for COVID-19

Manmeet Kaur, Mohan Singh, and Jaskanwar Singh

CONTENTS

14.1 INTRODUCTION

Technological innovation in industry has spanned several decades and now the Industry 4.0 era is coming. The Industry 4.0 approach was originally proposed in 2011 to develop the German economy. At the end of the eighteenth century, the first industrialization began and is typified by mechanical assembly plants premised on water and steam power; the second revolution began at the turn of the 20th century with the emblem of electrical energy-based mass labor production. In the 1970s, the third industrial revolution took place with heightened focus on electronics and internet technology; and the fourth industrial revolution, Industry 4.0, is currently under way, with the characteristics of the production of cyber-physical systems (CPS) based on clustered technologies and information unification. Several technologies and related paradigms are covered by Industry 4.0, including radio frequency identification (RFID), enterprise resource planning (ERP), Internet of Things (IoT), cloud-based manufacturing, and development of social products. The aims of Industry 4.0 are to reach a higher level of efficiency and productivity in operations, as well as a higher level of automation. Industry 4.0 concepts are interoperability, virtualization, decentralization, potential for real-time, service orientation, and modularity. Industry 4.0 will offer more versatility in terms of functionality, minimize lead times, configure small batch sizes, and decrease costs. Cloud/intranet, data integration, agile adaptation, intelligent self-organization, interoperability, production processes, optimization, safe connectivity, and service orientation are the main concepts of Industry 4.0. Cloud computing is a smart device used by artificial Intelligence (AI), the IoT, and other emerging technologies as a modular production line for almost entire production processes, delivering real-time information. Factories in Industry 4.0 have devices supported by wireless networking and sensors. These sensors are linked to a device that can envision and track the entire production line, and they can make their own choices as well. To meet the COVID-19 pandemic scarcity, Industry 4.0 uses smart manufacturing processes to produce critical disposable products. During this crisis, it offers a smart supply chain of medical disposables and supplies from which patients can get the required critical medical products in time.

14.2 SARS-CoV-2

COVID-19 is a novel type of pathogenic disease spread by a member of the 'Coronaviridae' family that was identified in Wuhan, China, at the end of 2019. The International Committee on Virus Taxonomy (ICTV) immediately named the causative virus as "Severe Acute Respiratory Syndrome Coronavirus-2" because the virus was linked to the SARS-CoV virus that had caused the 2003 SARS outbreak (ICTV, 2020; Yuen et al., 2020). On February 11, 2020, the World Health Organization (WHO) announced the description of the disease affected by the new virus was "COVID-19."

SARS-CoV-2 has been found to have more than 88% of the near sequence identity with that of two bat-derived SARS-like coronaviruses (bat-SL-CoVZC45 and bat-SL-CoVZXC21); other researchers have reported greater identity with previously published bat SARS-like CoVs, but found it to be less related to SARS-CoV (approximately 79%) and 'MERS-CoV' (approximately 50%) (Chan et al. 2020).

TABLE 14.1

Characterization of SARS-CoV-2 virus (ICTV)

Order	Nidovirales
Family	Coronaviridae
Subfamily	Orthocoronavirinae
Genus	α coronavirus, β coronavirus, γ coronavirus, δ coronavirus
Subgenus	Sarbecovirus

14.3 CLASSIFICATION OF SARS-CoV-2

Coronaviruses has been classified by the ICTV, as shown in Table 14.1. SARS-CoV-2 is a member of the family Coronaviridae, subfamily Orthocoronavirinae that has been classified into four genera named "alpha (α)-coronavirus," "beta (β)-coronavirus," "gamma (γ)-coronavirus," and "delta (δ)-coronavirus." The alpha-coronavirus and beta-coronavirus infect humans and other mammals, while the gamma-coronavirus affects the avian species, and the delta-coronavirus infects both mammals and avian species (Li, 2016). SARS-CoV-2, which is responsible for the pandemic COVID-19 has been identified as a member of β genus (Yang & Wang, 2020). Currently, six coronavirus strains that infect humans have been identified. The first four strains, 229E, OC43, NL63, and HKU1 were found to be broadly distributed among humans, and the other two strains, SARS-CoV and MERS-CoV, were of zoonotic origin (Zhu et al., 2020). Presently, SARS-CoV-2 strains are classified into two lineages, L and S, that are well defined by two different SNPs at nucleotide, i.e. L lineage: T28, 114 is in the codon of leucine; S lineage: C28, 144 is in the codon of serine that shows nearly complete linkage across the viral strains that have been sequenced to date (Dawood, 2020). Currently, SARS-CoV-2 isolates have revealed that the spike protein (S) containing D614G is predominant, and this change has enhanced the virus's transmission, as the SG614 is more stable and transmits more efficiently than the SD614 (Zhang et al., 2020)

14.4 CORONAVIRIDAE FAMILY, CORONAVIRUS GENOME ORGANIZATION & STRUCTURAL PROTEINS

Large, enveloped, single-stranded RNA viruses are the majority of the family Coronaviridae. They are the biggest single RNA viruses, with genome sequences ranging in diameter from 25 to 32 kb and a 118–136 nm virion. Nearly two-thirds of the genome embeds the necessary nonstructural proteins (nsps) for transcription and replication of the genome. Among these is nsp12, the RNA polymerase RNA-dependent large 930 amino acid RNA (RdRp). A multiprotein complex with other CoV nsps is formed by nsp12. CoV nsps, cleaved by virally encoded proteases, are synthesized as long precursor polypeptides. Particles of the coronavirus are enveloped with prominent spikes.

Virions are spherical and vary from 118 to 140 nm in size. A flexible nucleocapsid (N) consisting of genomic RNA linked to the nucleoprotein is within the envelope. The major glycoprotein that extends from the virion's surface is the spike (S) protein.

Membrane (M) and envelope (E) are other membrane-associated proteins. All CoVs encode four structural proteins: three membrane-associated proteins (S, M, and E) and a single protein called nucleocapsid (N). A prominent projection from the virus envelope forms the spike (S) protein and gives CoVs their characteristic appearance. S is glycosylated and is the protein for attachment and fusion. S is always associated and modified by N-linked glycosylation with the endoplasmic reticulum. In the endoplasmic reticulum (ER), S is partially or entirely split by host furin-like proteases. At the S1/S2 cleavage site, the extent of proteolysis correlates with the number of highly basic residues. Products S1 (N-terminal) and S2 (C-terminal) remain associated non-covalently. The protein of the membrane (M) is the most abundant protein in virions. M contains three hydrophobic domains and is thus closely associated with the envelope of the virus. In promoting membrane curvature, M plays a major role.

It has a short glycosylation-modified ectodomain (extracellular domain). M interacts with the proteins of N and E as well. When M is co-expressed with either N or E, virus-like particles are released. The protein envelope (E) is found in very small amounts in virions (approximately 20 molecules per virion), although there are larger amounts of E in infected cells. E assembles to form ion channels in membranes, so E is a viroporin. Viroporins in subcellular compartments affect the electrochemical balance. The single protein present in the ribonucleoprotein particle is the nucleocapsid (N) protein. N forms and binds genomic RNA to homodimers and homooligomers and bundles it into a long, versatile nucleocapsid.

N localizes to the cytoplasm in the infected cell and N is also present in the nucleus for certain CoVs. N associates with other structural proteins of CoV, thereby playing a role in budding and assembly. N also colonizes and is needed for RNA synthesis with replicase-transcriptase components. Other functions for N include cell cycle regulation (promoting cell cycle arrest) and inhibition of host cell translation (Tereda et al., 2014).

14.5 METHODS TO STUDY VIRUSES

14.5.1 PURIFYING VIRUSES

It is possible to purify, quantify, photograph, and biochemically analyze viruses grown in cultured cells. The greater the initial concentration of the virus, the simpler it is to purify the virus away from the components of cell debris and media. It is present among the cell debris if a virus is cytopathic. The virus is highly cell-related. To release the virus, the cells must be lysed. Virus and cell debris mixtures are subjected to low-speed pellet cell debris centrifugation, but not the much smaller virions. The supernatant is retained and the pellet is discarded. The supernatant containing the virus can be recentrifugated to pellet the virus at a higher velocity ($\sim$30,000-100,000 $\times$ g).

The supernatant is disposed after the centrifugation is over, and the virus pellet is retained. The virus can be further purified by a density gradient by centrifugation. A sucrose gradient ranges from 40% sucrose at the bottom of the tube to 5% sucrose at the top of the tube; components of sucrose and glycerol are commonly used. The gradient is prepared and the sample of the virus is carefully layered on the top.

Depending on their buoyant density, the components of the sample can break into layers during centrifugation. The visible bands can form and can easily be removed from the tube if the virus is present in the sample.

14.5.2 Chemical/Physical Methods of Virus Quantitation

The sum of a viral protein, genome, or enzyme in a sample is determined by chemical/physical methods of virus quantitation, but they are also convenient, rapid, and very reproducible. Infectious and inactivated virions are not differentiated by physical/chemical assays. The different kinds of chemical/physical techniques include:

- Direct visualization of virions by electron microscopy (EM)
- Hemagglutination assay (HA)
- Serological assays, i.e. enzyme-linked immunosorbent assay (ELISA), fluorescent-tagged antibody assays, and precipitation assay
- Genome quantification by polymerase chain reaction (PCR).

(a) **HA**: Hemagglutinating viruses fasten to the residues of sialic acid on the red blood cells (RBCs). Some viruses bind to the RBCs. A single virion can connect to several different RBCs, and multiple virions can link an RBC to form a wide, easily visualized network of cells and viruses. HA is simple and cheap and needs no system identification. The dilution factor of a virus sample is prepared and each dilution is applied to the RBCs in a well of a microtiter plate or test tube.

There are RBCs and saline (negative control) in one well and a positive reference virus sample in another. The specimens are mixed thoroughly and left at room temperature. The RBCs will move quickly to cover a tight button at the bottom of the tube in the negative wells. The RBCs and virions will join together in the positive wells to form a matrix of cells on the bottom of the channel. The HA titer is the reciprocal of the maximum virus dilution that gives a positive HA.

(b) **Serological techniques**:
- **Hemagglutination inhibition assay (HIA)**. This is a serological assay used to diagnose or recognize a suspected virus from the antibody to the virus. The HIA is done by first combining serum dilutions with the virus samples. Antibodies are allowed to bind to the virus and then add RBCs to the mix. Antibody-bound viruses would not be able to bind to RBCs. Therefore, the lack of hemagglutination in the HIA is a good outcome. If the known reagent is a hemagglutinating virus, then the HIA can be used to detect the antibody.
- **Virus neutralization assay**. In combination with an infectivity assay, such as the plaque test, a virus neutralization assay is used. This assay recognizes the antibody that is further able to prevent the replication of the virus. A special form of immunoassay is virus neutralization since it does not detect all reactions of the antigen-antibody. Only the antibody that can inhibit replication of the virus is found.

- **ELISA**. If the antibody is "tagged," the association of antibodies with their cognate antigens can be visualized. The tag can be an enzyme which cleaves a substrate, a fluorescent molecule, or a radioactive isotope. ELISAs require adsorption of either antigen or antibody onto a plate or tube. Either horseradish peroxidase (HRP) or alkaline phosphatase (AP) is often the enzyme used for identification. Relatively stable, cheap, and easy to purify, these enzymes can be chemically connected to an antibody to help detect an immune complex. ELISAs can be used for either antigen or antibody detection. In a cell culture supernatant, antigen detection ELISAs can be used to quantify the amount of virus.

(c) **Cell-based fluorescent antibody assays**. To detect the presence of viral antigens in cells, highly specific antibodies tagged with fluorescent molecules can be used. Assays can be direct (using primary antibody labeling) or indirect (using labeled secondary antibody). The techniques enable one to differentiate between infected and uninfected cells.

- **Western blots**. Western blots are labor-intensive and costly, but provide a means of confirming a reactive antigen's identity. Western blots are performed by sodium dodecyl sulfate (SDS) and polyacrylamide gel electrophoresis of antigen electrophoresis (PAGE) (i.e. purified virus). In a mixture, the proteins are separated by the size of the gel and transferred from the gel to a solid substrate often referred to as the membrane. The membrane is incubated with patient serum, and the presence of patient antibodies is detected by incubating the membrane with secondary antibodies labeled afterwards. The sensitivity of the western blot lies in its capability to demonstrate the protein molecular weight recognized by the serum of the patient. When the immunoreactive protein band does not correspond in size to known viral proteins, false positive reactions can be identified.

- **Immunohistochemistry**. The basis of immunohistochemistry is that enzyme-tagged antibodies are incubated by a tissue section. In the sample, a colorless substratum is added. If enzyme-tagged antibodies are present, a colored precipitate is produced by cleaving the substrate. As it makes it possible to examine specific virus-infected cells in a particular tissue, it is a powerful technique. Patient samples are often stored in formaldehyde or, at ultra-cold temperatures, stored frozen.

14.5.3   DETECTING VIRAL NUCLEIC ACIDS

a) **PCR**. PCR can be used to identify or quantify viral genomes in a sample. PCR is a very sensitive method and uses oligonucleotide primers designed to detect suspect viruses. Advantages of PCR include:

- The PCR product can be rapidly sequenced, providing genetic information about the viruses
- Primer sets can be designed to recognize sequences common among groups of related viruses or can be used to detect a specific member of a virus group

- Multiple primer sets can be used to look for more than one suspect virus in a sample
- PCR assays are sensitive, but sensitivity can be a disadvantage as well as an advantage. When performing PCR for diagnostic purposes, it is essential that every precaution be taken to avoid contaminating patient samples.

b) **High throughput sequencin**. While PCR amplification necessitates some prior understanding of a viral sequence, using high throughput and impartial sequencing techniques, all nucleic acid (DNA/RNA) must be sequenced in a sample. Technologies for sequencing can provide data worth of a genome in a day. In order to analyse the data and compare it to information stored in public databases, powerful algorithms are used. An explosion of new viruses from humans, animals, and environmental samples has resulted from the use of unbiased sequencing. The current challenge is to develop an understanding of which viruses may be threats to our normal viral flora and which are part of it.

While a significant number of methods for the detection of virus particles are available, there are a number of difficulties which limit the practical use of these methods. Such restrictions include:

1) Absorbs time
2) Not ideal for on-site, rapid anaylsis
3) Requirement for an individual with highly technical skills
4) The need for preparing and purifying samples
5) Lower precision and sensitivity
6) Higher instrument, accessories, and price maintenance
7) Dynamic operation of resources
8) Availability on a wide scale.

Therefore, new, effective methods for the rapid detection of analytes are needed, taking into account the flexibility of viruses and their replication niches. To test the mass population, implementation of these methods must ensure higher precision, ease of operation and portability, and large-scale availability. In addition, networked computers need to be able to operate more effectively, collaboratively, and resiliently by using advanced information analytics. This pattern is shaping the next generation of the manufacturing sector, namely Industry 4.0 (Samson et al., 2020).

14.6 CONSEQUENTIAL ADVANTAGES OF INDUSTRY 4.0 TECHNOLOGIES FOR COVID-19

Industry 4.0 innovations have the potential to provide our everyday lives with better digital solutions during this crisis. As envisaged by scientists for the mitigation effects of the COVID-19 pandemic, some advantages of Industry 4.0 technologies are as follows (Javaid et al., 2020):

- The planning of COVID-19 events
- Providing a healthier experience without putting healthcare and other staff at risk

- Manufacturing of virus-related precautionary items
- Providing a medical function in time, using the intelligent supply chain
- Using robotic-based care for infected patients to reduce the risk to doctors
- Digital reality being used for educational purposes
- Promoting a flexible working climate for therapy
- Using advanced manufacturing and emerging technologies, which provides many developments
- Researchers using these technologies to detect irregular data for social and media networks
- Improved risk management of this virus and a worldwide public health evacuation.

14.7 TECHNOLOGIES OF INDUSTRY 4.0 WHICH MAY HELP IN COVID-19 OUTBREAKS

The signs of COVID-19 are identified by Industry 4.0 technologies, which helps to prevent any uncertainty about this disease and can also predict the chances of contracting the disease. This helps track possible health conditions and anticipated chances of recovery. Important Industry 4.0 innovations that may help with outbreaks of COVID-19 are summarized briefly in Table 14.2.

14.8 INDUSTRY 4.0-BASED TECHNIQUES USED FOR DETECTION OF SARS-CoV-2

Due to its continued global spread and the unavailability of suitable treatment and diagnostic systems, the COVID-19 pandemic is becoming more severe. International health agencies are making serious efforts to manage the COVID-19 epidemic by exploring every aspect of the development of therapy with special attention to researching the smart diagnostic tools needed to detect the COVID-19 virus quickly and selectively. Innovative approaches have recorded the search for rapid testing of mass populations for COVID-19 in biosensors, deep learning (DL) or AI, CPS, and the IoT.

14.8.1 BIOSENSORS: FRONTIERS IN FIGHTING COVID-19

Sensors consist of receptors and transducers which are chemical or biological. The receptor communicates directly with the target analyte, and the transducer translates the process of identification into a quantitative signal (Ozer et al., 2020). Biosensors are analytical instruments in which a transducer and a detector that recognizes the interacted analyte and provides a digital output are coupled with biological detection molecules, such as enzymes, antibodies, or nucleic acids. Compared to conventional laboratory-based approaches, viral biosensors provide exciting alternatives to traditional diagnostic tests and can provide affordable, responsive, fast, miniaturized, and portable platforms (Souf, 2016). There are four types of biosensors, namely optical

TABLE 14.2
Delineation of Industry 4.0 technologies

S.No.	Advanced technology	Description	How is it efficacious?	References
1	Biosensor	For the conversion of a biological signal into an electrical signal, biosensors are used. Optical, thermal, piezoelectric, and electrochemical biosensors are some of the essential forms of biosensors. They are being noticed in a wide range of areas, applications such as medical research, food industry, and marine industry, etc. This biosensor technology, completely new to the market, has been used successfully as a wireless system in a multi-patient hospital environment.	Biosensors are capable of providing devices that can be easy to use, sensitive, and cost-saving in the current time of the COVID-19 pandemic. They can also provide high precision. A great example of a biosensor used in clinical analysis and disease diagnosis is the glucose meter. The 1AX patch for single-use wireless biosensors is under development. This biosensor patch is used for early detection of COVID-19. In the monitoring of COVID-19 symptoms, this patch will perform real-time recording of temperature, ECG trace, respiration rate, etc.	Pejcic et al., 2006
2	AI	AI is a versatile instrument that can be very effective in evaluating the risks of infection and population screening against the COVID-19 pandemic. Similar to machine learning (ML), computer vision, and natural language processing, it is an application that can teach computers to use big-data-based models to identify, describe, and predict patterns.	AI can predict the outbreak, and the spread of the virus can also be decreased or even stalled. Incorrect details about COVID-19 presented on social media affiliated with the application of AI can be detected and subsequently deleted. With the use of AI, clinical trials for drugs and vaccines against this virus can be streamlined. It can be used to build robots that can assist in the undertaking of sanitization jobs and the online medical evaluation of individuals. CT scans that are needed to diagnose pneumonia caused by a virus can be provided by this technology. In order to generate the equipment needed for the healthcare system, the application of this technology is advantageous.	Petropoulos, 2020

(Continued)

TABLE 14.2 (Continued)

S.No.	Advanced technology	Description	How is it efficacious?	References
3	3D printing	3D printing for manufacturing is already emerging in the medical industry. The customized part of the digital CAD file input can easily, in less time and for less cost, revise the previous product version. As a result of the necessary shortage, it helps in the design and production of ventilator components. Thus, the need for the global supply chain is met by producing the requisite precautionary components.	In certain important applications, 3D-printing technology may be used to contain the spread of the COVID-19 disease. A face mask is already under production to be developed using this technology. This face mask can be used to screen a large number of individuals for COVID-19 within 30 minutes. It is not safe for the environment to use surgical masks and N95 respirators, and it can be harmful to the ecosystem. On the other hand, the newly created NanoHack 3D-printed mask is believed to be recyclable and able to be reused.	Celia, 2020
4	Big data	Big data is an analytical methodology that is very suitable for tracking and managing the worldwide distribution of COVID-19. A significant number of patients infected by this virus can be saved by this technology. This technology provides the framework for decision making to be tested more easily and almost in real time. It will help save people's lives and find successful treatments quickly.	Big data can be very beneficial in assessing and predicting the human reach and effect of the coronavirus. The trackers of COVID-19 are able to collect the close real-time data from sources around the world and then equip scientists, physicians, epidemiologists, and policymakers with recent data that can be very useful in making better decisions to tackle the virus.	Bean, 2020
5	IoT	The IoT is an integrated solution that has led to immense growth in automated development, asset management, etc. It consists of data collection, transmission, analytics, and storage. Sets of collections with the support of sensors embedded in cell phones, robots, data, etc. are made. The collected data is then sent to analytics and decision-making department for analysis and server for the central cloud.	In the battle against COVID-19, the IoT is proving to be very helpful. For example, in order to ensure the deployment of drones, which are used for surveillance of mask-wearing and quarantine. It is possible to use this technology for tracing the route of an epidemic. For epidemiologists, it can be useful to scan for patient zero and also in the detection of individuals that come into contact with the patients. Patients should be confident of compliance with quarantine. It is possible to track down patients who violate quarantine. In addition, this technology may be helpful in providing medical personnel with relief by home patient remote control.	He, 2020

TABLE 14.3

Types of biosensors for respiratory virus detection

Virus	Type of portable biosensor	Recognition element	References
MERS-CoV (Middle East respiratory syndrome coronavirus)	Optical biosensor	Recombinant Spike protein S1	Ravina et al. (2020)
SARS-CoV	Piezoelectric immunosensor	Spike protein S1	Kizek et al. (2015)
Influenza A virus	Electrochemical biosensor	M 1 protein	Dziabowska et al. (2018)
SARS-CoV	Thermal biosensor	RNA-dependent RNA polymerase gene	Saylan et al. (2019)

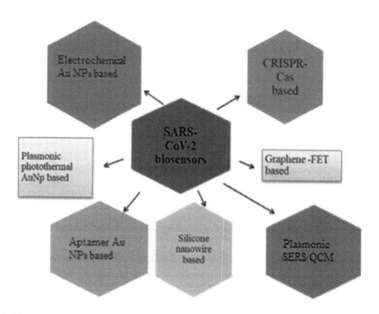

FIGURE 14.1 Biosensors used for the detection of COVID-19.

biosensors, piezoelectric biosensors, electrochemical biosensors, and thermal biosensors, depending on the technology involved (Saylan et al., 2019). Table 14.3 lists descriptions of the various biosensor platforms for the identification of respiratory viral infections. And the extremely precocious biosensors used for COVID-19 detection are shown in Figure 14.1. In addition, the use of the current clustered regularly interspaced short palindromic repeats (CRISPR)-Cas gene-editing method for virus detection has been studied (Zuo et al., 2017). Bacteria, microRNAs, and cancer mutations can also be identified by this process, in a simple and easily scalable way, merely by means of target-specifc crRNA/sgRNA. Nanoparticles (NPs) have recently gained considerable interest because of their biological activity and sensing properties (Holzinger et al., 2014). The gene-editing technique was updated using CRISPR-Chip as a biological sensor, coupled with a graphene-based field effect transistor (FET) that can detect up to 1.7 fM of nucleic acid within the period of 15 minutes without the need for amplification (Hajian et al., 2019).

It was also recently developed for the detection of infection with COVID-19 in less than 40 minutes. The lateral flow assay technique based on CRISPR-Cas12 is simple to implement and an effective and good substitute for diagnostics based on real-time RT-PCR. The FET-based biosensing devices use a monoclonal antibody against the SARS-CoV-2 spike protein to coat the graphene sheets of the FET. Using antigen protein, cultured virus, and nasopharyngeal swab specimens from COVID-19 patients, they determined its sensitivity. This FET biosensor system could detect 1 fg/mL concentration (conc.) of SARS-CoV-2 spike protein in phosphate-buffered saline (PBS) and 100 fg/Ml conc. in medical means of transport (Seo et al., 2020). Dual-functional, plasmonic biosensor plasmonic photothermal (PPT) and localized surface plasmon resonance (LSPR) have also been investigated for the current pandemic (SARS-CoV-2) detection (Qiu et al., 2020). Shen et al. (2012) suggested that biosensor devices based on silicon nanowire (SiNW) are also precise and sensitive to detect viral infections, so this could be used to detect recent SARS-Cov-2 virus pandemics.

14.8.1.1 Applications of Biosensors in Combating SARS-CoV-2

(a) **Virus detection:** Infectious virus outbreaks, such as HIV-1, hepatitis, influenza, dengue, zika, and COVID-19, have a difficult impact on the global survival of all living species. These viruses impact a lot of people from time to time and cause health issues or problems on an ongoing basis. Studies on the production of vaccines are still a major problem for the planet. Therefore, there is a need to efficiently and safely produce a vaccine for the above viruses. Biosensors may have a huge effect from this point of view (Saylan et al., 2019).

(b) **Predicting future disease:** Biosensors could become part of our daily lives and be capable of detecting potential diseases as well. In real time, this device can easily track our health at home. It will include applications for health and tracking patient status. Specific molecules and actionable insight into what is occurring in human bodies can be identified. This technology will be used in the future to better track patient wellbeing to avoid future diseases (Javaid et al., 2020).

(c) **Wireless medical biosensor patch for COVID-19 monitoring:** There is an underdevelopment of a new-to-market wireless medical biosensor patch to be used to track COVID-19 patients. This new biosensor technology, known as biosensor patch 1AX, can be attached to the patient's chest without any assistance and can be safely disposed of after use. The biosensor will prove to be highly useful for both patients and healthcare facilities, according to NS Medical ingevices (2020). It is capable of monitoring the body temperature, breathing rate, ECG trace, and heart rate in real-time patients.

(d) **Measure human body temperature:** Biosensors are now available to monitor body temperature and seem to be better equipment for detecting fever with the COVID-19 virus symptoms. They analyze the cause of illness during a shift in body temperature. By assessing body temperature, a physician can easily diagnose a specific illness (Liu et al., 2017).

(e) **Environmental monitoring and measurement of virus concentration in the air:** In situations such as accidental pesticide release or acute toxicity, biosensors are more effective than conventional approaches, such as chromatographic techniques for environmental monitoring of pollutants (Guo et al., 2017). The most important forms of biosensors for environmental monitoring have emerged from immunosensors and enzymatic biosensors. However, due to their unique features, such as being simple to alter and having thermal stability, aptasensors are also becoming a common option (Justino et al., 2015).

14.8.2 DL OR AI

Owing to its rapid spread, COVID-19 has now become a global pandemic. Detecting exposed individuals is very difficult since they do not immediately exhibit signs of the disease. Therefore, it is important to find a way of calculating on a regular basis the number of potentially infected persons in order to take effective steps. As an alternative to conventional time-consuming and costly approaches, AI may be used to test an individual for COVID-19. The key value of AI is that it can be applied to identify unseen images in a trained model. The science of teaching computers using mathematical models to learn and interpret data is known as machine learning (ML). The data is analyzed once ML is introduced into a system, and interesting patterns are observed. In various fields, including malware detection (Alazab et al., 2012), mobile malware detection (Batten et al., 2016), medicine (Xu et al., 2019), and data recovery, the use of ML has increased rapidly (Mesleh et al., 2012). A modern ML framework called deep learning (DL) was introduced in 2012, based on a convolutional neural network (CNN). DL algorithms allow data representation across several abstraction layers to be learned from computational models composed of multiple processing layers.

To perform classification tasks directly from images, texts, or sounds, computer models are trained. DL models have high accuracies (LeCun et al., 2015), and in some instances can enhance human performance. In 1972, Godfrey Hounsfield and Allan Cormack invented the computed tomography (CT) scan. It utilizes sophisticated X-ray technology in order to diagnose critical internal organs carefully. CT scanning can generate 3D images and is quick, painless, non-invasive, and accurate. Inner organ, muscle, soft tissue, and blood vessel CT scans provide greater clarity than standard X-rays, especially for soft tissues and blood vessels.

In order to automatically classify COVID-19 patients and analyse the disease burden quantification on CT scans using a dataset of CT scans from 157 international patients from China and the USA, Gozes et al. (2020) used a DL approach. Their proposed framework measures the CT scan at two distinct levels: subsystems A and B. Subsystem A performs a 3D analysis and subsystem B performs a 2D analysis to identify and locate wider diffuse opacities, including ground-glass infiltrates, for each section of the scan. The authors applied Resnet-50-2 to subsystem B to test their device and obtained an area of 99.6% under the curve. Sensitivity and specificity were, respectively, 98.2% and 92.2%.

In the early stage of COVID-19, Xu et al. (2020) stated that real-time RT-PCR has a low positive rate. They developed an early screening model using DL techniques to differentiate influenza (viral pneumonia) from COVID-19 pneumonia and stable cases using pulmonary CT images. For the study, a dataset of 618 CT samples was collected, and the images were categorized as COVID-19, influenza (viral pneumonia), and other cases, using methods based on ResNet-18 and ResNet.

14.8.3 CYBER-PHYSICAL SYSTEMS

CPS are characterized as disruptive technologies between their physical assets and computational capabilities for the management of interconnected systems (Baheti & Gill, 2011). The rising use of sensors and networked devices has also resulted in the continuous generation of high-volume data known as big data. CPS can be further built in such an environment to handle big data and exploit system interconnectivity to achieve the objective of intelligent, resilient, and self-adaptable machines (Lee et al., 2015).

14.8.3.1 CPS 5C-level Structure

There are two key functional components of the proposed 5-level CPS structure: (1) advanced networking that ensures real-time physical world data collection and cyberspace knowledge feedback; and (2) intelligent cyberspace data management, analytics, and computational capacity. Figure 14.2 outlines the detailed 5C architecture (Lee et al., 2015).

I.   **Smart connection:** The first phase in designing a cyber-physical device implementation is the collection of precise and accurate data from machines and their components. The data could be directly calculated by sensors or collected from manufacturing systems, such as ERP, MES, SCM, and CMM, from controllers or enterprises. Two significant considerations have to be taken into consideration at this stage. First, considering different types of data, a smooth and tether-free method is needed to handle the process of data acquisition and data transfer to the central server when specific protocols such as MT Connect are required (Vijayaraghavan et al., 2008). The second important consideration for the first level is the selection of proper sensors (type and specification).

II. **Data to information conversion:** From the results, meaningful information must be inferred. There are currently many tools and methodologies available for the level of data conversion to information. However, in order to build these algorithms specifically for prognostics and healthcare applications, intensive emphasis has been applied. The second stage of the CPS architecture gives self-awareness to machines by measuring health benefits, projected remaining useful life, etc.

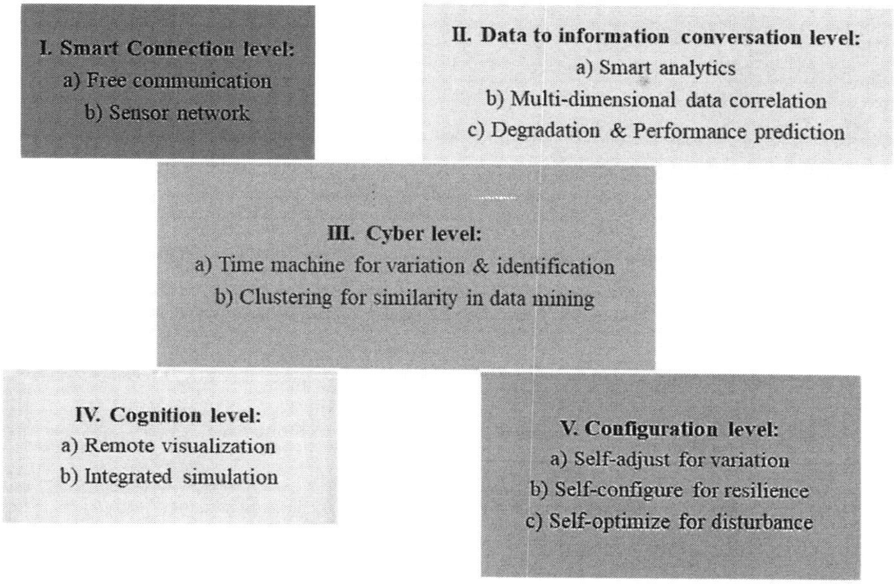

I. Smart Connection level:
a) Free communication
b) Sensor network

II. Data to information conversation level:
a) Smart analytics
b) Multi-dimensional data correlation
c) Degradation & Performance prediction

III. Cyber level:
a) Time machine for variation & identification
b) Clustering for similarity in data mining

IV. Cognition level:
a) Remote visualization
b) Integrated simulation

V. Configuration level:
a) Self-adjust for variation
b) Self-configure for resilience
c) Self-optimize for disturbance

FIGURE 14.2   5C schematic diagram showing implementation of cyber-physical system.

III. **Cyber:** The cyber level functions as a central center for knowledge. In order to form the system network, data is pushed to it from any linked machine. Relevant analytics must be used to extract additional data with vast data collected to provide greater insight into the condition of individual machines within the fleet. These analyses provide self-comparison capabilities for computers, where the performance of a single machine can be compared and graded among the fleet.

IV. **Cognition:** Implementing CPS at this stage provides a detailed understanding of the device being controlled. The right decision to be made is accompanied by a proper presentation of the gained information to expert users. Since comparative data as well as individual machine status is available, it is possible to make a decision on the priority of tasks to improve the maintenance method. In order to fully pass acquired knowledge to users, proper info-graphics are required for this stage.

V. **Configuration:** The level of configuration is the input from cyber space to physical space and acts to make machines self-configure and self-adaptive as supervisory control. This stage functions as a resilience control system to apply to the controlled system the corrective and preventive decisions taken at the cognition level.

14.8.4 Internet of Medical Things

The IoT is the arrangement of interrelated devices and operations compatible with all components of the network, such as software, hardware, network access, and any

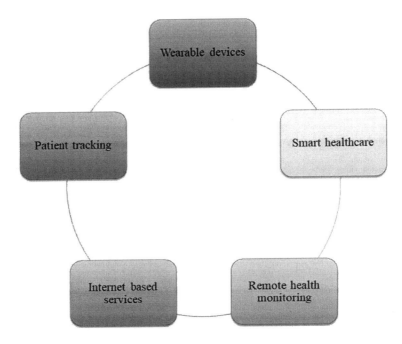

FIGURE 14.3   IoMT facilities for COVID-19 patients.

other electronic and computer requisites, which essentially makes them responsive by facilitating data exchange and compilation. The Internet of Medical Things (IoMT) can be described as the application, especially for the medical and healthcare sectors and fields, of the fundamentals, values, tools, techniques, and concepts of the well-recognized internet approach. All efforts were required to make the network of services possible so that the available healthcare resources and the different medical services could be linked via the ultimate internet-based computer applications (Haoyu et al., 2019). Figure 14.3 describes the facilities offered by the technology based on Industry 4.0 during COVID-19.

14.8.4.1   Working Process

The IoMT approach workflow method requires the integration of healthcare appliances, the infrastructure of medical treatment, the internet network, software applications, and services. The IoMT system helps patient facilities to more efficiently collect data, track records, patient databases, test images, and analysis, etc. The process links the main IoMT components, medical devices, and advanced technology-based devices that ultimately serve the intended functions to enhance patient care, particularly in remote areas (Dong et al., 2020). The proposed IoMT concept workflow method for serving orthopedic patients during the duration of COVID-19 is shown in Figure 14.4.

14.8.4.2   Applications of IoMT during COVID-19

In order to address the critical impacts of epidemics and pandemics, the IoMT offers many essential applications. Some of the major applications of IoMT are the facility

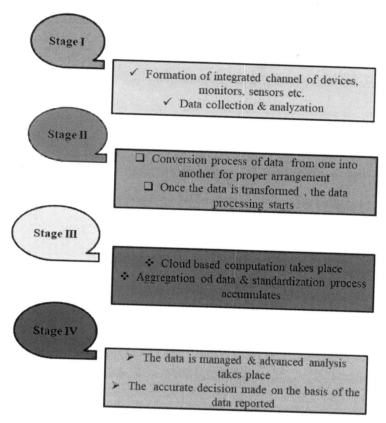

Stage I
- ✓ Formation of integrated channel of devices, monitors, sensors etc.
- ✓ Data collection & analyzation

Stage II
- ❑ Conversion process of data from one into another for proper arrangement
- ❑ Once the data is transformed , the data processing starts

Stage III
- ❖ Cloud based computation takes place
- ❖ Aggregation od data & standardization process accumulates

Stage IV
- ➤ The data is managed & advanced analysis takes place
- ➤ The accurate decision made on the basis of the data reported

FIGURE 14.4 Flow diagram showing processing of IoMT.

for providing medical services in a remote area, online and on-screen check-ups, report analysis, database sharing, knowledge computing, and overall tracking and monitoring of patients. The comprehensive applications of the IoT based on Industry 4.0 are listed in Table 14.4.

14.9 INTEROPERABILITY AND EMINENCE OF INDUSTRY 4.0

Industry 4.0 has two main variables: convergence and interoperability (Romero & Vernadat, 2016). Industry 4.0 can achieve smooth operations across organizational borders and realize networked enterprises embedded with malicious applications and software systems. One of the main benefits of Industry 4.0. is interoperability. Interoperability, according to Chen et al. (2008), is "the ability of two systems to understand each other and use each other's functionality." It reflects the ability of two systems to exchange data and share information and knowledge. Throughout the diversified, heterogeneous, and autonomous process, the interoperability of Industry 4.0 will synthesize software pieces, application solutions, business processes, and the

TABLE 14.4
Role of IoMT in COVID-19 (Joyia et al., 2017)

S.No.	Uses of IoMT	Description
1	Cost reduction	It is cost-effective, since it is possible to avoid the expenses borne by patients during regular medical visits, research, etc.
2	For drug management	Drug storage and consumption can be measured, as IoMT deals with the connected system channel throughout.
3	Proactive treatment	It opens the door to interrupted health monitoring and constructive treatment to be given.
4	Rapid disease detection	In diagnosing the disease at a very initial stage, the real-time database supports the data well controlled on the cloud base.
5	Emergency care	By using analytics and new digital devices, IoMT builds an innovative culture in the care process, such that any potential emergency can be anticipated and analyzed from a distance.
6	Remote care	Patients can be handled in a remote location by cloud-based software, telemedicine, etc.
7	For physicians	The IoMT helps physicians to monitor patients in a smart way. IoMT provides a super-class facility in the current COVID-19 case, where the pandemic does not allow face-to-face daily encounters. The analytical results given help doctors decide on more appropriate treatment protocols.
		IoMT provides innovative digital wearable systems designed to monitor patients properly. It is possible to check and monitor rhythms, blood pressure, etc.
8	Health monitoring & tracking	Using cloud-based data collection and report testing, health monitoring of patients can be performed.

business background. Industry 4.0's interoperability comprises four stages (Sowell, 2006):

a) operational (organizational)
b) systematical (applicable)
c) technical
d) semantic interoperability.

In particular, organizational interoperability reflects general frameworks within CPS and Industry 4.0 of principles, norms, languages, and relationships. Systematic interoperability describes methodology, specifications, domains, and model guidelines and principles. Technical interoperability articulates technical development tools and platforms, as well as IT systems, the ICT ecosystem, and related applications. Semantic interoperability ensures the sharing of information between various groups of individuals and different levels of institutions.

These four operating levels make it more efficient and cost-saving for Industry 4.0 and the cyber-physical infrastructure, IoT, and DL. Figure 14.5 illustrates the Industry 4.0 interoperability platform. Industry 4.0 increases the adaptability, resource utilization, and convergence of supply and demand systems, so plants, manufacturing,

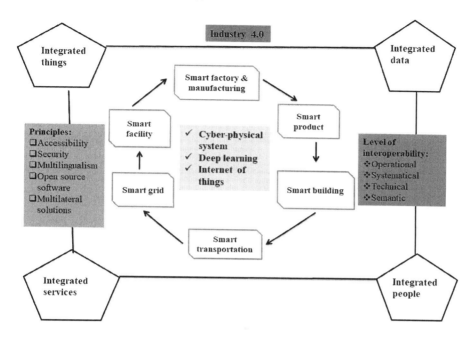

FIGURE 14.5   Framework of interoperability.

towns, and future intelligent equipment and artifacts are smart (Varghese & Tandur, 2014). The word "smart" is used to refer in the literature to applications of Industry 4.0, demonstrating intelligence and expertise. Smart factories and manufacturing, smart goods and smart cities are the core applications of Industry 4.0, according to Stock and Seliger (2016).

14.9.1   SMART FACTORY AND MANUFACTURING

By equipping production with sensors and autonomous systems, Industry 4.0 makes factories more intelligent, mobile, and dynamic. High levels of self-optimization and automation can also be accomplished by computers and equipment. In addition, the development process has the potential to meet more nuanced and professional product specifications and requirements (Roblek et al., 2016). Smart factories and smart manufacturing are also the key priorities of Industry 4.0. (Siddiqui et al., 2016). In particular, the horizontal process incorporates the value-creation modules from the material flow to the product life-cycle logistics, while the vertical process integrates the product, equipment, and human needs with the various aggregation stages of the value-creation and production processes. From the procurement of raw materials to the production method, product use, and the end of product life, knowledge and digitization are incorporated. In Industry 4.0, because of the rapid growth of technology, the production process would require more sensors, microchips, and autonomous systems.

14.9.2 SMART PRODUCT

Industry 4.0's architecture enables human beings to interact with goods, including ICT, IoT, CPS, cloud-formed data integration, standardized intelligent control, and visualized monitoring. In order to cooperate with Industry 4.0, the current production processes need to be incorporated. This involves advanced automated production processes, expertise, details, and real-time adoption. Thus, in interlinked digital and physical processes, smart products need to be produced with high technology. Big data, cloud computing, mass customization, IoT, and improvement of development time are drivers dominating the growth of Industry 4.0 (Schmidt et al., 2015). A cyber-physical framework offering intelligent user interfaces and context-sensitive user interfaces, as well as user-focused assistance systems, is provided by Gorecky et al. (2014).

14.9.3 SMART CITY

In its development strategy, Smart City is "a city that includes six variables: smart economy, smart mobility, smart environment, smart people, smart living and smart governance." (Roblek et al., 2016). IoT will facilitate the development of a new IT generation and knowledge-based economy by integrating the internet, a telephone network, a broadcast network, a wireless broadband network, and sensor networks (Tang, 2015). Lom et al. (2016) noted that technological discipline, fiscal, humanitarian, and legal aspects are included in Smart City. People are transitioning from consumers to main players in a smart city. High technology is the enabler of dynamics. Businesses become collaborators. Output is dependent upon the orientation of demand. During their life cycles, goods get smart. And with advanced planning, quality, and effectiveness, transport is a smart service. The goal of Smart City is to ensure urban sustainability, to enhance the quality of life and safety of people, and to ensure energy efficiency. Moreover, the transformation from a typical town to a smart city takes time.

14.10   FUTURE ASPECTS

Industry 4.0 technology can be used in the future to store confidential data from our healthcare system that can be used for another pandemic close to COVID-19. The practitioners, physicians, and workers who can impact the treatment line of COVID-19 and other similar pandemics or epidemics will quickly follow this revolution. It can be used to centralize all medical equipment, devices, and the method of treatment. In the future, to build a smart healthcare system, the medical industry will evolve and have to adapt to new technology, and so there is a need to update the software platforms and devices to the latest ones.

CONCLUSION

Industry 4.0 provides complex manufacturing sectors and other associated fields with an automatic solution. This includes the collection, transfer, storage, review, and proper monitoring of information systems through multiple manufacturing and digital information technologies. Digital technologies offer a revolutionary way to better isolate the infected patient, minimizing the high risk of mortality, speeding up the development of medications, and the process of treatment and care. People work from home with the use of these technologies; they experience a modern workplace community, work schedules, virtual offices, virtual meetings, and detailed written communications. Industry 4.0 uses smart technology to enable remote working, which is useful for COVID-19 outbreaks. With improved crowd control, transit management, and public security, this revolution will accelerate the digital transformation. Through the introduction of telemedicine consultation, these emerging innovations will build a virtual clinic. So, the physical crowding of patients in hospitals and clinics can be minimized. These systems monitor the patient records and reduce unnecessary medical consultations for the patient.

REFERENCES

Alazab, A., Hobbs, M., Abawajy, J., and Alazab, M. (2012). Using feature selection for intrusion detection system. *International Symposium on Communications and Information Technologies* (ISCIT), 27: 296–301.

Baheti, R., Gill, H. (2011). Cyber-physical systems. *Impact of Control Technology*, 15: 1–6

Batten, L. M., Moonsamy, V. and Alazab, M. (2016). Smartphone applications, malware and data theft, *in Computational Intelligence, Cyber Security and Computational Models*, ed: Springer, pp. 15–24.

Bean, R. (2020). *Big data in the time of coronavirus (COVID-19)*. CIO Netw [Internet]. Available from: https: //www.forbes.com/sites/ciocentral/2020/03/30/big-data-in-the-time-of-coronavirus-covid-19/#161ff87558fc.

Celia, S. (2020). *Covid-19: developing high tech protective masks* [Internet]. Medical Expo e-mag. Available from: http://emag.medicalexpo.com/covid19-development-of-high-tech-protective-masks.

Chan, J. F., Kok, K. H., Zhu, Z., Chu, H., To, K. K., Yuan, S., & Yuen, K. Y. (2020). Genomic characterization of the 2019 novel human-pathogenic coronavirus isolated from a patient with atypical pneumonia after visiting Wuhan. *Emerging Microbes & Infections*, 9: 221–236.

Chen, D., Doumeingts, G., Vernadat, F. (2008). Architectures for enterprise integration and interoperability: Past, present and future. *Computers in Industry*, 59: 647–659.

Dawood, A. A. (2020). Mutated COVID-19, May foretells mankind in a great risk in the future. *New Microbes and New Infections*, 35: 100673.

Dong, P. et al. (2020). Edge computing-based healthcare systems: Enabling decentralized health monitoring in Internet of medical Things. *IEEE Network*, 20: 0890–8044.

Dziabowska, K., Czaczyk, E., Nidzworski, D. (2018). Detection methods of human and animal infuenza virus—current trends. *Biosensors* 8: 94.

Gorecky, D., Schmitt, M., Loskyll, M., Zuhlke, D. (2014). *Human –machine interaction in the industry 4.0 era. 12th IEEE international conference on industrial informatics (INDIN)*, Porto Alegre, Brazil. pp, 289–294.

Gozes, O. et al. (2020). Rapid ai development cycle for the coronavirus (covid-19) pandemic. *Initial Results For Automated Detection & Patient Monitoring Using Deep Learning Ct Image Analysis*, arXiv preprint arXiv, 05037.

Guo, L. et al. (2017). Colorimetric biosensor for the assay of paraoxon in environmental water samples based on the iodine-starch color reaction. *Analytica Chimica Acta*, 967: 59–63.

Hajian, R. et al. (2019). Detection of unamplifed target genes via CRISPR–Cas 9 immobilized on a grapheme feld-efect transistor. *Nature Biomedical Engineering*, 3: 427–437.

Haoyu, L., Jianxing, L., Arunkumar, N., Hussein, A. F., Jaber, M. M. (2019). An IoMT cloud-based real time sleep apnea detection scheme by using the SpO2 estimation supported by heart rate variability. *Future Generation Computer System*, 98: 69e77.

He, S. (2020). Using the Internet of Things to fight virus outbreaks [Internet]. Available from: https://www.technologynetworks.com/immunology/articles/usingthe-internet-of-things-to-fight-virus-outbreaks-331992.

Holzinger, M., Le, G. A., Cosnier, S. (2014). Nanomaterials for biosensing applications: a review. *Frontiers in Chemistry*, 2: 63.

ICTV (2020) International Committee on Taxonomy of Viruses. Virus taxonomy 2020 report. Available at: https://talk.ictvonline.org/information/w/news/1300/page.

Javaid, M., Haleem, A., Vaishya, R., Bahl, S., Suman, R., Vaish, A. (2020). Industry 4.0 technologies and their applications in fighting COVID-19 pandemic. *Diabetes & Metabolic Syndrome: Clinical Research & Reviews*, 14: 419–422.

Joyia, G. J., Liaqat, R. M., Farooq, A., Rehman, S. (2017). Internet of Medical Things (IOMT): Applications, benefits and future challenges in healthcare domainz. *Journal of Community*, 12: 240–247.

Justino, C. I., Freitas, A. C., Pereira, R., Duarte, A. C., Rocha, T. A. (2015). Recent developments in recognition elements for chemical sensors and biosensors. *Trac Trends in Analytical Chemistry*, 68: 2–17.

Kizek, R. et al. (2015). Nanoscale virus biosensors: State of the art. *Nanobiosensors in Disease Diagnosis*, 4: 47.

LeCun, Y., Bengio, Y., Hinton, G. (2015). Deep learning, *Nature*, 521: 436–444.

Lee, J., Bagheri, B., Kao, A. H. (2015). A cyber physical systems architecture for industry 4.0 based manufacturing systems. *Manufacturing Letters*, 3: 18–23.

Li, F. (2016). Structure, function, and evolution of coronavirus spike proteins. *Annual Reviews of Virology*, 3: 237–261.

Liu, Y. et al. (2017). Development of a thermosensitive molecularly imprinted polymer resonance light scattering sensor for rapid and highly selective detection of hepatitis A virus in vitro. *Sensors and Actuators B Chemical*, 253: 1188–1193.

Lom, M., Pribyl, O., Svitek, M. (2016). *Industry 4.0 as a part of smart cities, in: Smart Cities Symposium Prague (SCSP)*, IEEE, pp. 1–6.

Mesleh, A., Skopin, D., Baglikov, S., Quteishat, A. (2012). Heart rate extraction from vowel speech signals. *Journal of Computer Science and Technology*, 27: 1243–1251.

NS Med Devices. *Life Signals to Roll Out Biosensor Patch for COVID-19 Monitoring*, 2020. Available from: https://www.nsmedicaldevices.com/news/lifesignals-biosensor-patch-covid-19/

Ozer, T., Geiss, B. J., Henry, C. S. (2020). Review—chemical and biological sensors for viral detection. *Journal of the Electrochemical Society*, 167: 037523.

Pejcic, B., De Marco, R., Parkinson, G. (2006). The role of biosensors in the detection of emerging infectious diseases. *Analyst [Internet]*, 131: 1079e90.

Petropoulos, G. (2020). *Artificial intelligence in the fight against COVID-19[Internet]*. Available from: https://www.bruegel.org/2020/03/artificialintelligence-in-the-fight-against-covid-19/.

Qiu, G. et al. (2020). Dual-functional plasmonic photothermal biosensors for highly accurate severe acute respiratory syndrome coronavirus 2 detection. *ACS Nano*, 14: 5.

Ravina, D. A., Mohan, H. et al. (2020) Detection methods for infuenza A H1N1 virus with special reference to biosensors: a review. *Bioscience Reports*, 40: 1–18.

Roblek, V., Mesko, M., Krapez, A. (2016). A complex view of Industry 4.0. *SAGE Open*, 6: 2158244016653987.

Romero, D., Vernadat, F. (2016). Enterprise information systems state of the art: Past, present and future trends. *Computers in Industry*, 79: 23–13.

Samson, R., Navale, R. G.,Dharne, M. S. (2020). Biosensors: Frontiers in rapid detection of COVID-19. *3 Biotech*, 10: 385.

Saylan, Y., Erdem, O., Unal, S., Denizli, A. (2019) An alternative medical diagnosis method: biosensors for virus detection. *Biosensors*, 9: 65.

Schmidt, R., Harting, C. R., Mohring, M., Neumairr, M., Jozinovic, P. (2015). Industry 4.0- potentials for creating smart products: Empirical research results. *Business Information System*, 18: 16–27.

Seo, G. et al. (2020). Rapid detection of COVID-19 causative virus (SARS-CoV-2) in human nasopharyngeal swab specimens using feld-efect transistor-based biosensor. *ACS Nano*, 14: 5135–5142.

Shen, F. et al. (2012). Rapid fu diagnosis using silicon nanowire sensor. *Nano Letters* 12: 3722–3730.

Siddiqui, M. S., Legarrea, A., Escalona, E., Parker, M. C., Koczian, C., Walker, M. (2016). Ulbricht, Hierarchical, virtualised and distributed intelligence 5G architecture for low-latency and secure applications. *Transactions on Emerging Telecommunications Technologies.*, 27: 1233–1241.

Souf, S. (2016). Recent advances in diagnostic testing for viral infections. *Bioscience Horizons: International Journal of Student Research*, 9: 1–16.

Sowell, P. K. (2006). The C4ISR architecture framework: history, status and plans for evolution. *Mitre corp, mclean VA*, 24: 1–12.

Stock, T., Seliger, G. (2016). Opportunities of sustainable manufacturing in Industry 4.0. *Procedia CIRP*, 40: 536–541.

Tang, Z. W. (2015). *The industrial robot is in conjunction with homework and system integration*. In *5th International Conference on Information Engineering for Mechanics & Materials (ICIMM)*. pp. 1679–1683.

Tereda, Y. et al. (2014). Emergence of pathogenic coronavirus in cats by homologous recombination between feline and canine coronaviruses. *Plos One* 9: e106534.

Varghese, A., Tandur, D. (2014). *Wireless requirements and challenges in industry 4.0*, In: *International Conference on Contemporary Computing and Informatics (IC3I)*, IEEE, pp. 634–638.

Vijayaraghavan, A., Sobel, W., Fox, A., Dornfeld, D., Warndorf, P. (2008). Improving machine tool interoperability using standardized interface protocols: MTConnect. in: *Proceedings of the 2008 International Symposium on Flexible Automation (ISFA)*, Atlanta, GA, USA.

Xu, X. et al. (2020). Deep learning system to screen coronavirus disease 2019 pneumonia. arXiv preprint arXiv: 2002.09334.

Xu, Y., Wang, Y., Yuan, J., Cheng, Q., Wang, X., Carson, P. L. (2019). Medical breast ultrasound image segmentation by machine learning. *Ultrasonics*, 91: 1–9.

Yang, P., Wang, X. (2020). COVID-19: a new challenge for human beings. *Cellular and Molecular Immunology*, 17: 555–557.

Yuen, K. S., Ye, Z. W., Fung, S. Y., Chan, C. P., Jin, D. Y. (2020). SARS-CoV-2 and COVID-19: The most important research questions. *Cell & Bioscience*, 10: 40.

Zhang, L., Jackson, C. B., Mou, H., Ojha, A., Rangarajan, E. S., Izard, T., Farzan, M., Choe, H. (2020). The D614G mutation in the SARS-CoV2 spike protein reduces S1 shedding and increases infectivity. *Bio Rxiv*, 10: 1–25.

Zhu, N., Zhang, D., Wang, W., Li, X., Yang, B., Song, J., Zhao, X., Huang, B., Shi, W., Lu, R., Niu, P., Zhan, F., Ma, X., Wang, D., Xu, W., Wu, G., Gao, G. F., Tan, W (2020). China novel coronavirus investigating and research team. a novel coronavirus from patients with pneumonia in China, 2019. *New England Journal of Medicine*, 382: 727–733.

Zuo, X., Fan, C., Chen, H. Y. (2017). Biosensing: CRISPR-powered diagnostics. *Nature Biomedical Engineering*, 1: 1–2.

Index

Page numbers in **bold** indicate tables and page numbers in *italic* indicate figures.

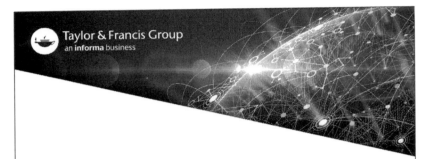